Solution Models Based on Symmetric and Asymmetric Information

Solution Models Based on Symmetric and Asymmetric Information

Special Issue Editors

Edmundas Kazimieras Zavadskas
Zenonas Turskis
Jurgita Antuchevičienė

MDPI • Basel • Beijing • Wuhan • Barcelona • Belgrade

Special Issue Editors
Edmundas Kazimieras Zavadskas
Vilnius Gediminas Technical University
Lithuania

Zenonas Turskis
Vilnius Gediminas Technical University
Lithuania

Jurgita Antuchevičienė
Vilnius Gediminas Technical University
Lithuania

Editorial Office
MDPI
St. Alban-Anlage 66
4052 Basel, Switzerland

This is a reprint of articles from the Special Issue published online in the open access journal *Symmetry* (ISSN 2073-8994) from 2018 to 2019 (available at: https://www.mdpi.com/journal/symmetry/special_issues/Solution_models_based_symmetric_asymmetric_information)

For citation purposes, cite each article independently as indicated on the article page online and as indicated below:

LastName, A.A.; LastName, B.B.; LastName, C.C. Article Title. *Journal Name* **Year**, *Article Number*, Page Range.

ISBN 978-3-03921-006-0 (Pbk)
ISBN 978-3-03921-007-7 (PDF)

Contents

About the Special Issue Editors

Edmundas Kazimieras Zavadskas, PhD, DSc, Professor at the Department of Construction Management and Real Estate, Chief Research Fellow at the Laboratory of Operational Research, Research Institute of Sustainable Construction, Vilnius Gediminas Technical University, Lithuania. He received his PhD in Building Structures (1973), Dr. Sc. (1987) in Building Technology and Management. He is a member of Lithuanian and several foreign Academies of Sciences, Doctore Honoris Causa from Poznan, Saint Petersburg and Kiev Universities, the Honorary International Chair Professor in the National Taipei University of Technology. Awarded by International Association of Grey System and Uncertain Analysis (GSUA) for huge input in Grey System field, has been elected to Honorary Fellowship of International Association of Grey System and Uncertain Analysis, a part of IEEE (2016), awarded by "Neutrosophic Science—International Association" for distinguished achievement in neutrosophics and has been conferred an honorary membership (2016), awarded of the Thomson Reuters certificate for access to the list of the most highly cited scientists (2014). Highly Cited Researcher in the field of Cross-Field (2018), recognized for exceptional research performance demonstrated by production of multiple highly cited papers that rank in the top 1% by citations for field and year in Web of Science. Main research interests: multi-criteria decision-making, operations research, decision support systems, multiple-criteria optimization in construction technology and management. Over 460 publications in Clarivate Analytic Web of Science, h = 55, a number of monographs in Lithuanian, English, German and Russian. Editor in Chief of journals Technological and Economic Development of Economy and *Journal of Civil Engineering and Management*, Guest Editor of over twenty Special Issues related to decision making in engineering and management.

Zenonas Turskis, PhD, Professor at the Department of Construction Management and Real Estate, Chief Researcher at the Laboratory of Operational Research, Research Institute of Sustainable Construction, Vilnius Gediminas Technical University, Lithuania. Education and training: 1979 Civil engineer at Vilnius Civil Engineering Institute (now Vilnius Gediminas Technical University), Vilnius (Lithuania); 1990 Applied mathematics at Kaunas Polytechnic Institute (now Kaunas Technological University), Kaunas (Lithuania); 1993 PhD degree, Technical sciences at Vilnius Technical University (now Vilnius Gediminas Technical University), Vilnius (Lithuania); 2009 Habilitation procedure, Technological Sciences, Civil Engineering at Vilnius Gediminas Technical University. Work experience: Manager at different Joint Stock Companies more than 15 years, Computer programmer—Manager at different companies more than 5 years, Senior Engineer, master, chief dispatcher at construction companies more than 5 years. Particular areas of research interest are: sustainability in architecture, design, and construction; infrastructure and transport projects; life cycle analysis; innovations in construction and management; decision support models to solve abovementioned (but not only) problems. Over 140 publications in Clarivate Analytic Web of Science, h = 41.

Jurgita Antuchevičienė, PhD, Professor at the Department of Construction Management and Real Estate at Vilnius Gediminas Technical University, Lithuania. She received her PhD in Civil Engineering in 2005. Her research interests include multiple-criteria decision-making theory and applications, sustainable development, construction technology, and management. Over 90 publications in Clarivate Analytic Web of Science, h = 23. A member of IEEE SMC, Systems Science and Engineering Technical Committee: Grey Systems and of two EURO Working Groups: Multicriteria Decision Aiding (EWG-MCDA) and Operations Research in Sustainable Development and Civil Engineering (EWG–ORSDCE). Deputy Editor in Chief of Journal of Civil Engineering and Management, Editorial Board member of Applied Soft Computing and Sustainability journals. Guest Editor of several Special Issues: "Decision Making Methods and Applications in Civil Engineering" (2015) and "Mathematical Models for Dealing with Risk in Engineering" (2016) in *Mathematical Problems in Engineering*, "Managing Information Uncertainty and Complexity in Decision-Making" (2017) in *Complexity*, "Civil Engineering and Symmetry" and "Solution Models based on Symmetric and Asymmetric Information" (2018) in *Symmetry*, "Sustainability in Construction Engineering" (2018) in *Sustainability*, and "Multiple-Criteria Decision-Making (MCDM) Techniques for Business Processes Information Management" (2018) in *Information*.

symmetry

MDPI

Editorial

Solution Models Based on Symmetric and Asymmetric Information

Edmundas Kazimieras Zavadskas [1,2], Zenonas Turskis [1,2] and Jurgita Antucheviciene [1,*]

1 Department of Construction Management and Real Estate, Vilnius Gediminas Technical University, Sauletekio al. 11, LT-10223 Vilnius, Lithuania; edmundas.zavadskas@vgtu.lt (E.K.Z.); zenonas.turskis@vgtu.lt (Z.T.)

2 Laboratory of Operations Research, Institute of Sustainable Construction, Vilnius Gediminas Technical University, Sauletekio al. 11, LT-10223 Vilnius, Lithuania

* Correspondence: jurgita.antucheviciene@vgtu.lt; Tel.: +370-5-274-5233

Received: 2 April 2019; Accepted: 2 April 2019; Published: 5 April 2019

Abstract: This Special Issue covers symmetry and asymmetry phenomena occurring in real-life problems. We invited authors to submit their theoretical or experimental research presenting engineering and economic problem solution models dealing with the symmetry or asymmetry of different types of information. The issue gained interest in the research community and received many submissions. After rigorous scientific evaluation by editors and reviewers, nine papers were accepted and published. The authors proposed different solution models as integrated tools to find a balance between the components of sustainable global development, i.e., to find a symmetry axis concerning goals, risks, and constraints to cope with the complicated problems. We hope that a summary of the Special Issue as provided in this editorial will encourage a detailed analysis of the papers.

Keywords: hybrid problem solution models; multiple-criteria decision-making (MCDM); hybrid MCDM; criteria weight assessment; fuzzy sets; rough sets; Z-numbers; neutrosophic numbers; Bonferroni mean (BM) operator; engineering problems; economic decisions

---

1. Introduction

An integral part of contemporary human activities is choosing the most efficient solutions and justifying the selected alternatives and judgments of selected justifying procedures. All objective measurement involves subjective judgments. Firstly, developers of plans decide which problems must be solved and which not. Model development consists of the definition of model objectives, conceptualization of the problem, translation into a computational model, and model testing, revision, and application. Theory, prior knowledge, and other inputs determine which features of a given process to highlight and which to leave out under a given set of conditions that will dictate the specification of the model. Symmetry and asymmetry phenomena occur in real-life problems. Structural symmetry and structural regularity are essential concepts in many natural and human-made objects and play a crucial role in problem solutions. Real (accurate) balance in the real world is an exceptional case [1]. It is an essential feature that facilitates model description and the decision-making process itself.

Decision-makers need to be clear and explicit about the objectives of the problem and the importance of multiple goals, benchmarking values and acceptable compromises. The existence of information asymmetry causes difficulties when achieving an optimal solution. As the asymmetric information is more important, its role is more crucial. Therefore, various solution models propose integrated tools to find a balance between components of global development, i.e., to find symmetry axes concerning goals, risks, and constraints to cope with complicated problems. When confronted with complex problems, a solution's problem is divided into smaller issues. The analyst then uses a method to integrate the results so that the action can be selected temporarily.

Other stakeholders should align the decision on complex and strategic issues. Moreover, decision-makers should strike a balance between objectivity and subjectivity of data [2].

Objectivity is often considered the basis for the evaluation of the knowledge society. Objectivity is a value. The objectivity, balance, and symmetry of decision-making emphasize paradoxes [3] in terms of groups and outcomes. Science is objective when setting and summarizing facts. It is an obvious way of dealing with the requirements of scientific realism.

Confirmation of objectivity and induction problem; choice of theory and exact change; realism; scientific explanation; to experiment; measurement and quantification; evidence and basis for statistics; science based on actual data; experimental values are the central, fundamental debates in the philosophy of science. Understanding scientific objectivity is, therefore, essential to understanding the nature of science and its role in society. Under the concept of product objectivity, science is objective, or to such an extent that its products—theories, laws, experimental results, and observations—represent an accurate representation of the outside world. According to the understanding of the objectivity of the process, science is objective, or to such an extent that its necessary procedures and methods depend on the associated social and ethical values, the bias of the individual scientist. In particular, this second understanding is independently multi-faceted; and it includes explanations related to measurement procedures, self-justification processes, or socio-scientific scales.

The latter projects are characterized by high investment, long construction, and sophisticated technology. Many decision-making problems arise from imperfect information. This means that not all the information needed to create a reasonable solution is known [4]. In a market where customers reach balance, and product developers should have detailed information about product features, it is necessary to understand the importance of asymmetric information so that nobility, whether this inefficiency should cause concern, and when the degree of asymmetry is economically essential. Information asymmetry is usually greatest in areas where information is complex, difficult to obtain, or both [5]. Besides, asymmetric information is typical of a problem where the party has more information than the other and this is quite problematic. Insufficient information makes market problems more difficult. However, stakeholders also have incentives to create mechanisms that allow them to form mutually beneficial decisions even in the face of imperfect information [6–17]. The degree of asymmetry is different, yielding testable implications for the prevalence of asymmetric learning. In such a personal situation, decision-making is optional, using compensation data [18]. People practice multifaceted engineering solutions. Therefore, they should acknowledge a critical parameter corresponding to the degree to which the information is asymmetric. Humans implement multi-faceted decisions of engineers in practice [19–25]. Humans necessarily fill all measurement in science and technology with subjective elements, whether in selecting measures or in collecting, analyzing or interpreting data. Symmetric and asymmetric information play a critical role in engineering problems.

In Kant's view, all knowledge begins with human experience and is concurrent with the experience. The need for qualitative multi-criteria evaluation caused this—information content is determined by by the inexact scale of measurement [26]. The main problem, however, is dealing with qualitative information. Many methods consider qualitative data as pseudo-metric data, but officially forbid it as a way to consider qualitative details. Qualitative multi-criteria methods, in general, have to be survivable from the classification of the actual data. The lack of information in a multi-criteria analysis may emerge from two sources: 1) an imprecise definition of alternatives, evaluation criteria and preferences (or preference scenarios); and 2) an inaccurate measurement of the effects of other options on evaluation criteria and preference weights. One symmetry description is to say that it is the result of a balanced proportion harmony. There is a symmetrical balance when all the parts of the objects are well-balanced [27]. The perfect Yin Yang symbol is a sign of balance, harmony, and moderation. It is all about finding unity amidst duality (Figure 1).

Figure 1. The Yin Yang symbol.

Scientists have proposed many strategies to improve the profitability of industries and apply sustainable production methods [28]. The evolution of design has highlighted the advantages of the principle of symmetry [29]. The balance in humans' duty affects such product conditions as structural efficiency, attractiveness, and economic, and functional or aesthetic requirements. It includes compliance with standardization requirements, production of repeat elements and mass production that reduces production costs [30]. Therefore, symmetry and regularity are generally reliable and symmetrical shapes are preferred but not asymmetric [31].

Besides the methodological developments, there are a large number of successful applications of multiple-criteria decision-making (MCDM) methods to real-world problems that have made MCDM a domain of great interest both for academics and for industry practitioners [32]. Often, different MCDM techniques do not lead to the same results. Multi-criteria utility models are models designed to obtain the utility of items or alternatives that are evaluated according to more than one criterion.

The most popular hybrid MCDM methods demonstrate the advantages over traditional ones for solving complicated problems, which involve stakeholder preferences, interconnected or contradictory criteria, uncertain environment. Decision-makers could use MCDM methods [33] such as the analytic hierarchy process [34], fuzzy analytic hierarchy process [35], fuzzy Delphi [36], analytic network process under intuitionistic fuzzy set [37], additive ratio assessment (ARAS) [38], simple additive weighting, and game theory [39], Discrete two persons' zero-sum matrix game theory [40], evaluation based on distance from average solution (EDAS), complex proportional assessment (COPRAS), technique for order preference by similarity to ideal solution (TOPSIS) [41], as well as develop original models [42]. Decisions made in complex contexts need these methods for practical solutions. Many studies proved the fact that construction materials contribute to sustainable building management [43,44].

The primary features on which depend the effectiveness of a project's life cycle [45] are a selection of proper place [46] and time to implement a plan [47], and to select a decent contractor [48].

The researchers directed to the hybrid MCDM approaches. The right knowledge for supporting systematic improvements evolution of the hybrid MCDM approaches can be characterized by [49,50].

When decision-makers disagree, analysis of decisions can help to understand the situation of each person better, raise awareness of the issues involved and the cause of any conflict. Such improved communication and understanding can be of particular value when a team of professionals from different disciplines meets to make a decision. The analysis of decisions allows various stakeholders to participate in the decision-making process. It is the basis of a common understanding of the problem and makes is more likely that there will be a commitment to ultimately chosen action.

Keeney [51] pointed out that modern decision analysis does not create an optimal solution to the problem; the results of the study can be considered relatively prescriptive. The report shows the decision-maker what he should do, based on the decisions made during his analysis [52]. The central premise is rationality. When the decision-maker adopts rules or axioms that most people consider reasonable, he should give preference to the way they choose alternatives. The actions prescribed in the analysis may contradict the intuitive feelings of the decision-maker. He can then analyze this conflict of analysis and intuition. The study allows the decision-maker to understand the problem better so that his or her preference changes match the analysis priorities. This explains why the reasoned opportunity presented in the analysis is different from the natural choice of the decision-maker.

2. Contributions

Nine original research articles are published in the current Special Issue. Authors from four continents contribute to the papers: Europe, Asia, South America and Africa (Figure 2). Three intercontinental papers are published: two articles co-authored by European and Asian researchers and one document involving European and African co-authors.

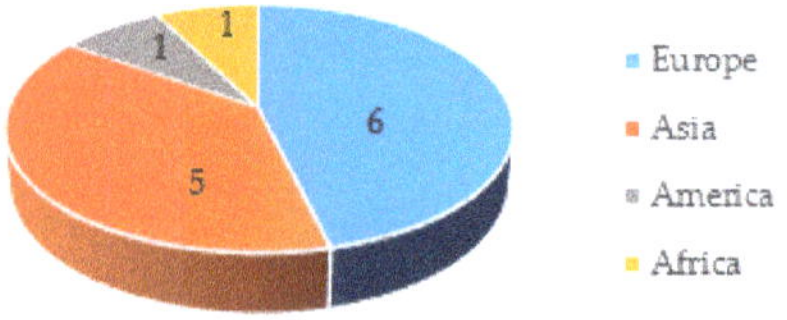

Figure 2. Distribution of papers by countries.

Thirty-seven authors from eight countries contributed to the Issue (Figure 3). The most numerous contributions are from Lithuania, China, Iran, and Romania. Moreover, we received submissions contributed by authors from Bosnia and Herzegovina, Serbia, Brazil, and Libya.

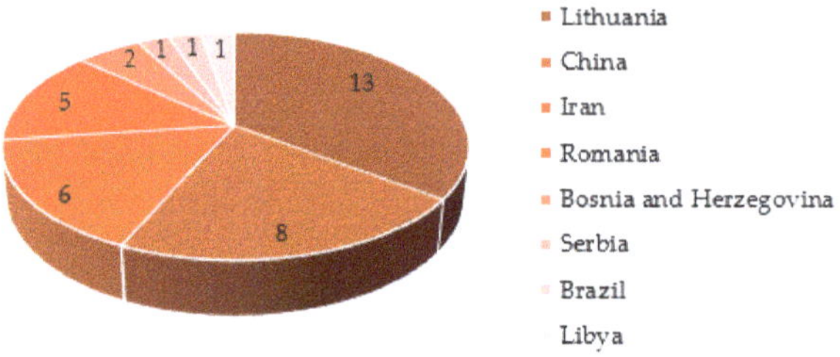

Figure 3. Distribution of authors by countries.

The delivery of papers according to authors' affiliations is presented in Table 1. Co-authors from Lithuania contribute to two papers together with co-authors from China and by one document with Iran, also with Serbia, Bosnia, and Herzegovina, and Libya. The other research teams are not international, and they involve authors from Brazil, Romania, China, Iran, and Lithuania.

Table 1. Publications by countries.

Countries	Number of Papers
Brazil	1
Romania	1
China	1
Iran	1
Lithuania	1
China–Lithuania	2
Iran–Lithuania	1
Bosnia and Herzegovina–Serbia–Libya–Lithuania	1

All the papers suggest solution models based on symmetric or asymmetric information and they contribute to decision-making in various fields of engineering, economy or management. Most of the proposed models include novel or extended MCDM methods under uncertainty. Usual MCDM methods are combined with interval-valued fuzzy sets, rough numbers or Z-numbers. Only one-third of papers published in the current issue does not apply MCDM methods. They contribute to problems related to symmetry by offering other solution models like Bernoulli's binary sequences, repeated experiments or financial models (Figure 4).

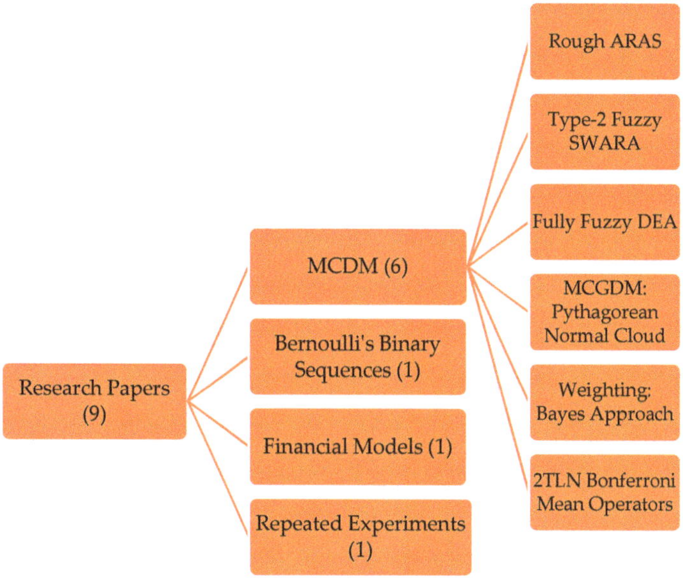

Figure 4. Applied decision-making approaches.

The presented case studies applying the proposed solution models dealing with symmetric or asymmetric information in the technological, economy or managerial problems are grouped into three research areas consisting of 2–4 papers each (Figure 5).

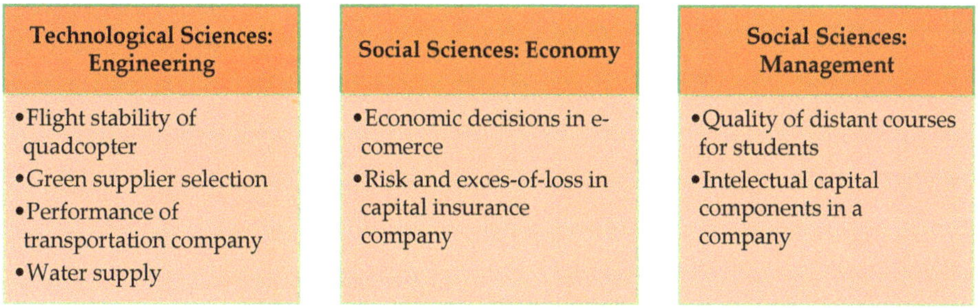

Figure 5. Research areas of the presented case studies.

Grouping of the papers in three research areas as presented in Figure 5 is rather conditional. In many of the research works, the fields are interrelated. The first paper explores water usage by analyzing Bernoulli's binary sequences in the representation of empirical events [53]. The analysis is also related to the economic problem of water usage–expenditure systems.

The next paper analyses the performance of transportation companies [54]. A novel multi-criteria rough ARAS model is developed in the paper. It is applied to companies' evaluation in developing countries. Sensitivity analysis is performed as well as comparison with other methods based on rough numbers is provided. The suggested approach will be further applicable for solving different problems.

Solving the efficiency evaluation with fuzzy data is also analyzed in another paper. The paper presents a new method for solving the fully fuzzy DEA (data envelopment analysis) model where all parameters are Z-numbers [55].

The topic of data fuzziness is continued in the paper aimed at the weighting of criteria in multi-criteria decision models [56]. An extended SWARA (step-wise weight assessment ratio analysis) method with symmetric interval type-2 fuzzy sets for determining the weights of criteria is developed. In the current paper, the suggested approach is applied for importance evaluation of intellectual capital components in a company.

One more paper aimed at the evaluation of weights of criteria proposes use of a Bayes approach for weight recalculation [2]. The core idea of the article is to suggest a plan for combining of criteria weights obtained by different subjective and objective criteria weight assessment methods.

Continuing a topic of data fuzziness, an emerging tool for uncertain data processing, that is known as neutrosophic sets, is applied. Several 2-Tuple linguistic neutrosophic number Bonferroni mean operators are developed [57]. They are applied in models for a currently topical issue of green supplier selection.

The approach partly resembling the TOPSIS (technique for order preference by similarity to ideal solution) method because of considering the symmetry of distances to the positive and the negative ideal solutions, and based on the Pythagorean normal cloud is proposed [58]. Moreover, some cloud aggregation operators are presented. The proposed approach is designated to economic decisions, and an example from e-commerce is presented.

The next paper related to economic decisions does not apply MCDM methods. It suggests financial models for optimal dividend and capital gains problem [59]. A reinsurance case with excessive losses based on risk information is presented.

The last paper representing the field of technological sciences and engineering, analyses symmetrically structured quadcopter and its flight stability [60]. The research focuses on developing a data logger and then applying repeated experiments.

After the above short presentation of research, we encourage the readers to undertake a detailed analysis of the papers published in the Special Issue.

3. Conclusions

The Guest Editors are very happy that the topics of the Special Issue generated interest among researchers from four Continents: Europa, Asia, South America, and Africa. Researchers from eight countries, including three international collectives, contributed to the papers published in the issue.

As could be expected concerning the aforementioned topics, multiple-criteria decision-making models are suggested in two-thirds of the papers. The authors of six articles (from nine articles published) apply MCDM methods in their research. Therefore, we can conclude that multiple-criteria decision-making techniques proved to be well applicable to symmetric information modeling.

Most approaches suggested decision models under uncertainty, combining the usual MCDM methods with interval-valued fuzzy or rough sets theory, also Z numbers.

The application fields of the proposed models involved both problems of technological sciences and social sciences. The papers cover three essential areas: engineering, economy, and management.

Author Contributions: All authors contributed equally to this work.

Acknowledgments: The authors express their gratitude to the journal Symmetry for offering an academic platform for researchers to contribute and exchange their recent findings in civil engineering and symmetry.

Conflicts of Interest: The authors declare no conflict of interest.

References

1. Turskis, Z.; Urbonas, K.; Daniūnas, A. A Hybrid Fuzzy Group Multi-Criteria Assessment of Structural Solutions of the Symmetric Frame Alternatives. *Symmetry* **2019**, *11*, 261. [CrossRef]
2. Vinogradova, I.; Podvezko, V.; Zavadskas, E.K. The Recalculation of the Weights of Criteria in MCDM Methods Using the Bayes Approach. *Symmetry* **2018**, *10*, 205. [CrossRef]

3. Schad, J.; Lewis, M.W.; Raisch, S.; Smith, W.K. Paradox research in management science: Looking back to move forward. *Acad. Manag. Ann.* **2016**, *10*, 5–64. [CrossRef]
4. Zavadskas, E.K.; Turskis, Z.; Antucheviciene, J. Selecting a contractor by using a novel method for multiple attribute analysis: Weighted Aggregated Sum Product Assessment with grey values (WASPAS-G). *Stud. Inform. Control* **2015**, *24*, 141–150. [CrossRef]
5. Chalekaee, A.; Turskis, Z.; Khanzadi, M.; Ghodrati Amiri, G.; Keršulienė, V. A New Hybrid MCDM Model with Grey Numbers for the Construction Delay Change Response Problem. *Sustainability* **2019**, *11*, 776. [CrossRef]
6. Mardani, A.; Zavadskas, E.K.; Khalifah, Z.; Jusoh, A.; Nor, K. Multiple criteria decision-making techniques in transportation systems: A systematic review of the state of the art literature. *Transport* **2016**, *31*, 359–385. [CrossRef]
7. Mardani, A.; Jusoh, A.; Zavadskas, E.K. Fuzzy multiple criteria decision-making techniques and applications—Two decades review from 1994 to 2014. *Expert Syst. Appl.* **2015**, *42*, 4126–4148. [CrossRef]
8. Kahraman, C.; Onar, S.C.; Oztaysi, B. Fuzzy Multicriteria Decision-Making: A Literature Review. *Int. J. Comput. Intell. Syst.* **2015**, *8*, 637–666. [CrossRef]
9. Mardani, A.; Jusoh, A.; Nor, K.M.D.; Khalifah, Z.; Zakwan, N.; Valipour, A. Multiple criteria decision-making techniques and their applications—A review of the literature from 2000 to 2014. *Econ. Res.-Ekon. Istraz.* **2015**, *28*, 516–571. [CrossRef]
10. Antucheviciene, J.; Kala, Z.; Marzouk, M.; Vaidogas, E.R. Solving Civil Engineering Problems by Means of Fuzzy and Stochastic MCDM Methods: Current State and Future Research. *Math. Probl. Eng.* **2015**, *2015*, 362579. [CrossRef]
11. Keshavarz Ghorabaee, M.; Amiri, M.; Zavadskas, E.K.; Antucheviciene, J. Supplier evaluation and selection in fuzzy environments: A review of MADM approaches. *Econ. Res.-Ekon. Istraz.* **2017**, *30*, 1073–1118. [CrossRef]
12. Mardani, A.; Nilashi, M.; Antucheviciene, J.; Tavana, M.; Bausys, R.; Ibrahim, O. Recent fuzzy generalisations of rough sets theory: A systematic review and methodological critique of the literature. *Complexity* **2017**, 1608147. [CrossRef]
13. Yazdanbakhsh, O.; Dick, S. A systematic review of complex fuzzy sets and logic. *Fuzzy Sets Syst.* **2018**, *338*, 1–22. [CrossRef]
14. Rajab, S.; Sharma, V. A review on the applications of neuro-fuzzy systems in business. *Artif. Intell. Rev.* **2018**, *49*, 481–510. [CrossRef]
15. Khan, M.; Son, L.H.; Ali, M.; Chau, H.T.M.; Na, N.T.N.; Smarandache, F. Systematic review of decision making algorithms in extended neutrosophic sets. *Symmetry* **2018**, *10*, 314. [CrossRef]
16. Mardani, A.; Nilashi, M.; Zavadskas, E.K.; Awang, S.R.; Zare, H.; Jamal, N.M. Decision making methods based on fuzzy aggregation operators: Three decades review from 1986 to 2017. *Int. J. Inf. Technol. Decis. Mak.* **2018**, *17*, 391–466. [CrossRef]
17. Turskis, Z.; Dzitac, S.; Stankiuviene, A.; Šukys, R. A Fuzzy Group Decision-making Model for Determining the Most Influential Persons in the Sustainable Prevention of Accidents in the Construction SMEs. *Int. J. Comput. Commun. Control* **2019**, *14*, 90–106. [CrossRef]
18. Govindan, K.; Rajendran, S.; Sarkis, J.; Murugesan, P. Multi criteria decision making approaches for green supplier evaluation and selection: A literature review. *J. Clean Prod.* **2015**, *98*, 66–83. [CrossRef]
19. Mardani, A.; Jusoh, A.; Zavadskas, E.K.; Khalifah, Z.; Nor, K.M. Application of multiple-criteria decision-making techniques and approaches to evaluating of service quality: A systematic review of the literature. *J. Bus. Econ. Manag.* **2015**, *16*, 1034–1068. [CrossRef]
20. Shen, K.Y.; Zavadskas, E.K.; Tzeng, G.H. Updated discussions on "Hybrid multiple criteria decision-making methods: A review of applications for sustainability issues". *Econ. Res.-Ekon. Istraz.* **2018**, *31*, 1437–1452. [CrossRef]
21. Govindan, K.; Hasanagic, M. A systematic review on drivers, barriers, and practices towards circular economy: A supply chain perspective. *Int. J. Prod. Res.* **2018**, *56*, 278–311. [CrossRef]
22. Govindan, K.; Soleimani, H. A review of reverse logistics and closed-loop supply chains: A Journal of Cleaner Production focus. *J. Clean. Prod.* **2017**, *142*, 371–384. [CrossRef]
23. Correia, E.; Carvalho, H.; Azevedo, S.G.; Govindan, K. Maturity models in supply chain sustainability: A systematic literature review. *Sustainability* **2017**, *9*, 64. [CrossRef]

24. Mardani, A.; Jusoh, A.; Zavadskas, E.K.; Kazemilari, M.; Ahmad, U.N.U.; Khalifah, Z. Application of multiple criteria decision making techniques in tourism and hospitality industry: A systematic review. *Transform. Bus. Econ.* **2016**, *15*, 192–213.
25. Keshavarz Ghorabaee, M.K.; Zavadskas, E.K.; Amiri, M.; Turskis, Z. Extended EDAS method for fuzzy multi-criteria decision-making: An application to supplier selection. *Int. J. Comput. Commun. Control* **2016**, *11*, 358–371. [CrossRef]
26. Khanzadi, M.; Turskis, Z.; Ghodrati Amiri, G.; Chalekaee, A. A model of discrete zero-sum two-person matrix games with grey numbers to solve dispute resolution problems in construction. *J. Civ. Eng. Manag.* **2017**, *23*, 824–835. [CrossRef]
27. Hashemkhani Zolfani, S.; Zavadskas, E.K.; Turskis, Z. Design of products with both International and Local perspectives based on Yin-Yang balance theory and SWARA method. *Econ. Res. -Ekon. Istraživanja* **2013**, *26*, 153–166. [CrossRef]
28. Ruzgys, A.; Volvačiovas, R.; Ignatavičius, Č.; Turskis, Z. Integrated evaluation of external wall insulation in residential buildings using SWARA-TODIM MCDM method. *J. Civ. Eng. Manag.* **2014**, *20*, 103–110. [CrossRef]
29. Sousa, J.P.; Xavier, J.P. Symmetry-based generative design and fabrication: A teaching experiment. *Autom. Constr.* **2015**, *51*, 113–123. [CrossRef]
30. Jaganathan, S.; Nesan, L.J.; Ibrahim, R.; Mohammad, A.H. Integrated design approach for improving architectural forms in industrialized building systems. *Front. Archit. Res.* **2013**, *2*, 377–386. [CrossRef]
31. Banginwar, R.S.; Vyawahare, M.R.; Modani, P.O. Effect of plans configurations on the seismic behaviour of the structure by response spectrum method. *Int. J. Eng. Res. Appl.* **2012**, *2*, 1439–1443.
32. Balali, V.; Zahraie, B.; Roozbahani, A. A Comparison of AHP and PROMETHEE Family Decision Making Methods for Selection of Building Structural System. *Am. J. Civ. Eng. Archit.* **2014**, *2*, 149–159. [CrossRef]
33. Ye, K.; Zeng, D.; Wong, J. Competition rule of the multi-criteria approach: What contractors in China really want? *J. Civ. Eng. Manag.* **2018**, *24*, 155–166. [CrossRef]
34. De la Fuente, A.; Armengou, J.; Pons, O.; Aguado, A. Multi-criteria decision-making model for assessing the sustainability index of wind-turbine support systems: Application to a new precast concrete alternative. *J. Civ. Eng. Manag.* **2017**, *23*, 194–203. [CrossRef]
35. Prascevic, N.; Prascevic, Z. Application of fuzzy AHP for ranking and selection of alternatives in construction project management. *J. Civ. Eng. Manag.* **2017**, *23*, 1123–1135. [CrossRef]
36. Chen, C.J.; Juan, Y.K.; Hsu, Y.H. Developing a systematic approach to evaluate and predict building service life. *J. Civ. Eng. Manag.* **2017**, *23*, 890–901. [CrossRef]
37. Shariati, S.; Abedi, M.; Saedi, A.; Yazdani-Chamzini, A.; Tamošaitienė, J.; Šaparauskas, J.; Stupak, S. Critical factors of the application of nanotechnology in construction industry by using ANP technique under fuzzy intuitionistic environment. *J. Civ. Eng. Manag.* **2017**, *23*, 914–925. [CrossRef]
38. Štreimikienė, D.; Šliogerienė, J.; Turskis, Z. Multi-criteria analysis of electricity generation technologies in Lithuania. *Renew. Energy* **2016**, *85*, 148–156. [CrossRef]
39. Kalibatas, D.; Kovaitis, V. Selecting the most effective alternative of waterproofing membranes for multifunctional inverted flat roofs. *J. Civ. Eng. Manag.* **2017**, *23*, 650–660. [CrossRef]
40. Gardziejczyk, W.; Zabicki, P. Normalization and variant assessment methods in selection of road alignment variants–case study. *J. Civ. Eng. Manag.* **2017**, *23*, 510–523. [CrossRef]
41. Bielinskas, V.; Burinskienė, M.; Podviezko, A. Choice of abandoned territories conversion scenario according to MCDA methods. *J. Civ. Eng. Manag.* **2018**, *24*, 79–92. [CrossRef]
42. Keshavarz Ghorabaee, M.; Zavadskas, E.K.; Olfat, L.; Turskis, Z. Multi-criteria inventory classification using a new method of evaluation based on distance from average solution (EDAS). *Informatica* **2015**, *26*, 435–451. [CrossRef]
43. Giama, E.; Papadopoulos, A.M. Assessment tools for the environmental evaluation of concrete, plaster and brick elements production. *J. Clean. Prod.* **2015**, *99*, 75–85. [CrossRef]
44. Van Kesteren, I.E.H. Product designers' information needs in materials selection. *Mater. Des.* **2008**, *29*, 133–145. [CrossRef]
45. Shen, L.; Tam, V.W.Y.; Tam, L.; Ji, Y. Project feasibility study: The key to successful implementation of sustainable and socially responsible construction management practice. *J. Clean. Prod.* **2010**, *18*, 254–259. [CrossRef]

46. Zavadskas, E.K.; Turskis, Z.; Bagočius, V. Multi-criteria selection of a deep-water port in the Eastern Baltic Sea. *Appl. Soft Comput.* **2015**, *26*, 180–192. [CrossRef]
47. Dahooie, J.H.; Zavadskas, E.K.; Abolhasani, M.; Vanaki, A.; Turskis, Z. A novel approach for evaluation of projects using an Interval-Valued Fuzzy Additive Ratio Assessment (ARAS) Method: A case study of oil and gas well drilling projects. *Symmetry* **2018**, *10*, 45. [CrossRef]
48. Hashemi, H.; Mousavi, S.M.; Zavadskas, E.K.; Chalekaee, A.; Turskis, Z. A New group decision model based on grey-intuitionistic fuzzy-ELECTRE and VIKOR for contractor assessment problem. *Sustainability* **2018**, *10*, 1635. [CrossRef]
49. Zavadskas, E.K.; Antucheviciene, J.; Turskis, Z.; Adeli, H. Hybrid multiple-criteria decision-making methods: A review of applications in engineering. *Sci. Iran.* **2016**, *23*, 1–20.
50. Zavadskas, E.K.; Govindan, K.; Antucheviciene, J.; Turskis, Z. Hybrid multiple criteria decision-making methods: A review of applications for sustainability issues. *Econ. Res.-Ekon. Istraz.* **2016**, *29*, 857–887. [CrossRef]
51. Keeney, R.L. Decision Analysis: An Overview. *Oper. Res.* **1982**, *30*, 803–838.
52. Phillips, L.D. Decision Analysis in the 1990's. In *Tutorial Papers in Operational Research*; Shahini, A., Stainton, R., Eds.; Operational Research Society: Birmingham, UK, 1989.
53. Telles, R.C. Geometrical Information Flow Regulated by Time Lengths: An Initial Approach. *Symmetry* **2018**, *10*, 645. [CrossRef]
54. Radović, D.; Stević, Ž.; Pamučar, D.; Zavadskas, E.K.; Badi, I.; Antuchevičiene, J.; Turskis, Z. Measuring Performance in Transportation Companies in Developing Countries: A Novel Rough ARAS Model. *Symmetry* **2018**, *10*, 434. [CrossRef]
55. Namakin, A.; Najafi, S.E.; Fallah, M.; Javadi, M. A New Evaluation for Solving the Fully Fuzzy Data Envelopment Analysis with Z-Numbers. *Symmetry* **2018**, *10*, 384. [CrossRef]
56. Keshavarz-Ghorabaee, M.; Amiri, M.; Zavadskas, E.K.; Turskis, Z.; Antucheviciene, J. An Extended Step-Wise Weight Assessment Ratio Analysis with Symmetric Interval Type-2 Fuzzy Sets for Determining the Subjective Weights of Criteria in Multi-Criteria Decision-Making Problems. *Symmetry* **2018**, *10*, 91. [CrossRef]
57. Wang, J.; Wei, G.; Wei, Y. Models for Green Supplier Selection with Some 2-Tuple Linguistic Neutrosophic Number Bonferroni Mean Operators. *Symmetry* **2018**, *10*, 131. [CrossRef]
58. Zhou, J.; Su, W.; Baležentis, T.; Streimikiene, D. Multiple Criteria Group Decision-Making Considering Symmetry with Regards to the Positive and Negative Ideal Solutions via the Pythagorean Normal Cloud Model for Application to Economic Decisions. *Symmetry* **2018**, *10*, 140. [CrossRef]
59. Yan, Q.; Yang, L.; Baležentis, T.; Streimikiene, D.; Qin, C. Optimal Dividend and Capital Injection Problem with Transaction Cost and Salvage Value: The Case of Excess-of-Loss Reinsurance Based on the Symmetry of Risk Information. *Symmetry* **2018**, *10*, 276. [CrossRef]
60. Kuantama, E.; Tarca, I.; Dzitac, S.; Dzitac, I.; Tarca, R. Flight Stability Analysis of a Symmetrically-Structured Quadcopter Based on Thrust Data Logger Information. *Symmetry* **2018**, *10*, 291. [CrossRef]

symmetry

MDPI

Article

Geometrical Information Flow Regulated by Time Lengths: An Initial Approach

Charles Roberto Telles

Administrative Sectorial Group, Continuous Services Sector, Secretary of State for Education of Paraná, Curitiba 80240-070, Brazil; charlestelles@seed.pr.gov.br

Received: 26 September 2018; Accepted: 6 November 2018; Published: 16 November 2018

Abstract: The article analyzes Bernoulli's binary sequences in the representation of empirical events about water usage and continuous expenditure systems. The main purpose is to identify among variables that constitute water resources consumption at public schools, the link between consumption and expenditures oscillations. It was obtained a theoretical model of how oscillations patterns are originated and how time lengths have an important role over expenditures oscillations ergodicity and non-ergodicity.

Keywords: probabilistic systems analysis; nonlinear dynamics; public management; pattern formation; resources distribution; population sizes; information theory; oscillations

---

1. Introduction

When considering a large number of administrative agents within a public institution, several aspects of how those agents execute public services must be considered to establish a proper analysis of public expenditures and public budgets [1]. Those aspects can be attributed, for example, to how public services are affected by variables set in a complex environment. Policy making, expenditure provision, unstable budgets, and government attempt to organize a big and very complex system, commonly, found limitations due to the multilevel of aspects that influence the provision of services to society in any public sector [2]. In this article, it will be considered one example of the public provision of service of water, and its correlation between the usage by public schools and the random behavior of expenditures presented with this continuous expenditure.

A public institution, due to the multiple relations with distinct sectors of society, assumes a complex organization of how expenditure is produced [1]. To control water expenditure, it is necessary to understand the variables under this context and starting from it, understand how information unit flow among variables that compose this service. This article points an approach that analyzes variables maximum oscillations inputs and outputs in their possible interactions, iterations, frequency of iterations, and time regulations. These analysis outputs can be used as a tool for predictive management towards a massive number of administrative units that are commonly found in big countries, cities, or states.

The main positive result of this type of analysis is the variety of paths in which it is possible to interact with a public expenditure. For example, in order to reduce the expenditure with water usage at schools, uncommonly, the problem goes beyond the administrative scope, flowing into docent activities, students, parents, local community, and society as a whole. For this solution, it is required, for example, to start a water saving campaign in the schools that constitute one of many options that compose the great variable's scenario of how water is used by nowadays life. But how much the frequency in which students consume or use water is relevant to the public expenditure? Or which variables affect this event to be considered relevant in order to have the best optimum control of public provision of service and expenditures? Is there a method that makes it possible to predict risks about messing with one of those variables? For these questions, this article addresses the possibility of adopting information flow as a measure of the expenditure system level of randomness by a theoretical view a methodological proposal.

A very similar approach to this article's scope of investigation can be found in Buchberger and Wu [3] where time-dependent Markovian queuing system is used to estimate temporal and spatial variations of the flow regime of water demands at one block of heterogeneous homes. However, this model is not suitable for the large number of variables and sample size when analyzing a public system of a state, city or country. This same argument can be addressed and confirmed by own author's studies, Guercio, Magini and Pallavicini [4] where, the basis method of Buchberger and Wu are carried out adapted to larger sample sizes. In this research [4], 85 residences of four blocks were analyzed extending the scope and methodology to be applied in larger samples sizes. Thou, heteroscedasticity was not considered since only homogeneous samples of users for the estimation of water usage were used as parameter.

In analyzing the article "Methodology for Analysis of Water Consumption in School Buildings" [5], it can be seen that the flows of oscillations present in continuous expenditures were related directly with water usage at public schools. Using coefficient of determination-R^2 to verify water usage characteristics among 149 public educational establishments of Secretary of State for education at Paraná, it was hypothetically confirmed that the water usage is influenced by different variables that can assume different intensities and categories at the different regions of the State with 199.315 km^2 of area, 51 water utilities, and 2274 water distribution points (schools and administrative units). In this way, public schools present a large variety of features (variables, Figure 1) [2] that make difficult for managers to set a unique method to determine how to manage water consumption or other natural resource types that is of use in other public provision services. Therefore, the proposed methodology in the article [5] of knowing in large samples how water consumption in schools occurs through linear time series of analysis and coefficient of determination-R^2, trying to extract universal indicators that can serve as a reference for the whole State, was shown to be limited, due to impossibility of evaluating and predict for future time how consumption behavior will be expressed.

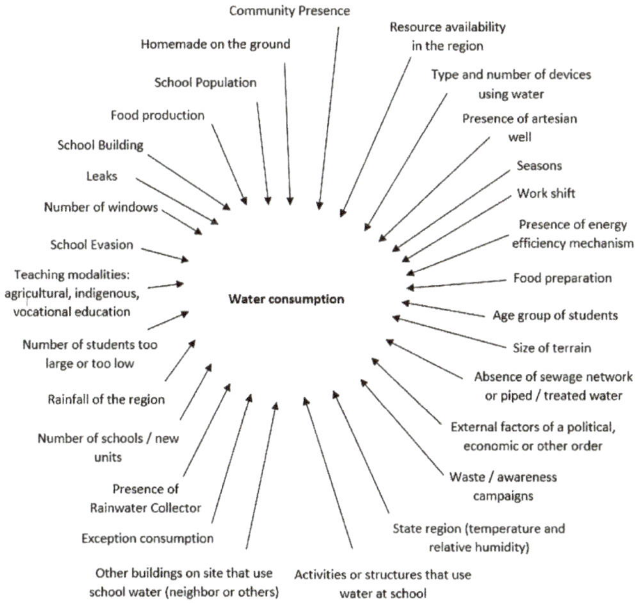

Figure 1. Variables affecting water consumption at the public schools of Secretary of State for Education of Paraná, Brazil [2].

Generally, indicators, such as school population size (per capita model), are used to estimate urban public provision of services, however, according to Figure 2, population factor itself can affect the consumption of water in a very smooth way, sustaining a continuous growth of consumption that accompanies the population, but does not indicate a direct ratio between population and consumption of the resource in a directly proportional order (Figure 2). The intensity in which the school population influences the consumption of water is not proportional in a quantitative aspect, and thus it is assumed that other variables exert their own internal mechanics in the event and generates modifications in the dynamics of the system [5,6].

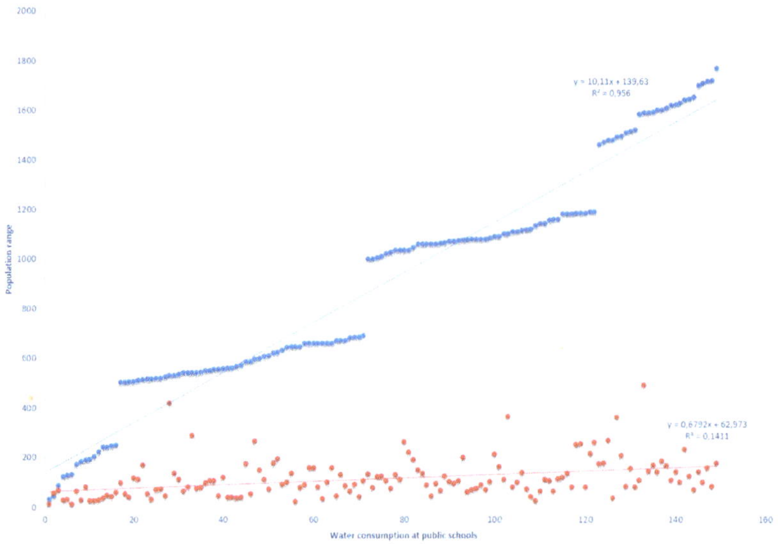

Figure 2. Water usage at public schools. The data consist of 149 schools at different regions of the Paraná State with a population of $u = 133.783$ individuals. Using R^2 (determination coefficient) for both data, the linear function presented for the chosen population data can't be equally found in the water usage behavior.

2. Methodology

In more recent studies, linear time series and cross-checking of variable's categories were performed [6,7], in a sequence of analysis started with Guercio, Magini, and Pallavicini [4] and Buchberger and Wu [3] in order to identify intensities and coupling effects of variables among them, or through coefficient of determination-R^2 [5], but results, subjectively understood in the graph lines, may not be sensitive enough [8–11] to a management analysis with a view to controlling the system by producing intervention actions on the variables with each other, or in isolation, or estimating what results will be possible with precision when interfering in certain processes of the event.

The per capita model, in which quantitative indicators, for example, % of the population, are used as objective parameters whose purpose is to describe the behavior of dependent variables in a system. For example, the relation between population variable and water usage per individual. Therefore, per capita model states that the larger the population, the greater the water consumption, and if not observable by Figure 2, this statement could remain as the best optimal approximation to measurement of water usage by population, assuming the cross-checking of variables [6,7] already under consideration. Although, there are other variables that influence the system so that the population factor does not have enough force to produce high vibration in the system, enough to generate the observed maximum and minimum oscillations in the consumption (Figure 2) [12]. One empirical experiment [13] for predator-prey dynamics, understanding it as resources-consumption by analogy, observations lead to the conclusion that large population and no controlled dynamics have not a per capita model explanation, enough to sustain oscillations in the ratio-dependence between resources availability and consumption. Also, the per capita model, according to Dahl, Bhattacharyya, and Timilsina [9,11], is a reduced method of investigation when the variables in the system present dynamics, therefore, assuming heteroscedasticity form.

From this brief historical point of view, methodologies to estimate water usage in buildings have shared main concepts, evolving towards new knowledge, and mainly serving as empirical methods for specific analysis. This feature allows for water usage estimation to be investigated only by real situations, not giving a glance of a possible theoretical view of the problem. In other words, a method in which oscillations of water usage can be predicted by a theoretical view, composing administrative knowledge of private and public organizations. This knowledge could cause positive effects for planetary distribution of resources at every dimension of human organization, being it residences, commerce, industry, cities, states, and countries [11].

In Figure 2, it is possible to see ranges of population in blue (μ = 0–200; 500–700; 1000–1200; 1400–1700) from public schools at State of Paraná, Brazil. Those ranges are compared to water consumption in red and the objective of these data was to arrange a discrepant rising for population variable intending to see if water consumption follows population increases proportionally. The observable result was an influence of population sizes at water consumption in a very smooth way. In this sense, using demographic bases with per capita methodology as management indicators might be imprecise and controversial criteria [14], since, in fact, there may be hidden variables [15] that influence the final set of the event.

For better methodological results and risk analysis in this article, it would be more accurate to analyze the variables of a system in ideal sense rather than a realistic data as Figure 2. The reasons to opt for it is to investigate if there is a prior organization of the event influenced by other aspects, in which it becomes possible to make approaches on the quantitative aspect of frequency [16] with which the variable population and water consumption interact, excluding other traditional data correlation. Though, the frequency aspect of consuming water by population was not considered in an exact sense (real time for each trial), but an ideal model reflecting the binary mode system of Bernoulli's method of analysis. Following this way, it is possible to indicate possible sequences of interactions between variables more than the set of intensities in which the variables express themselves [8,17], trying to see if organization of elements in the event through time assumes more relevant outputs regarding expenditure oscillations, than only considering the quantitative aspects of how much water each individual consumed. This approach is not suitable for Granger causality techniques due to particularities of variables and samples range of variance that can reach a broad output and heteroscedasticity [9,11,18] (Figure 2—red data).

The analysis of this article brings not to trivial ways in calculating by the method of Bernoulli the probabilities of an event in occurrence at first, but the analysis of static parameters of information (deterministic starting condition) inside a system of linear binary sequences, being this last characteristic investigated relative to the number of iterations, frequency, and time of which can result in many possible expressions when the function of time distorts variable's expressions. Binary trajectories do not express probabilistic modifications through time regarding the presence or absence of variables, but in the model given, express frequencies distortions, which may lead to new properties of information flow and probabilistic time dependent variables. In this way any binary system with the same mathematical starting conditions already reflects the same methodology that was developed in this article and the main objective for this that is to analyze oscillations of systems by a multivariate and intuitively stochastic model based on numerical information that was extracted from Bernoulli sequence method [19,20].

The main approach to this method for dynamical systems will be shown as how oscillations of panel data (Figure 2) can be caused by time regulation flows at small events (microstructure) that compose the entire system (macrostructure). Leading to posterior pathways regarding entanglements of other small events [16]. In this way, chaotic behavior could achieve equilibrium and a freeze phase state of patterns in variables that promotes oscillations [16]. It means also in other words, to regulate flows of information by time, understanding it as to regulate the order of iterations in its expression's frequencies. For this reason, linear time series method is useful only for checking data variance on time, but do not constitute as a method for problem issued by this article, since the phenomenon is not an expression of iteration of time to be measurable, but a frequency in which iterations assume on time, a specific order (pathways) [13,16,21].

Presume for the non-ergodicity of data in Figure 2 for water consumption, as well it follows analogously for the expenditures at a public institution, as caused by the smallest time flow of information for each pathway (variables affecting the system in their coupling functions) and its frequency of iteration along it. In this sense, the flow of information in the system assumes geometrical or non-geometrical properties possibly periodic by time lengths. For the observer, consume seems to be random, but for the internal movements of sequences, the geometrical properties of variables can be extracted when considering an information theory approach. Population and consume is one example to be issued in this article, but not limited to it. The aim of analyzing oscillations by information theory and Bernoulli sequences is to indicate whether the numerical information can be used as a tool for predicting the behavior of stability or instability of expenditure systems rather than probability density functions or queue theory. This way it becomes possible for the manager the decision making process and management purposes.

2.1. Theoretical Framework of Experiment

As a model of analysis, the microstructure of events that compose water expenditures will be represented by a single framework, in which concepts can be analyzed regarding how modifications at the microstructure can affect the macrostructure as well. Considering a main concept of the theory of information, in which low probabilities of events contain more information than high probabilities [22], the methodology demonstrates how in an ideal model, variables assume a Bernoulli binary entropy information modifying its possible probabilities and ergodicity due to the time aspect of the event and the flow of information. The theoretical experiment can result in the following situations according to Figure 3.

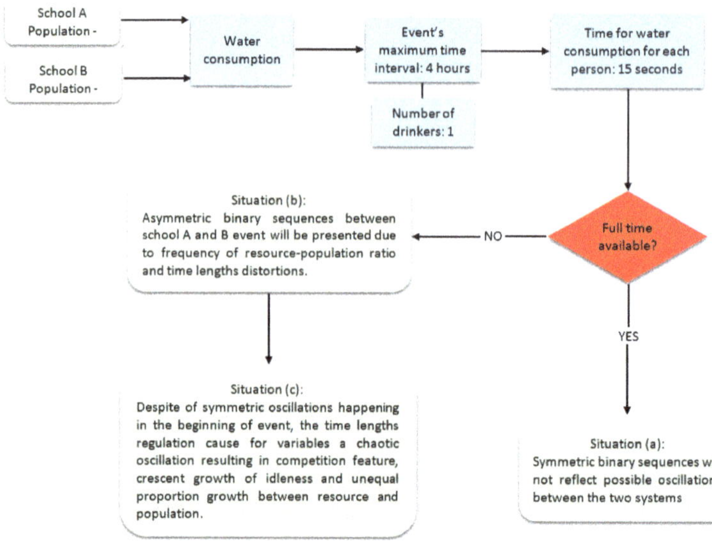

Figure 3. Flowchart of Section 2. Methodology, showing theoretical experiment of information regulated by time length's effects on water consumption and school's population.

In order to problematize the effect of maximum entropy of information [23] in two systems with equal starting conditions (Figure 3), when considering the model of analysis in which two schools work in only one shift with time schedule of 4 h. Which difference exists between the two schools relating to the size of their population μ (number of students) as a function of water consumption, for every 15 s? This theoretical experiment will describe different analyzes, considering interactions between one by one individual and resource availability through time. Time is considered as the main factor in which iterations (frequency of interaction between variables in quantitative aspect) exert more influence on the behavior of the variables than their probabilities from binary sequences [23]. To start the theoretical investigation, it was selected two sample sizes for analysis. They present five times unequal proportions: school A, $\mu = 200$ and school B, $\mu = 1000$, and the number of drinkers in each school: 1.

2.1.1. Information Flow and Ergodic Properties

Theorem 1. *Considering the systems as ideal and not possible of having non-observable variables that directly influence consumption behavior. For both systems (school A and B) for the first 15 s (Y), there is the possibility of a student consuming the resource* (x_1)*, and the opposite time and space effect that none will consume it* (x_2)*, (without possibility of the same individual consuming again). Indicating the consumption of resources by an individual as "1" and not consumption, "0", in a given number K of dependent iterations (a function of time), the constant flow of variables is contained in the time, as it is indicated in the following Equation (1):*

$$P^{y \to k}(x_1(1,0) \cap x_2(0,1)) \tag{1}$$

After the 15 s end, there is the second expression of the system as a potential possibility of another student repeat the starting condition. Resulting as,

Lemma 1. *Given the probabilities, the events (variables) of success p and failure q, are considered as* $\boldsymbol{p = x_1}$ *and* $\boldsymbol{q = x_2}$*, where the variables p and q are dependent and not identically distributed (not iid), and we get the following probability of the event:* $\boldsymbol{q = 1 - p}$*. The odds of the event following the given probability can be set as:*

$$x_1 - 1 = x_2 \text{ with 50\% chance and } x_2 - 1 = x_1 \text{ with 50\% chance} \tag{2}$$

Proof of Lemma 1. Since, the system has a geometrical property, for all n binary vectors $(x_1, \ldots, x_n)$, it obtains 100% odd for any time length as Equation (3):

$$\begin{gathered} P[x_1 = i_1, \ldots, x_n = i_n] = p_{i,\ \ldots,\ i_n} \\ \sum_{i_1=0}^{1} \cdots \sum_{i_n=0}^{1} p_{i_1,\ \ldots,\ i_n} = 1 \end{gathered} \tag{3}$$

□

However, the main concern of this article is not to calculate the probability of the event on time (Y), but to verify the behavior of the event from the point of view of the binary sequences and information entropy as a method that makes it possible to visualize the event in its information characteristics, such as oscillation properties. When the probabilities between the two systems are identical and not observable in Bernoulli's method in terms of probability density function regarding the behavior of the two systems as compared to each other in the function Y, it can be of use considering the frequency in which variables x_1 and x_2 have their iterated behavior regulated by the time. Therefore, to conceive the analysis as information entropy [22] of the system, the variables considered assume an evolution on time in bits [24].

As time T_k passes, there is a growth of the variable x_1 and x_2 revealing binary sequences that repeat cumulatively and asymmetrically on time length $(T_k \to \infty)$, according to Table 1.

Table 1. Bits distribution over time.

Time. Variables	T_1 X_1	T_2 X_2	T_3 X_1	T_4 X_2	T_5 X_1	T_6 X_2	T_7 X_1	T_8 X_2	...
Bits	1	0	1	0	1	0	1	0	...
	0	1	0	1	0	1	0	1	...

The sequences originated by bits of information and distributed by variables with geometrical properties can be described by the next Equation (5) and Figure 4:

$$\lim_{y \to \infty} (x_{1t} + x_{2t}) = x_1 + x_2 \quad (4)$$

where, in other way it can be represented as a combination of variables defined as: $0_n C_r(T_k \to \infty) = (p) + (p-q) + (p-q+q) + (p-q+p-q) + (p-q+p-q+p)$, and so on.

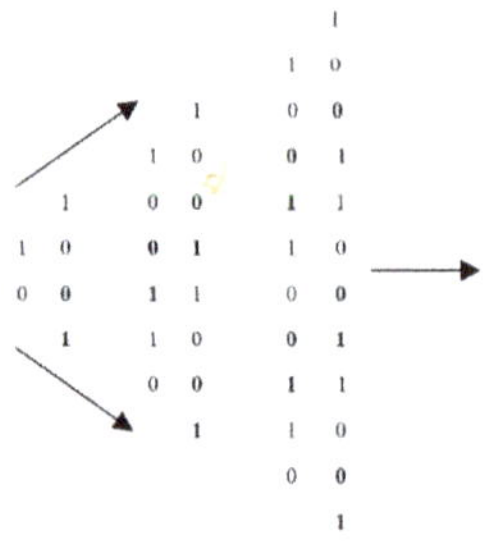

Figure 4. Schematic of bit evolution over time.

Proof of Theorem 1. The probability of the infinite sequences of iterations set to happen in values 0 and 1 is constant as k becomes infinite and remains always in the given proportion of 50% [25]. where,

$$\begin{gathered} P\ (Y = 0 \text{ and } 1,\ \forall Y) \\ \leq P\left(T_1 = \cdots = Y^k = 0 \text{ and } 1\right) = P^k \\ \lim_{k \to \infty} P^k = 0 \text{ and } 1 \end{gathered} \quad (5)$$

□

Figure 4 can be represented by the Figure 5 where both variables assumes in ideal condition an evolution on time presenting constant probability and oscillation. Regarding the odds of 50% for the sequence evolution, for realistic conditions, it is not correct to assume that value neither for the theoretical investigation of this article, in which it was considered a constant flow of variables performing 100% constant sequencing. For that point, the article describes the event for the most ideal behavior of variables (deterministic) when considering it as the parameter, a model, in which for real life situations, analogously, it is possible to exclude the interference of multiple other variable's effects from event, as seen in Figure 2, whose presence in it have the punctual potential of affecting the final results and it will be shown in a later Section 3. Results, that frequency, iteration, and time can be more deterministic in organizing the event than other external variables of it.

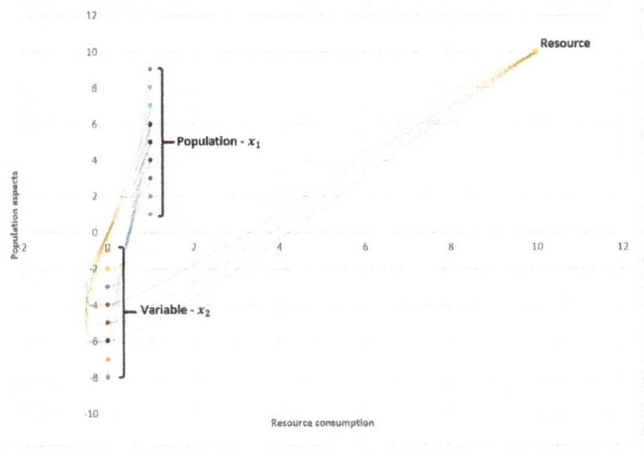

Figure 5. Deterministic evolution of water consumption by population over time.

In Figure 5, the x axis represents the resource consumption trajectory while y positive axis represent population interaction (x_1) with resource and negative y axis, the variable x_2. x_2 is expressed in the same manner as x_1 in the time function Y and as a necessary effect caused by resource consumption and individual interaction. The information sequence, from the left to the right, in which the variable x_1 and x_2 are expressed reveal common periodic oscillations (constant patterns of event starting conditions), in which regardless of the size of the sequence, the results will always be the same. Thus, the maximum information entropy [23] of the system is finite or infinitely constant, but asymmetric on time length regarding the alternated presence of variables x_1 and x_2 set to happen. The evolution of iterations (Y) perform a periodic oscillation or as a geometric variable L, consisting of a constant odd of events on time, whatever the time length chosen from the sequence.

Theorem 2. *Following this path, the ratio of x_1 to x_2 is shown as increasing, but asymmetric in the length of Y function as a geometric variable L:*

$$x_2\,(0,1) \propto\, x_1\,(1,0) \qquad \forall_{x1,x2}\, \in\, Y(x_1 R x_2 \rightarrow\, \neg\,(x_2 R x_1)) \tag{6}$$

In this sense, it is possible to affirm that both the population of $\mu = 200$ and $\mu = 1000$ will have consumption and idle time, defined by Y, in which the variable x_1 and x_2 will not have different expressions of probability and maximum entropy of information for both populations. Otherwise, a result is obtained where time affects event influencing number of sequences to happen defined by the following Equation of Cauchy [26]:

$$Y = f(x_1 +\, x_2)$$

where,

$$\propto\, \cap Y$$

Produces dynamical properties in the event as:

$$f(Y\!+\propto) - f(Y)$$

Resulting in,

$$Y = f(x_1 +\, x_2) \text{ and } x_1 +\, x_2 = f(Y) \tag{7}$$

The geometric variable L affects infinitely every time $T_k \rightarrow \infty$, expressing turn shifts between variable x_1 and x_2. The geometric variable L expresses no probability functions, except if determined by Y, in which this article aims to associate with resources management within a system as a simulator of resources distribution management. This is represented as:

$$L + 1 = geometric\ p$$

where L is equal to,

$$Y = f(x_1 + x_2) + (x_1 + x_2) = f(Y) = 1 \tag{8}$$

The probability of L_k is equal to the variables x_1 and x_2 in its probabilistic expressions as follows:

$$\begin{array}{c} +1 = x_1 - 1 = x_2 \text{ with 50\% chance and} \\ L - 1 = x_2 - 1 = x_1 \text{ with 50\% chance} \end{array} \tag{9}$$

Proof of Theorem 2. For variable L with geometric distribution of $p = 1/2$, where, $P(L = n) = pq^{n-1}$, $n \in \{1, 2, \ldots\}$; the entropy of L in bits is: [27]

$$\begin{array}{rl} H(L) = & -\sum_{n=1}^{\infty} pq^{n-1} \log(pq^{n-1}) \\ -\left[\sum_{n=1}^{\infty} pq^n \log p + \sum_{n=0}^{\infty} npq^n \log q\right] & \\ = & \frac{-p\log p}{1-q} - \frac{pq\log q}{p^2} \\ = & \frac{-p\log p - q\log q}{p} \\ = & H(p)/p \text{ bits.} \end{array} \tag{10}$$

If $p = 1/2$, then $H(L) = 2$ bits. □

Some patterns are produced as constant features of the system. They are the probability and entropy of x_1, x_2 and geometric p. But, other variables influence the event, the number of iterations (Y). This effect, as represented in Figure 6, counts towards amount of resources available and population size. Y can be represented for managerial purposes as a controlling tool in which types of resources management can be achieved.

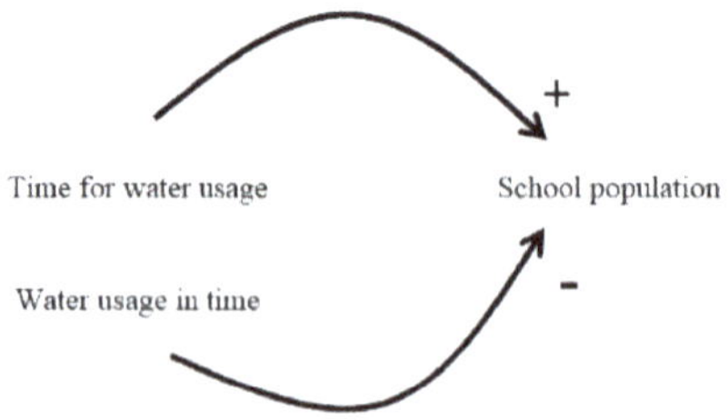

Figure 6. Nonlinearity effects caused by time according to Figure 7.

2.1.2. Information Flow and Time as the Cause of Oscillations

Considering the event in the starting condition as a linear system, and consisting of two dependent random variables, with memory and probabilities in maximum finite or infinite lengths, constant and equal to 1/2 for both variables (stationary process). The expressions of the possible trajectories remain constant in sequences that are repeated alternating the presence and absence of one of the variables in each iteration (asymmetric). There are constant oscillations in the event (geometrical variable), except if the variables x_1 and x_2 are regulated as a function of time Y. It is observed that the variables x_1 and x_2 assume on time $T_k \to \infty$ specific behaviors (non-ergodicity) that can be used as management tools for random systems (nonlinearity). In this way, a complex model for population sizes and natural resource distribution was obtained, sustained by concepts of iteration, frequency, and time regulations.

Analyzing the non-oscillation properties of any event by the theoretical framework, the management of resources and population can assume distinct effects on time, types I and II of information flow, according to Figure 7. Variable A = resources; Variable B = population.

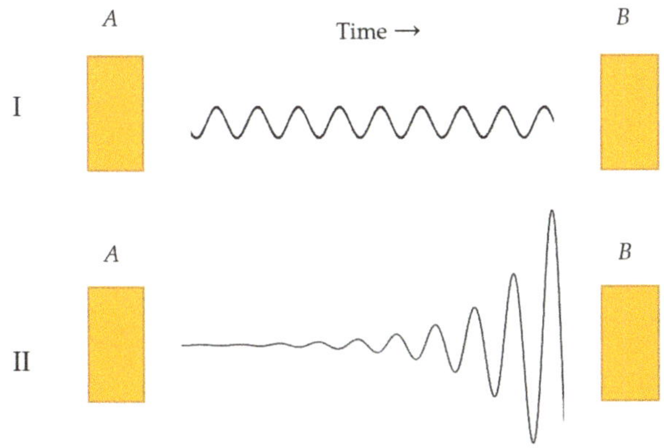

Figure 7. Information flow at geometric variables (I) and not geometric variables (II). A: Resources and B: Population.

Coupling effects of variables x_1 and x_2 towards variables A and B, regulated by time Y and with the geometrical properties of the axiomatic conception of variable x_1 and x_2.

It means that the two bits of information, in ideal condition, can be controlled by an external variable (time) without changing the maximum output of oscillations due to the geometrical property of variables (Figure 7). For other types of entropy information, different behaviors will be observed. Two bits are easier to regulate with time than, for example, 15 bits due to low variance and length of variable's distributions. Entropy information flow in the example given remains constant as time passes. However, it is possible to control the distribution of information (resources flow among interactions) in the given system for arbitrary inputs and outputs [16]. The effects of regulating the event through time can cause specific effects for the phenomena. Therefore, for managerial purposes of this article, the amount of information distribution can be influenced by coupling conditions (time length or other dependent variable) with small or big intensities, making possible to obtain low risks and optimal control concerning the flow of resources in a given set of elements that constitute the event, being this flow understood by how variables increase in interaction's frequencies as time of phenomena goes infinite (see Figure 3). The effect of it for real systems is observed for the coupling functions regarding bit distribution and real system frequency quantitative aspects [8].

In contrast to the nonlinearity properties, if the time be considered in terms of short or long duration (Y, the number of iterations), it is possible to affirm that the larger the school population, the lower the water consumption on time (Figure 8), in an effect of increasing the frequency with which the variable x_2 will be present in the system. Larger population (considering interaction process active, not counting idle population) generate more void spaces (variable x_2) and soon extend the resource consumption over time when compared to smaller populations. In the problem in question, a variable x_3, defined as the number of drinkers in the place, will affect the dynamics of the event, however, although there were two or 10 drinkers in the place, still the system has its behavior, as already described. The difference in the increase in the number of drinkers is at the rate of frequency with which the water resource is consumed and the increase of frequency with which the variable x_2 of the system also expresses itself reducing its effect due to the large number of drinkers. For management purposes, it is possible to reduce the number of drinking troughs to reduce resource consumption or increase as needed. It is to be considered that a large number of drinking fountains are inefficient, as idleness in the system is a constant and not very large quantities of drinking fountains would be required to provide water for an entire population of, say, 1000 students for a better distribution of the resource.

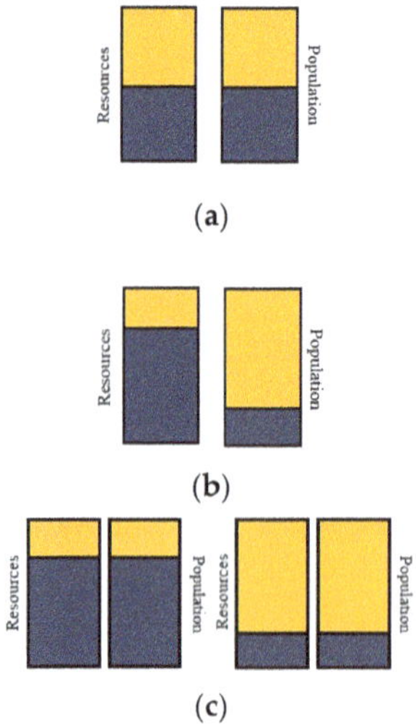

Figure 8. Oscillation's quantitative aspects due to information distribution regulated by time lengths. (**a**) Equal amount of variables A and B $\therefore Y \rightarrow \infty$, where $\uparrow Y \uparrow B \uparrow X_1 \uparrow X_2 \downarrow A$ for optimal resources distribution. (**b**) Time length for event where amount of variable A is not used fully by B caused by finite time $\therefore Y < \infty$, where in this case, $< Y < B < X_1 < X_2$ AND $\uparrow A$, being A not consumed in time given. Resources wrongly distributed. (**c**) Time length for event where amount of variable A is limited for use caused by B variable X_2 presence $\therefore Y < \infty$ where $\uparrow B \uparrow X_2 \uparrow A \uparrow X_1 \downarrow A \downarrow X_1 \downarrow B \downarrow X_2$. Resources containment and population-resource chaotic regulation.

The main objective for the manager is to work with the risks and uncertainty of the system in order to analyze how the system expresses itself and to have the best decision making [14]. The example that is described in this article illustrates situations that are present when large numbers of public management agents regarding the administrative and financial scope for several types of provisioning services are considered to achieve the best optimal solution for distribution, containment, or reduction of resource consumed. However, methods considering the linearity constant among variables that compose the expenditure systems might fall into false results, due to oscillations present in the system and information flow's frequency and time aspects [28,29]. In this way it is possible to use the results of analysis in maximum entropies of information on identical systems as an indicator of how to operate the system's variables.

The Figure 6 represents the behavior of variables x_1 and x_2 as a function of Y and not as a function of probabilities.

The Figure 6 indicates that the larger the population, the longer the time for water consumption and the Y function, in the opposite direction, the larger the population, the lower the use of water on time. This conclusion will be explained in Section 3 of article.

Consider now the use of the resource by individuals, with the possibility of repetition, in other words, it is possible that an individual will ingest water again. Thus, it is concluded that the higher the school population, the lower the water consumption, in an effect of increasing the frequency with which the variable x_2 will be present in the system in a certain time of analysis and not proportional to the number of different individuals who will have access to the resource (y negative axis beyond value $x_2 = -10$, see Figure 9). Following this situation, Figure 9 shows how projections of population-resource ratio will be increased, also expressing saturation in the system towards the time available for all the individuals. It is an example of suboptimal ratio among variables if compared with Figure 5 resource trajectory lines, which may cause initial oscillations regarding binary sequences flow.

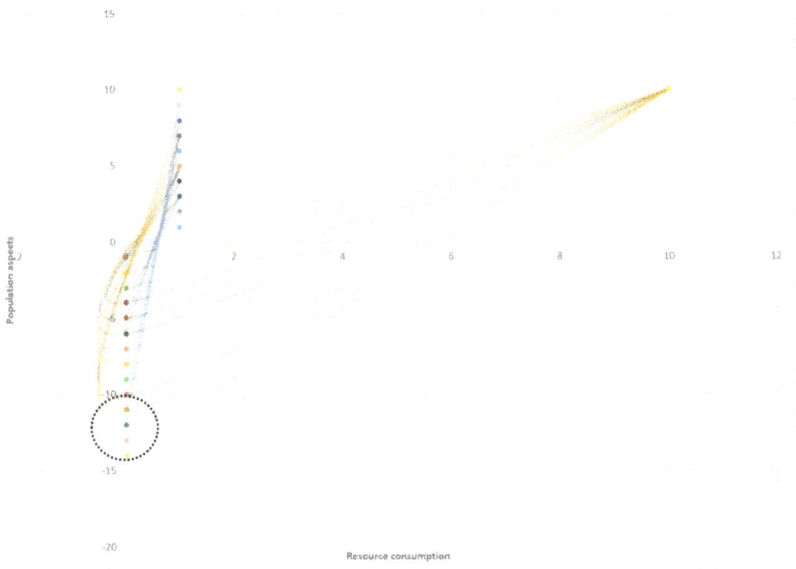

Figure 9. Water consumption and individual repetition. Observation: the lines are colored for the benefit of graph visualization.

When it is defined that oscillations can be caused by the frequency with which the variables in a system interact, it means that, in addition to the example of Figure 9, there are other more complex interactions that promote great fluctuation in the continuous expenses of a public institution. One important interpretation of the Theorems 1 and 2 is that the amount of water consumed in buildings, population or else, cannot be interpreted as a final expenditure value, but it reflects more to a budget. In other words, if consume is caused by variables affecting the system, it is not possible to assume an expenditure (continuous and previously planned financial resource) as a direct reflection of water usage, but rather, as a reflection of how iterations, interactions, frequency, and time are leading the system's network. If manager does not see internal features of the system, water, or other sort of resource distribution will not be appropriately achieved and on the contrary, budgets are annually being generated.

3. Results

For this section, there are items, ordered as (a), (b), and (c), and are represented also by Figure 5 (methodology section).

(a) No matter the size of sample, if time has short length, systems have no influence of order towards elements in a given set.

In real situations, short time intervals that are available for students, of about 15 min at our hypothetical schools A and B, have no influence on water usage if comparing each other, no matter how much population it has. This can be caused by the low amount of information flowing in the system for the entire sample. Since time interval is short and variables require naturally some time to be expressed (individual-resource interaction), low amount of interactions will be obtained due to time maximum interval, reflecting very tenuous oscillations to occur. It means the oscillation's effects of the two systems are nearly zero if compared against a long time run that would cause enough time to variables x_1 and x_2 express frequency features. Also, another feature of population sizes is that, since the binary sequences were shown to be constant for both populations of $\mu = 200$ and $\mu = 1000$, in the beginning of event, and, relatively, the binary sequences of each group will not express variations between them caused by order invariance effect, in which, as showed in Figure 10, repeated iterations don't differ from itself in ordinal response. But, express high oscillations properties if iteration assumes frequencies regulated by time lengths. In consideration of a real approximation of the event, in which there are different values in the time of water consumption and idle time of x_2, and after it, the results will be redirected to a nonlinear system in which properties will be demonstrated in item (b) and (c).

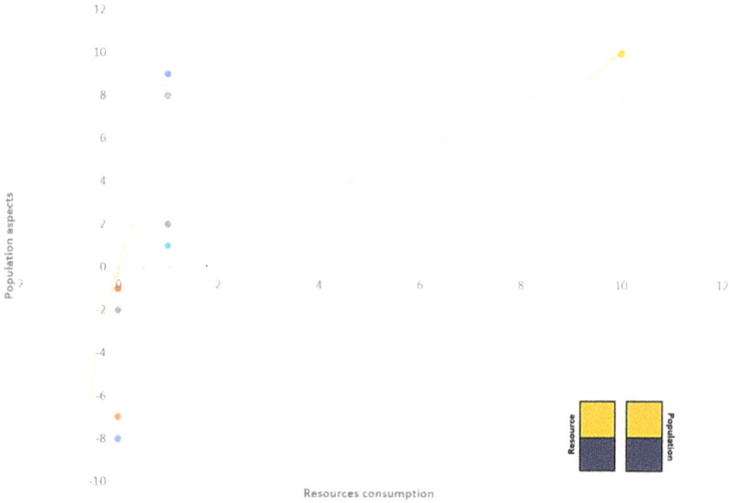

Figure 10. Representation of population-resource ratio analyzed by continuous time length. Iterations order can't affect the system if time is continuous. Another effect would be expected if a time interval interrupts the flow of variables, leading the system to an insufficient distribution of resources as described in item (b) and (c). Observation: the lines are colored for the benefit of graph visualization.

However, it is possible to assume that not all phenomena have any influence of order in the entropy outputs. In this particular case, the article explores this effect, in an ideal simulation. On the other hand, for all possible sequences of expression between variables in a system, there are possible paths in which each variable assumes specific aspects in ordered repeated iterations [16].

Non-Ergodicity

(b) In a given event in which manager needs to make a proper distribution of resources for all elements in a set, the proportion of resources needs to be equal to the number of elements of the set, however, if the time of event is relatively short to provide enough length to the number of element's correspondences, the resources are not going to be equally distributed among all elements. Incorrect management of resource distribution will probably be reported.

Phase space formation: the provision of a given resource for the population will have lower quality than expected if the objective is to provide a resource for the largest number of participants.

The assertion can be understood by the aspect of the analysis of the binary sequences of the event, in which, as there is s frequency of resource consumption, there is also the frequency of idleness for the consumption [21], which are added together and generate the impossibility that in short time length, large populations can consume a certain amount of resource.

To correct this effect, it is necessary to extend the maximum time of the event or to increase the speed of the variables that participate in the logistics of distribution and supply of the resource to the individuals.

The next Figure 11 represents the limiting situation described before, in which at a given time limit, the population consumes a certain resource and part of the population will not have access to the resource (interaction effect and not per capita aspects) due to lack of time or variables that influence the logistics of the system.

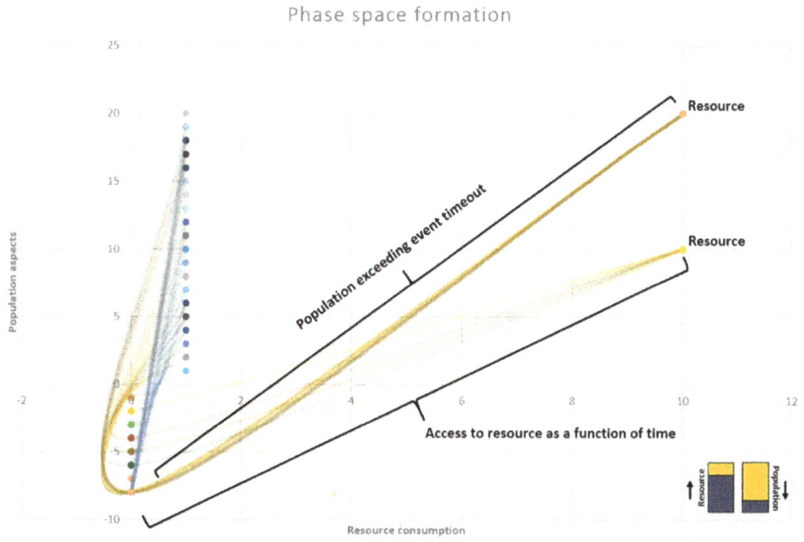

Figure 11. Flow of water consumption and population size regulated by finite time interval. Consider in the graph a maximum time of permanence in the place (lower resource trajectories) in which the population size demands more time to obtain full correspondence between population and resource ratio. The flow of resources is not reached to full system size due to the time given. Observation: the lines are colored for the benefit of graph visualization.

(c) In a given event in which manager needs to make a cut in the resource's distribution for all elements in a set, the proportion of resources can be lower than the number of elements of the set, however, it needs to adequate time of events in a way all elements and their expressions are not going to be able to supply themselves in the given time. Competition or lack of supplies for system' elements will probably be reported.

Phase space and time influence: Following the previous analysis, the difference lies in the logical proposition that instead of being necessary for the participants to consume the resources, the aim is non-consumption. Thus, the larger the population in a given location, the less time available for everyone to consume the resource equally, when considering for this the non-modification in the variables that provide the logistics of distribution and supply of the resource in the system. This example can be seen in Figures 12 and 13 where the situation was simulated for observation. Note that, in this system, there are flows of resources and population in continuous growth. Despite harmonic oscillations happening in the beginning of the event, the time regulation cause for variables a chaotic oscillation due to competition feature, crescent growth of variable x_2 and unequal proportion between resource and population in general. This scheme can be seen in Figure 8.

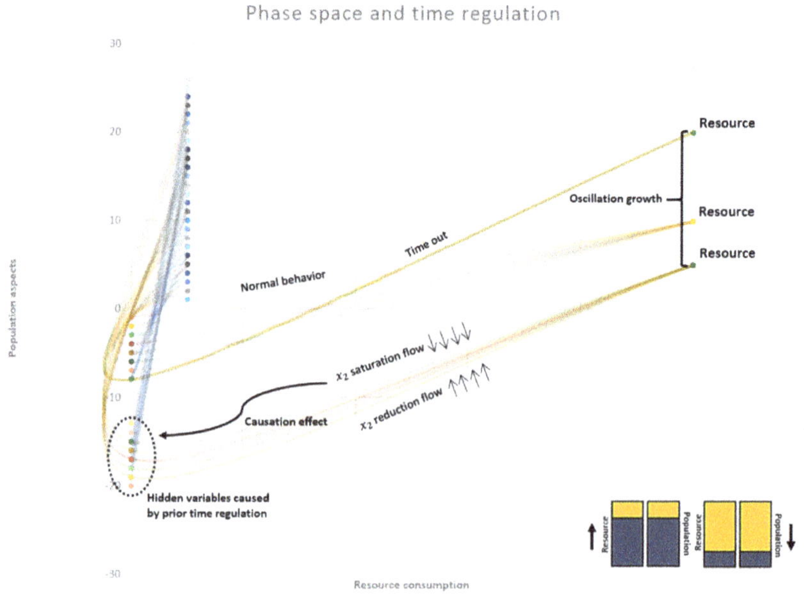

Figure 12. Deterministic to chaotic behavior of variables regulated by time lengths. Observation: the lines are colored for the benefit of graph visualization.

Note that every type of phenomenon has its own hidden variables, leading to specific causation effects. In the example given, organisms present life related characteristics to be expressed, such as competition, mutualism, commensalism, predation, parasitism, and other multidimensional features related to physical, chemical and biological properties.

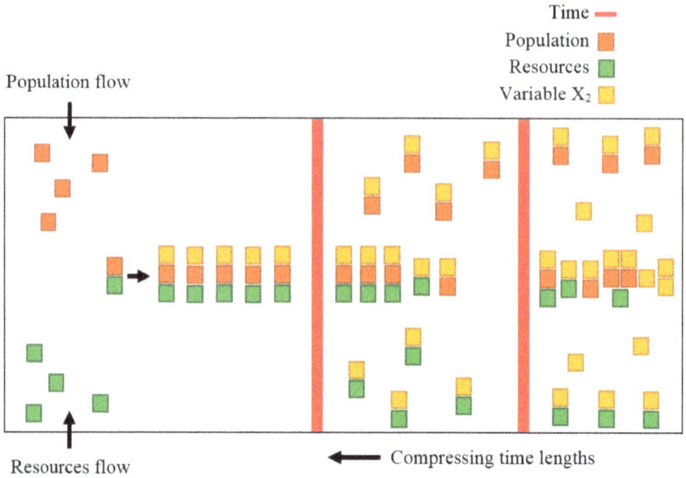

Figure 13. Scheme for resources and population dynamics regulated by time lengths. Possible results obtained through iteration, frequency and time over variables x_1 and x_2. It is theoretically postulated that time lengths have specific effects over the event, causing specific phase space's trajectories.

The analysis of information by means of maximum information input and output in binary Bernoulli sequences in this sense exposed reveals that the ratio between variable x_1 and x_2 increases as time passes, but not necessarily in the same proportion due to external time regulation and other coupling effects (hidden variables). For realistic conditions, time is not determined for the variable x_2. Therefore, it is concluded that there is no direct proportion of water consumption, the number of students and the time available for water consumption, since the non-consumption idle time (beyond the own existence of variable x_2 void effect) exists and is expressed indefinitely in the system, removing from the final result of the system possibilities of prediction on the previously treated question that the population directly affects the consumption of water in its quantitative aspect only. This count not only for repetition feature, but other characteristics mentioned for Figure 12.

When it is taken during observation the effect of time lengths within the investigated event, much more than the quantitative aspect of the population, it is the frequency (length of event-time regulation) with which the variables interact that generate distortions. That is for management purposes a possibility to modify in nondeterministic flows, the ideal geometrical property, turning the system into a nonlinear event, which is observable to the manager in a theoretical and realistic way, leading him to decide which ways to opt for intervention effects on managing risks regarding optimization for resource distribution or containment in a system [14,18].

For better visualization of phase portraits and theoretical description of the experiment, it will be represented in Figures 14 and 15, for example, a population-resource ratio situated in ideal condition of a restaurant. In Figure 14, the ratio is relatively constant, and in Figure 15, time regulation takes effect, modifying the event organization.

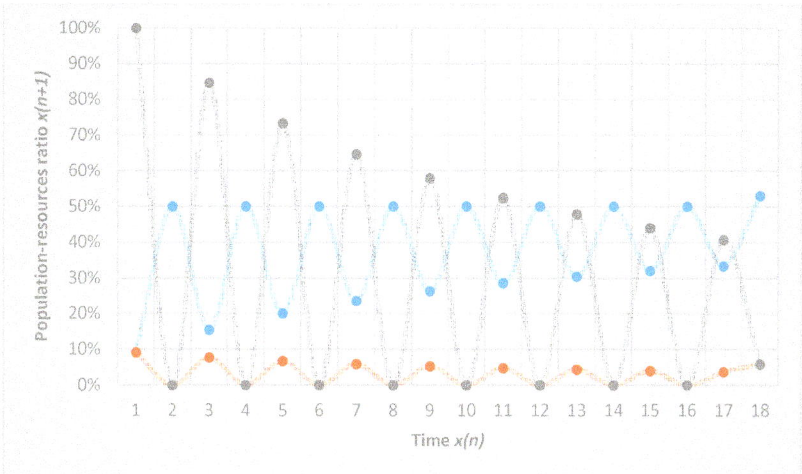

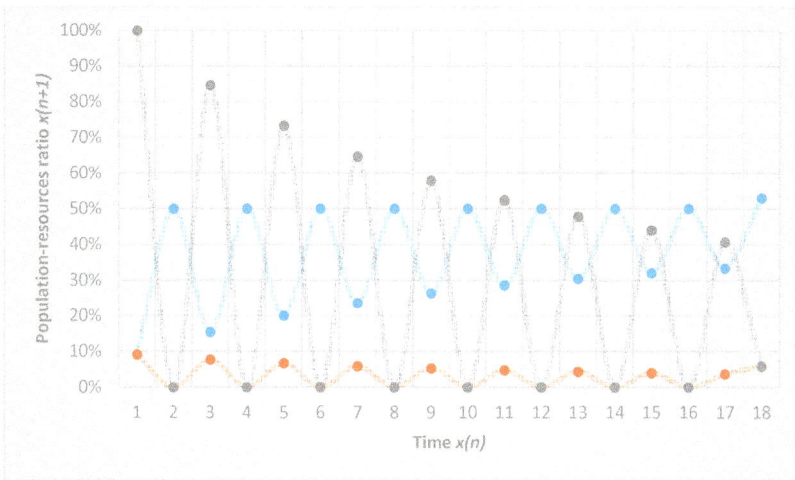

Figure 14. Constant binary distribution of population-resource ratios. Imagine a restaurant where the brown line represents the queue of individuals. Blue line, the variable x_2 and orange line, resources. As time passes, individuals at queue start getting access to resources, as waiting time stay relatively constant and resources are consumed in the same proportion of individuals in the queue.

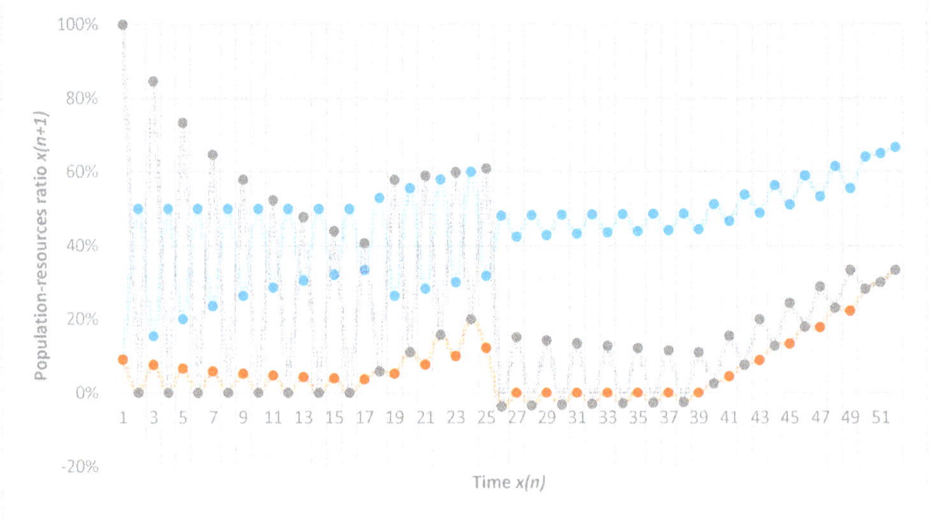

Figure 15. Time regulated dynamics of population-resource ratio. It is possible to observe the oscillations of variables in the system as time passes. Blue line, the variable x_2, orange line, resources and light brown line, population.

This Figure 15, with the same situation described in Figure 14, at the time of 17, time runs over to access resources, and it is observed an increase of individuals at queue as well variable x_2 and resource availability. In this point 27, hidden variables start expressing through the system, such as competition, mutualism, or other sorts of social behaviors. This effect causes the individual queue to be reduced by stress condition or positive association between individuals. As queue line reduces, resources are over consumed at a given short time proportionally by the number of individuals feeding at the same time. Variables x_2 remain relatively constant due to the new form of "queue organization". Starting from 39, time axis, another state occurs in the event. As resources start running out, individuals have to wait for new provision, raising variable x_2 in the same proportion as queue organization starts to be formed again.

As variables of the system start oscillating, resources seem to follow this direction of influence. The next Figures 16 and 17 are presenting two distinct situations of population-resources ratio. First, Figure 16, the proportionality of variables is stable, assuming the same values of 1 and 0 for the same y axis as each unit of time passes. There is an increase of proportion between the two variables x_1 and x_2 at the same rate.

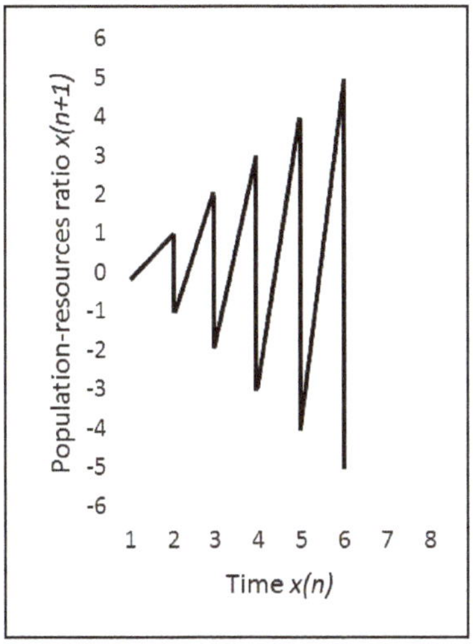

Figure 16. Time series of population variables x_1 and x_2 expressing proportionality for population-resource ratio.

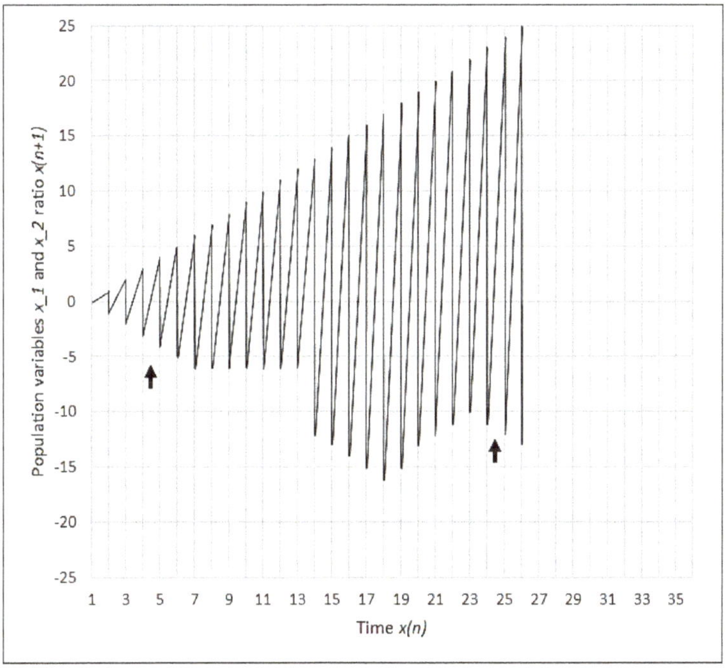

Figure 17. Time series of population variables x_1 and x_2 expressing disproportionality for population-resource ratio. The asymmetrical pattern shows recurrence at original state (indicated arrows) and time regulation equilibrium.

The Figure 17 shows the opposite situation where deterministic to chaotic behavior starts forming. It is possible to see at Figure 17 the oscillations starts similar to Figure 16, and when time intervention starts, population keep increasing with time, but variable x_2 ceases to grow. This effect is attributed to the interruption of interaction between x_1 and resource. As soon as hidden variables are triggered by time regulation, the flow of population keeps increasing in the same ratio, but is accompanied by an abnormal growth of variable x_2, caused by the saturation of individuals in the locality. This new configuration points out to the expressive reduction of resources due to a large number of individuals consuming as well cease of the state due to resource scarcity. After resource attractor ends activity, the system gets back to the original state (indicated arrows at Figure 17) and it can possibly keep flowing with the same features as far as all elements of the system are present as starting conditions set.

In Figure 18, exploring the view of Figures 16 and 17, binary values are displayed in Cartesian graph. Circles in black represent the raising of binary values at y axis starting from 0 to 1 and -1 and present continuous growth. This view is a little coarse, but it can give a glance of event phases and evolution.

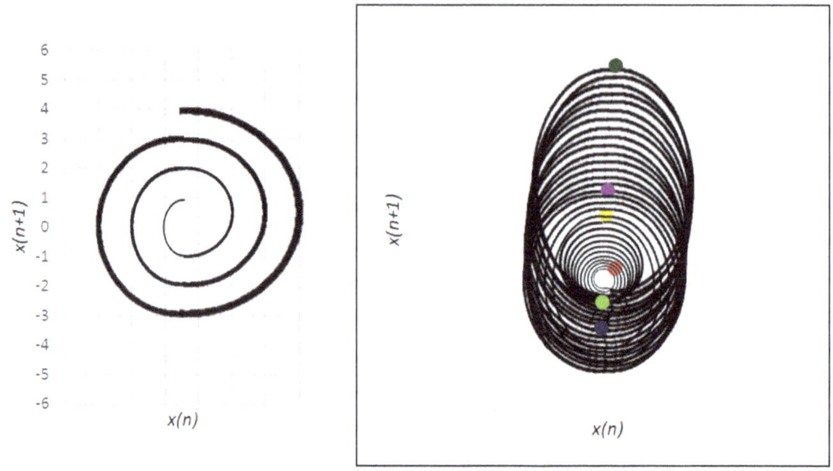

Figure 18. To the left, the representation of Figure 16 and to the right, Figure 17. In this view, it is possible to see specific phases of event of Figure 17 that are marked with color circles. Red circle, event start. Light green, time regulation. Blue circle, x_2 saturation. Yellow, chaotic phase. Purple, population growth and variable variable x_2 reduction. Dark green, recurrence of variable x_2.

Note that at second arrow in Figure 17, as resources are continuously available, flowing keeps repeating the same configuration. In the case of resources or population goes decreasing this second arrow state keeps reducing as previous state until it finds the zero point plot. Another consideration about variables distribution is about the recurrence of binary distribution when only population decreases. Consider this recurrence effect as instead of population goes increasing, its number after variable x_2 start reducing, and it decreases with the same proportion of variable x_2. In this situation, the chaotic event formed before will be dissolving into the population and variable x_2 reduction. These descriptions can be observed in Figure 19.

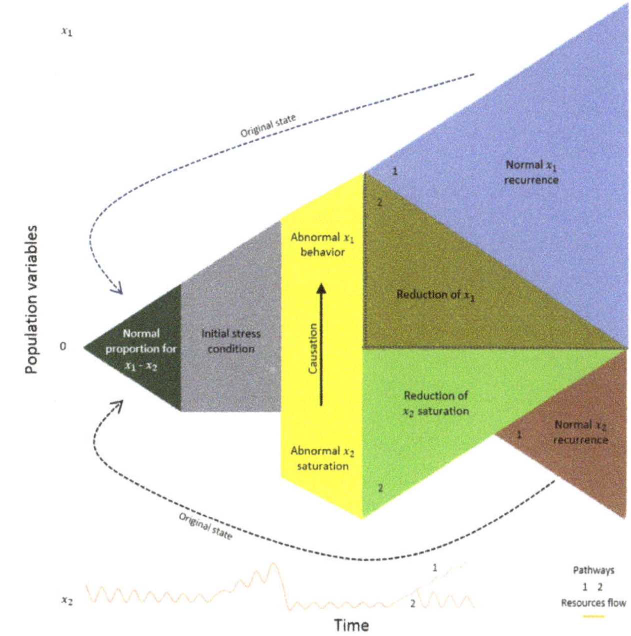

Figure 19. Evolution of system dynamics. Population and resource ratio are represented in two possible pathways. (1) Variables and resources recurrence to the original state. The amount of resource available at pathway 1 is proportional to the population previous aspects. In this case, resource amount is higher than the original state amount. (2) Pathway 2 leads to the end of the event. It is expected for the resource time series at bottom of Figure 19, a constant reduction of information flow until it reaches 0 (zero for both variables interaction).

4. Discussion

The scope of the article relies not only on resources consumption, managerial or risk assessment for administrative, or financial aspects of public administration but other issues in which analysis of information is set by the conditions exposed in the methodology section. For this analysis, the main proposal for future research is to deal with the flow of information in a nonlinear model system regulated by time aspect. Also, different views about the issue can be addressed, not only by information theory, but other disciplines with a variety of possible dimensions of analysis. If possible to control the flow of information by creating chaos and deterministic features in the same event, as the number of bits remains static or can have patterns of formation, deformation, defined phase spaces, it is possible to adjust the entropy through time lengths resulting in many possibilities of control and assessment of a sample towards physical, chemical and biological dimensions, considering it as a broader suggestion of analysis for the specific conditions stated for solving article issue problem.

The main proposition of considering time as a tool to regulate entropy flows is addressed to the aspect of how oscillation's behavior of variables can express in a phenomenon, and it reveals how entropy flows within the system and how the evolution of process can be forwarded or having a reversible state, to the containment or better distribution rules desired to be achieved. What if a duality based phenomena or other chaotic systems can be sustained by binary based events and have regulations caused by time lengths? [3,4,9,11,13,16,22,28,30,31] The intervention at binary based thermodynamics scope is obtained as far as its expressions can be time regulated and dependent on the specific internal logic of variables interacting within the system, and it means by axiomatic reasons, making it possible to dismantle (control) the binary valued orientation output of interactions in terms of flow directionality for multiple variable based events.

Regarding duality based phenomena, the possibility of reversibility of a particle state of energy can be attributed to time effects as well [30], where the amount of energy accumulated through sequences is delivered into partitions due to time limits at sequences formation (event internal logic and elements). Those approximations lead to a conception of starting and ceasing the particle-wave function sustained by energy flows, oscillation maximum output (entropy), and time regulations for pathways optimizations [21].

It is observable, for example, an empirical research about information theory for Risk-based Water System Operation [31], where forecasts for rainfall and water level measurement are time dependent. However, the results points for a perfect prediction for small periods of time, it confirms time as a factor that disrupts predictability regarding behavior of variables rather than potential information entropy [16]. When considering this thesis as a confirmation of the fact that thermodynamics concept given by Shannon [32] assumes, for these results of forecasting, the small effect in which each binary based interaction event tend to give several unpredictable outcomes due to diversity of pathways, but, nonetheless, those features are time regulated [3,4,9,11,13,16,19,22,28,30,31,33–38], and phenomena of this kind are promoted by the flow of sequences among one unit to another one, composing the whole system as far as it has enough time to express its potential phase spaces.

One empirical example that is very similar to this article, but does not offer a mathematical and a theoretical model of the problem presented, is the experiment [33] where resources that are consumed by bacteria are regulated and it modifies aspects of bacteria competition, evolution and survival. When experiment makes resources a case of availability and flows, it acts as a time regulation of events, giving results that are expressed in this article as found in "Nonlinearity from deterministic conditions" section. However, variable's properties (internal logic as mentioned in the last paragraph) differ between considering human water consumption or bacteria resources interaction, but the results and the event as a whole have time regulated dynamics analogously. Results overview can be observed in discussion section [33] and by the theoretical framework obtained through time regulation method, it is possible to anticipate future discoveries that could be derived from the experiment with bacteria [33] or with other investigations. It means iteration (number of organisms and variables expression), frequency (of iteration), and time are capable of disrupting any results obtained regarding survivability, evolution and competition of a life form. This statement is a suggestion for researchers of this field.

Another study very related to the approach that is presented in this article is about nutrition and scarlet fever mortality correlation in the 19th century [34], where it was observed wheat prices oscillations causing consequent mortality oscillations in the same ratio. According to information sequences that are regulated by time lengths, the correlation between variables can be predicted by investigating the maximum oscillation output in which frequency of iterations exhibit caused by time influence in the micro event of the disease-nutrition ratio. Possible other variables of biological origin will count towards the results making oscillation output be wider than expected.

The main objectives for future researches in sequence to this article results should be focused on control of oscillations in a given flow of information. One main aspect of oscillation observations can be attributed by the concept of transfer entropy and asymmetric information flow at different scales [20], where, when comparing different samples of resources distribution and population sizes, the effect of a time interval as described in this article will cease the constant probability distribution of variables x_1 and x_2, resulting in the chaotic behavior of variables, as was analyzed in Figures 12 and 13. The transfer of entropy among different samples could be described by appropriate methods in the previous condition adopted in this article example. Also, the proposed model analyzed gives a general view of binary system functioning for realistic events, system main concepts that organizes the events (iteration, frequency, and time regulations), a model for information theory analysis, probabilistic system analysis, empirical experiments analogous to the example given in this article, information entropy flows, and specific researches to be carried out in future time, such as:

- microorganism's reproduction flow [33], infection speed [34];
- vector based diseases [35];
- expenditures management and analysis of massive administrative units (agents) [2,28];
- water and electrical energy distribution over massive administration units commonly found in public administration of cities or countries [11,28,31];
- theoretical and empirical model for resource distribution considering containment, reduction, or regulation of resources production Vs consumption [11,13];
- logistics aspects of administration and commerce flow [36];
- mathematical model for entropy, geometry, and time regulations for nonlinear dynamic events [16,21];
- engineering of materials [37];
- civil engineering or related areas [3,4,6,7];
- language processing regarding space and time accelerations or other biological aspects [22];
- metaphysical concepts for physical and mathematical investigations regarding iteration, frequency and time [15,20];
- deterministic to chaos event's investigations [21,37];
- duality based and quantum based phenomena, such as economics and particles and other fields of knowledge related to the issue [30]; and,
- computing, networking, and communications [38].

The use of the Poisson distribution is not possible due to logical requirements of method and the ideal condition in which the method is presented, that is the opposite conception adopted by Guercio, Magini, and Pallavicini [4]. The consumption of water resources in the given example of this article is orientated by the continuous trial of events that are dependent of the previous trial, resulting in the failure of the method proposed if observed from the view of Poisson's method. It could be possible to describe the same phenomenon researched in this article considering a realistic system for the event using Poisson distribution and check if the results go to proximity of characteristics described before, since in realistic analysis the trials are not given by the ideal condition used in this article that considers every trial in the sequence of happening without delays or null time-valued sequence. However, the main task of the article is to show how event even if perfectly scaled in proportion (geometrical properties) assumes nonlinearity that is caused by time, originating a higher flow of information, no geometrical properties, diverse information patterns, probabilistic densities function, and as a result, high oscillation of phenomena expressions. These features are very similar to realistic systems involving natural resources consumption and population sizes or others nature events. In this way, it is possible to confirm realistic system's equilibrium in chaotic randomness from previous conditions that is achieved when the same properties of ideal condition (deterministic) analyzed in this article express specific behaviors regarding variable's distribution under time effects of length. This way, geometrical or not geometrical variables can be controlled in possible expressions by analyzing trajectories of each variable among themselves along finite or infinite iterations (time regulation).

It is possible to observe, for example, the use of queue theory that is associated with time regulations, in which, the pathways are designed with probabilistic density functions after time regulations oriented the intended information flow process to be achieved. This confirms the hypothetical framework of this article relating the basis potential entropy of systems and remodeling methods through time in order to extract queue theory best approximations [3,38].

For an open event, it means that there are flows of variables that increases with time and for a closed event, where variables in the initial condition are the same in the end for any given time (fixed number of samples), there are two frameworks of analysis considering the issue the article investigates:

(a) proportionality between/among variable's distribution (for open/closed system)—linearity

    - General management of deterministic flows methods; and,

(b) disproportionality between/among variable's distribution (open system)—nonlinearity

- Time regulated dynamics.

5. Conclusions

The analysis of information flow by binary sequences of Bernoulli regulated by time lengths can be of use to predict the behavior of nonlinear systems in a continuous way, allowing for the manager to visualize the events in their particularities of oscillations behaviors. This way, low amount of information in a non-linear system, as Shannon already stated, gives high probability and high amount of information, gives low probability. This statement can be forwarded accordingly to time regulations to investigate oscillation properties of systems of which in low amount of information, the time effects are unable to produce high oscillation properties. On the contrary, high amount of information that is associated with time regulations, can result in high output of oscillations. The analysis does not take place as an exhaustive methodology for understanding these types of events. However, logical propositions arising from such analyzes can be useful for planning, monitoring, and controlling complex systems in order to reduce the estimated risks for these events. Phase space of event investigated in this article was no performed. Though it was mentioned in results section, further studies are required. It is suggested to study behavior of event in its recurrence plots and establish more specific relations with subject fields of knowledge.

Another conclusion arising from article experiment is to demonstrate how non-ergodicity of variables that compose water consumption (microstructure) at schools is responsible for non-ergodic oscillations in water expenditures (macrostructure). The information flow found in water consumption at schools is very high in terms of interactions, iterations, frequency of iterations, and time expressions. These properties lead to a high level of oscillations in water usage, leading as well to an unstable degree of water consume from each one of the 2000 public schools. As a symmetric result, the instability that is produced in water usage reflects with instability of expenditures observed for all the public provision of service. This relation was discussed as a possibility of using these type of analysis as a tool (indicator) for managers in order to be able to infer in complex systems of expenditures, how to reduce unwanted effects, such as continuous growth of expenditures, chaotic behavior of the system, uncertainty and summarizing it, very suboptimal public administration, and the use of tax money.

Also the article scope of investigation brings a glance of how time is related to phenomena where diverse variables that are located in a chain of events, in which, probabilistic functions can assume an evolution of density and retrocede in its own properties regulated by time lengths. These densities evolutions promote together with other variables, increasing margins of possible outcomes in a complex chain of events, giving the whole view of deterministic to the chaotic control of events in duality transformations for each pathway of the event. In this way, it is possible to regulate through time the frequencies in which iterations assume the main role of possible oscillations in the event, thus reducing the non-ergodicity of the system as a whole. This statement was performed in theoretical view, using water-population ratio of interaction, iteration, frequency of iteration, and time regulations. It is considered as an approach for future researches with the same characteristics of this study. The objective to finish this approach is related to not only use the last paragraph statement as an indicator for a manager, but also have the possibility of measure the dynamics of all variable's information flow leading to the capability of manipulating events and therefore, increasing the precision and margin of possible outcomes.

The last, but inferred conclusion, at first can be considered as an unobservable factor, the analysis of conditions of which a system is, when the number of iterations through time, affects the expression of possible outcomes (pathways) of the event changing its internal probabilities or entropy flows. But, in this sense, the flow of information is a function of interaction, iteration, frequency of iterations, and time regulations effect rather than potential entropy of information presented in the observed event or variables at their starting conditions. This view can lead to the new possibilities of controlling deterministic to chaotic behavior in nonlinear phenomena. Following this line of phenomena observation, simulation methods can describe how resources demands arise from complex systems, being able to predict by simulation the oscillation levels of each dynamic variables that compose an investigated event. These observations results are priority actions needed for administrative provision of services and other related areas, giving the possibility to the manager to avoid uncertainty and aim at the best optimal outputs in the deep complexity environment.

Funding: This research received no external funding.

Conflicts of Interest: The author declares no conflict of interest.

References

1. Eppel, E.A.; Rhodes, M.L. Complexity theory and public management: A 'becoming' field. *Public Manag. Rev.* **2018**, *20*, 949–959. [CrossRef]
2. Telles, C.R.; Cunha, A.R.B.D.; Chueiri, A.M.S.; Kuromiya, K. Analysis of oscillations in continuous expenditures and their multiple causalities: A case study. *J. Econ. Adm. Sci.* **2018**, 34. [CrossRef]
3. Buchberger, S.G.; Wu, L. Model for instantaneous residential water demands. *J. Hydraul. Eng.* **1995**, *121*, 232–246. [CrossRef]
4. Guercio, R.; Magini, R.; Pallavicini, I. Instantaneous residential water demand as stochastic point process. *WIT Trans. Ecol. Environ.* **2001**, *48*, 10. [CrossRef]
5. Telles, C.R. Metodologia para Análise do Consumo de Água em Edificações Escolares. *Parana J. Sci. Educ.* **2017**, *3*, 1–10. (In Portuguese)
6. Blokker, E.J.M.; Vreeburg, J.H.G.; Van Dijk, J.C. Simulating residential water demand with a stochastic end-use model. *J. Water Resour. Plan. Manag.* **2009**, *136*, 19–26. [CrossRef]
7. Balacco, G.; Carbonara, A.; Gioia, A.; Iacobellis, V.; Piccinni, A.F. Investigation of peak water consumption variability at local scale in Puglia (Southern Italy). *Proceedings* **2018**, *2*, 674. [CrossRef]
8. Yarnold, P.R. Minimize usage of binary measurement scales in rigorous classical research. *Optim. Data Anal.* **2018**, *7*, 3–9.
9. Dahl, C. A survey of energy demand elasticities for the developing world. *J. Energy Dev.* **1994**, *18*, 1–47.
10. Bertsimas, D.; Thiele, A. Robust and data-driven optimization: Modern decision-making under uncertainty. *Model. Methods Appl. Innov. Decis. Mak.* **2006**, *3*, 95–122.
11. Bhattacharyya, S.C.; Timilsina, G.R. *Energy Demand Models for Policy Formulation: A Comparative Study of Energy Demand Models*; The World Bank: Washington, DC, USA, 2009.
12. Secretaria de Estado da Educação do Paraná (PARANÁ). *Excel, Work Sheet about Water Consumption at 149 Public Schools at Paraná State*; Continuous Service Sector: Curitiba, Brazil, 2013. [CrossRef]
13. Arditi, R.; Ginzburg, L.R. Coupling in predator-prey dynamics: Ratio-dependence. *J. Theor. Biol.* **1989**, *139*, 311–326. [CrossRef]
14. Aven, T. Risk assessment and risk management: Review of recent advances on their foundation. *Eur. J. Oper. Res.* **2016**, *253*, 1–13. [CrossRef]
15. Gibbins, P. *Particles and Paradoxes: The Limits of Quantum Logic*; Cambridge University Press: Cambridge, UK, 1987.
16. Harush, U.; Barzel, B. Dynamic patterns of information flow in complex networks. *Nat. Commun.* **2017**, *8*, 2181. [CrossRef] [PubMed]
17. Butucea, C.; Delmas, J.F.; Dutfoy, A.; Fischer, R. Maximum entropy distribution of order statistics with given marginals. *Bernoulli* **2018**, *24*, 115–155. [CrossRef]
18. Fried, R.; Didelez, V.; Lanius, V. Partial correlation graphs and dynamic latent variables for physiological time series. In *Innovations in Classification, Data Science, and Information Systems*; Springer: Berlin/Heidelberg, Germany, 2004; pp. 259–266.
19. Martin-Löf, P. Complexity oscillations in infinite binary sequences. *Zeitschrift für Wahrscheinlichkeitstheorie und Verwandte Gebiete* **1971**, *19*, 225–230. [CrossRef]
20. Gencaga, D.; Knuth, K.H.; Rossow, W.B. A recipe for the estimation of information flow in a dynamical system. *Entropy* **2015**, *17*, 438–470. [CrossRef]
21. Donner, R.V.; Small, M.; Donges, J.F.; Marwan, N.; Zou, Y.; Xiang, R.; Kurths, J. Recurrence-based time series analysis by means of complex network methods. *Int. J. Bifurc. Chaos* **2011**, *21*, 1019–1046. [CrossRef]

22. Li, M.; Xie, K.; Kuang, H.; Liu, J.; Wang, D.H.; Fox, G.E.; Shi, Z.F.; Chen, L.; Zhao, F.; Mao, Y.; et al. Neural coding of cell assemblies via spike-timing self-information. *Cereb. Cortex* **2018**, *28*, 2563–2576. [CrossRef] [PubMed]
23. Baran, T.; Harmancioglu, N.B.; Cetinkaya, C.P.; Barbaros, F. An extension to the revised approach in the assessment of informational entropy. *Entropy* **2017**, *19*, 634. [CrossRef]
24. Mézard, M.; Montanari, A. *Information, Physics and Computation*; Oxford University Press: Corby, UK, 2009.
25. Tsitsiklis, J. Bernoulli Process: Probabilistic Systems Analysis and Applied Probability, Lecture 13. MIT Opencourseware. 2011. Available online: https://ocw.mit.edu/courses/electrical-engineering-and-computer-science/6-041-probabilistic-systems-analysis-and-applied-probability-fall-2010/video-lectures/lecture-13-bernoulli-process/ (accessed on 25 September 2018).
26. Cauchy, A.L.B. *Cours d'Analyse de l'École Royale Polytechnique*; Debure: De L' Imprimerie Royale, France, 1821.
27. Cover, T.M.; Thomas, J.A. *Elements of Information Theory*, 2nd ed.; Willey-Interscience: Hoboken, NJ, USA, 2006.
28. Ha, J.; Tan, P.P.; Goh, K.L. Linear and nonlinear causal relationship between energy consumption and economic growth in China: New evidence based on wavelet analysis. *PLoS ONE* **2018**, *13*, e0197785. [CrossRef] [PubMed]
29. Telles, C.R.; Chueiri, A.M.S.; Cunha, A.R.B. *Pesquisa Operacional do Setor de Serviços Contínuos: Economia, Meio Ambiente e Educação*, 2nd ed.; Secretaria de Estado da Educação do Paraná: Curitiba, Brazil, 2018; 146p. [CrossRef]
30. Licata, I.; Chiatti, L. Timeless approach to quantum jumps. *Quanta* **2015**, *4*, 10–26. [CrossRef]
31. Weijs, S. Information Theory for Risk-based Water System Operation. Ph.D Thesis, Water Resources Management, Faculty of Civil Engineering & Geosciences of Technische Universiteit, Delft, The Netherlands, 2011.
32. Shannon, C.E. A mathematical theory of communication. *ACM SIGMOBILE Mob. Comput. Commun. Rev.* **2011**, *5*, 3–55. [CrossRef]
33. Pekkonen, M.; Ketola, T.; Laakso, J.T. Resource availability and competition shape the evolution of survival and growth ability in a bacterial community. *PLoS ONE* **2013**, *8*, e76471. [CrossRef] [PubMed]
34. Duncan, C.J.; Duncan, S.R.; Scott, S. The dynamics of scarlet fever epidemics in England and Wales in the 19th century. *Epidemiol. Infect.* **1996**, *117*, 493–499. [CrossRef] [PubMed]
35. Giannoula, A.; Gutierrez-Sacristán, A.; Bravo, Á.; Sanz, F.; Furlong, L.I. Identifying temporal patterns in patient disease trajectories using dynamic time warping: A population-based study. *Sci. Rep.* **2018**, *8*, 4216. [CrossRef] [PubMed]
36. Afy-Shararah, M.; Rich, N. Operations flow effectiveness: A systems approach to measuring flow performance. *Int. J. Oper. Prod. Manag.* **2018**. [CrossRef]
37. Klug, M.J.; Scheurer, M.S.; Schmalian, J. Hierarchy of information scrambling, thermalization, and hydrodynamic flow in graphene. *Phys. Rev. B* **2018**, *98*, 045102. [CrossRef]
38. Tao, Y.; Yu, S.; Zhou, J. Information Flow Queue Optimization in EC Cloud. In Proceedings of the International Conference on Computing, Networking and Communications (ICNC), Maui, HI, USA, 5–8 March 2018; pp. 888–892.

symmetry

Article

Measuring Performance in Transportation Companies in Developing Countries: A Novel Rough ARAS Model

Dunja Radović [1], Željko Stević [1,*], Dragan Pamučar [2], Edmundas Kazimieras Zavadskas [3], Ibrahim Badi [4], Jurgita Antuchevičiene [5] and Zenonas Turskis [6]

1 Faculty of Transport and Traffic Engineering Doboj, University of East Sarajevo, Vojvode Mišića 52, 74000 Doboj, Bosnia and Herzegovina; dunja.radovic@sf.ues.rs.ba

2 Department of Logistics, University of Defence in Belgrade, Pavla Jurisica Sturma 33, 11000 Belgrade, Serbia; dpamucar@gmail.com

3 Institute of Sustainable Construction, Vilnius Gediminas Technical University, Sauletekio al. 11, LT-10223 Vilnius, Lithuania; edmundas.zavadskas@vgtu.lt

4 Faculty of Engineering, Misurata University, 2429 Misurata, Libya; i.badi@eng.misuratau.edu.ly

5 Department of Construction Management and Real Estate, Vilnius Gediminas Technical University, LT-10223 Vilnius, Lithuania; jurgita.antucheviciene@vgtu.lt

6 Department of Construction Technology and Management, Vilnius Gediminas Technical University, Sauletekio al. 11, LT-10223 Vilnius, Lithuania; zenonas.turskis@vgtu.lt

* Correspondence: zeljkostevic88@yahoo.com or zeljko.stevic@sf.ues.rs.ba

Received: 6 September 2018; Accepted: 20 September 2018; Published: 25 September 2018

Abstract: The success of any business depends fundamentally on the possibility of balancing (symmetry) needs and their satisfaction, that is, the ability to properly define a set of success indicators. It is necessary to continuously monitor and measure the indicators that have the greatest impact on the achievement of previously set goals. Regarding transportation companies, the rationalization of transportation activities and processes plays an important role in ensuring business efficiency. Therefore, in this paper, a model for evaluating performance indicators has been developed and implemented in three different countries: Bosnia and Herzegovina, Libya and Serbia. The model consists of five phases, of which the greatest contribution is the development of a novel rough additive ratio assessment (ARAS) approach for evaluating measured performance indicators in transportation companies. The evaluation was carried out in the territories of the aforementioned countries in a total of nine companies that were evaluated on the basis of 20 performance indicators. The results obtained were verified throughout a three-phase procedure of a sensitivity analysis. The significance of the performance indicators was simulated throughout the formation of 10 scenarios in the sensitivity analysis. In addition, the following approaches were applied: rough WASPAS (weighted aggregated sum product assessment), rough SAW (simple additive weighting), rough MABAC (multi-attributive border approximation area comparison) and rough EDAS (evaluation based on distance from average solution), which showed high correlation of ranks by applying Spearman's correlation coefficient (SCC).

Keywords: rough ARAS; transport; performance; logistics; MCDM

---

1. Introduction

Transportation, according to Grabara et al. [1], represents the most important logistics activity and in order to enable the smooth running of a transportation process, a series of activities has to be integrated into one coherent whole. Transportation is, according to Borzacchiello et al. [2], a natural and dynamic part of any modern space-economy, offers great economic benefits and improves economic development, as confirmed by research [3] emphasizing that freight transportation plays a key role in today's economies. Mobility in this century is an everyday aspect of life, and it is necessary to take into account all trends in commodity flows, measure their performance and optimize them from economic, functional, technological and organizational aspects. The trend of urbanization is constantly increasing in the world, which increases the delivery frequency of goods and demands for transportation, which makes it an even more important factor of a complete supply chain. Adding to this the fact that it is necessary to connect production and consumption that are thousands of kilometers away from each other in many cases, understanding and measuring one's own performance is a prerequisite for the efficient operation of transportation companies. It has been recognized that the large number of criteria that contributed to the problem-solution process must be considered embody duality, paradox, unity in diversity, change, and harmony, offering a holistic approach to problem-solving. An effective integration of cross-functional processes has a considerable influence on the success [4] any company. According to Stević et al. [5], transportation represents the most expensive logistics subsystem, the subsystem that causes the largest percentage of logistics costs, as confirmed by Guasch [6], who states that transportation costs represent the most significant item in total logistics costs. Considering the structure of total logistics costs, the transportation costs, according to Karri and Ojala [7], account for about 40%, and in total costs account for up to 20%. According to Stević et al. [5], it is necessary to work day-to-day to reduce these costs, especially regarding large companies that have a large volume of transportation movements on a daily basis.

Taking into account the aforementioned, the procedure for measuring and monitoring performance indicators is of great importance for the operation of every transportation company. Determining the performance that has the greatest impact on the efficiency of the operation of transportation companies enables optimization of logistics processes and shorter time to complete them. In developed countries around the world, monitoring key performance indicators has been recognized as a necessity, both due to the improvement of the operation of companies as well as the increasing competition on the market, and it is an inevitable part of the logistics processes of every company. However, developing countries have not fully recognized the importance of identifying key performance yet or are at very beginning of recognizing it.

Throughout the research and development of the model in this paper, several goals have been presented. The first goal relates to the improvement and enrichment of the methodology for dealing with uncertainty in the field of group multi-criteria decision-making throughout the development and presentation of a novel rough additive ratio assessment (ARAS) approach. The second goal of this paper is to bridge the gap that currently exists in the methodology for measuring and monitoring performance indicators in the logistics subsystem of transportation. The second goal is achieved throughout a newly developed approach to dealing with imprecision based on rough numbers. The third goal of the paper is a possibility of improving the efficiency of the operation of transportation companies in developing countries and encouraging the development of competitiveness as an important factor in the success of every company. The fourth goal of the paper refers to a possibility of post-analysis in transportation companies that are evaluated in this paper, where the best-ranked ones can serve others as a benchmark. The last, fifth goal of the paper is to popularize and affirm the idea of rough numbers throughout a detailed calculation of operations with rough numbers, which are characteristic of the field of multi-criteria decision-making.

This paper contains six sections in total, the first of which relates to the importance of measuring and monitoring the performance of transportation companies. In the second section, which is related to a literature review, there is a review of studies on similar issues in which multi-criteria decision-making methods are applied. The third section presents the applied methods. Basic operations with rough numbers and a detailed algorithm of a newly developed rough ARAS approach are presented. The fourth section presents a detailed study on performance measurement in Bosnia and Herzegovina, Libya and Serbia based on the developed approach. The fifth section is a sensitivity analysis that involves checking the stability of the results using a three-phase procedure and the discussion of the results. The sixth section presents the key contributions of the developed model and the research, as well as proposals for future research.

2. Literature Review

In order to evaluate and improve the performance of transportation companies, it is necessary to know and analyse key performance indicators (KPIs) [8,9]. KPIs are the measures that organizations can use to assess their own performance. In other words, KPIs help organizations determine the degree of their success in achieving their goals [10,11]. KPIs help managers of transportation companies determine which components of the company require more care and continuous monitoring in order to achieve values that will satisfy service users. In general, the development of key performance indicators for every organization has always been one of the most stressful tasks [12,13]. In that sense, although today many organizations around the world use key performance indicators to find out how successful they are, a very small number of organizations use the most appropriate KPIs to assess their own performance. The reason for that is a lack of understanding of the key performance indicators by business leaders [14]. Performance measurement, in various logistics domains, using performance indicators was carried out in several studies to assess the logistics situation in countries such as Malaysia [15], Finland [16] and Turkey [17,18]. These studies show the acceptance of a logistics performance index as a measure for assessing the logistics performance in the country and linking the logistics performance to trade and transportation policies. Other studies have used indicators of logistics performance for research purposes. Hoekman and Nicita [19] consider various World Bank indices that affect the limitation and facilitation of trade and consider their application in developing countries.

In some studies, the estimation of KPIs is also used as a reflection of logistics performance that may be affected by certain policy measures [20–22]. Cemberci et al. [22] considered the effect of global competitiveness on the KPIs of transportation operators and concluded that better competitiveness could be achieved by timely improvement of the KPI components as well as through active monitoring of international shipments. Kim and Min [20] combined the KPIs and the environmental performance index to create an index of green logistics performance, which significantly influenced the final results of the research. Marti et al. [21] examined the importance of each component of KPIs in relation to trade in developing economies using a gravity model. They concluded that all components of KPIs have a positive connection with the scope of international trade. Civelek et al. [23] applied a hierarchical regression analysis to determine the correlation between the logistics performance index, the competitive index and the gross domestic product. Another study that applied the hierarchical regression analysis was by Uca et al. [24] who, similar to Civelek et al. [23], considered the correlation between transportation performance, a corruption perception index and the extent of external trade. The conclusion of the study is that there is a significant correlation and influence between the logistical capacity of a country, corruption, and the extent of external trade.

Leading practitioners who are researching the performance of transportation companies look for answers to several questions, including the question of which indicators they should use and when they need to use them to measure the performance of transportation companies [25]. Therefore, in the transportation industry there is a need to establish a framework for the implementation of a strategic system for measuring the performance of companies [26]. Such a system implies a choice of balanced indicators with a holistic approach [27]. However, the importance of identifying indicators for performance measurement in transportation companies has been recognized only by a small number of researchers and practitioners [26]. Several studies focus on evaluating the logistics performance from several perspectives [28] and logistics throughout performance measurement [29]. Consequently, in order to identify key indicators in logistics performance measurement, the balanced scorecard (BSC) concept is a widely accepted approach and it has been used in several studies [30–32].

Norin [33] investigated logistics in air transportation. The study proposed a set of key performance indicators to assess the performance of logistics components using a conceptual model. In recent studies, Kosanke and Schultz [9] have proposed a KPI set to assess the performance of air transportation operations at an airport. Humphreys and Francis [34] have come to the conclusion that increasing demand in air transportation and changes in airport ownership led to the introduction of new financial and environmental measures. The authors have provided a good discussion on a wide range of past, present and future key performance indicators of airports. Regarding the issue of the environment, Morrell and Lu [35] and Ignaccolo [36] studied noise related to transportation activities and its impact on communities near transportation terminals.

In a limited number of studies, models for multi-criteria decision-making are used. The research shown in [37] has indicated that KPIs are necessary to improve internal organization, relationship with customers, competitiveness, and strategy planning in the case of tanker companies in Greece. Nathnail et al. [38] used the analytic hierarchy process (AHP) to assess the significance of each criterion in KPIs and the overall performance of two terminals. Bentaleb et al. [39] identified and analyzed key performance indicators for a port using Measuring Attractiveness by a Categorical-Based Evaluation Technique (MACBETH) tools. The methodology based on the previous analysis helps managers in such companies to make decisions and increase global performance. In [40], the decision-making trial and evaluation laboratory (DEMATEL) method for analyzing the importance and relation among the criteria for evaluating intermodal transportation was used. Moreover, there are many other models used to evaluate and compare key performance indicators [41–45]. In addition to the presented multi-criteria decision-making models (MCDM) for the analysis of KPIs of transportation companies, methods of strategic management in combination with MCDM were also used, e.g., a benchmarking technique that is very useful for assessing the impact of various factors such as services, safety, environment, costs and profit indicators [46–50].

This study extends the existing knowledge of the applicability of KPI measurement in the transportation industry by introducing a comprehensive, balanced set of performance indicators. In complex real-life scenarios, there may be a need for modeling a hierarchical structure, as well as a need to determine the prioritization of different indicators. This presents a challenging and still insufficiently considered issue in the domain of transportation [51]. It is therefore necessary for transportation companies to explore the relations among their different abilities [52]. Therefore, managers of transportation companies should try to answer several questions, such as how to determine the priorities of indicators and how to build a hierarchical relation in order to identify the impacts among indicators [53]. In such cases, MCDM methods offer practical solutions. However, designing the MCDM framework for performance measurement is a complex process that is further elaborated in order to improve the area discussed in this paper [54].

Consequently, in order to face the above challenges, it is necessary to develop a model for identifying the key performance indicators of transportation companies and determining their relations. It is precisely this purpose that is the goal of this study, in order to provide a comprehensive decision-making model that identifies key performance indicators for the transportation industry and evaluate the relations among these indicators from the perspective of logistics providers using MCDM models. In order to achieve this goal, the main research question of this study is: how do you form a decision-making model in which key transportation performance indicators are implemented and which enables the setting of priorities for these indicators, taking into account mutual relations? To solve this problem, the study proposes the measurement of KPIs of transportation companies using step-wise weight assessment ratio analysis (SWARA) and ARAS methods, throughout the analysis of dependence among performance indicators and a proposal of their priorities. Although there are several studies focusing on the application of the MCDM concept in the transportation industry [55], the implementation of a rough approach in MCDM models, which are applied in the field of logistics, has received very limited attention. In particular, there are no studies that consider the integration of a rough approach in the SWARA-ARAS hybrid model, not only in the field of performance measurement of logistics companies, but in general in the MCDM literature. The SWARA-rough ARAS model is a new comprehensive decision-making model that allows management of transportation companies to measure performance even in situations where there are inaccuracies and uncertainties in data.

3. A Novel Rough Additive Ratio Assessment (ARAS) Method

The ARAS method belongs to a group of multi-criteria decision-making methods that have been developed over the past decade and it has found a wide application in different areas. A classic form of this method was developed in 2010 [56], while its fuzzy [57] and grey forms [58] were developed in the same year. Bearing in mind all the advantages of using rough theory [59,60] in the MCDM to represent ambiguity, vagueness and uncertainty, the authors have decided in this paper to modify the ARAS algorithm using rough numbers, which is an original contribution. The fuzzy form of ARAS method offers certain benefits in terms of uncertainty, and due to a possibility of reducing subjectivity to a minimum, reducing uncertainty, and obtaining clearer results, a new rough ARAS consisting of seven steps has been developed in this paper.

Step 1. Forming a multi-criteria model. In this step, it is necessary to define the problem that needs to be solved, to form a set of m criteria on the basis of which n alternatives will be evaluated. In addition, it is necessary to define a set of k decision-makers who will participate in a process of group decision-making.

Step 2. Forming a group rough matrix. In this step using Equations (A1)–(A6) it is necessary to aggregate individual matrices into a group rough matrix represented by Equation (1). An integral part of this initial matrix, unlike in other approaches, is an additional row that represents optimal values of the alternatives according to the criteria.

$$IRM = \begin{bmatrix} [x_{11}{}^L, x_{11}{}^U] & [x_{12}{}^L, x_{12}{}^U] & \cdots & [x_{1n}{}^L, x_{1n}{}^U] \\ [x_{21}{}^L, x_{21}{}^U] & [x_{22}{}^L, x_{22}{}^U] & \cdots & [x_{2n}{}^L, x_{2n}{}^U] \\ \vdots & \vdots & \ddots & \vdots \\ [x_{m1}{}^L, x_{m1}{}^U] & [x_{m2}{}^L, x_{m2}{}^U] & \cdots & [x_{mn}{}^L, x_{mn}{}^U] \\ [x_{O1}{}^L, x_{O1}{}^U] & [x_{O2}{}^L, x_{O2}{}^U] & \cdots & [x_{On}{}^L, x_{On}{}^U] \end{bmatrix} \tag{1}$$

Optimal values of alternatives according to criteria are formed by taking the highest or lowest values depending on whether the criterion belongs to a cost or benefit type. If it is a benefit-type criterion, then the maximum value is taken, and if it is a cost-type criterion, the minimum value is taken.

Step 3. Normalization of an initial rough matrix that involves three phases depending on the type of criteria. The first phase relates to the normalization of benefit-type criteria by applying Equation (2).

$$(\mathrm{a})\; n_{ij} = \frac{[x^L{}_{ij}; x^U{}_{ij}]}{\sum\limits_{i=1}^{m} [x^L{}_{ij}; x^U{}_{ij}]} \quad for \quad c_1, c_2, c_3 \ldots c_n \in B. \tag{2}$$

If we take into account the basic operations with rough numbers and the sum calculated by columns for each criterion in particular, the normalization process obtains the following form Equation (3)

$$n_{ij} = \left[\frac{x_{ij}{}^L}{x^{*U}{}_{ij}}; \frac{x_{ij}{}^U}{x^{*L}{}_{ij}} \right]. \tag{3}$$

where $x^{*U}{}_{ij}$ indicates the upper sum limit by the criterion, and $x^{*L}{}_{ij}$ the lower sum limit.

The second and third phase of normalization involve the application of Equations (4) and (5): for cost criteria; Equation (4) should be first applied in order to obtain inverse values for $\left[x_{ij}{}^{L}; x_{ij}^{U}\right]$.

$$(\mathrm{b})\ \overline{x_{ij}} = \left[\frac{1}{x_{ij}{}^{L}; x_{ij}^{U}}\right] = \left[\frac{1}{x_{ij}^{U}}; \frac{1}{x_{ij}^{L}}\right]. \tag{4}$$

and then the Equation (5) in order to complete the normalization process:

$$(\mathrm{c})\ n_{ij} = \frac{\left[\overline{x}_{ij}{}^{L}; \overline{x}_{ij}{}^{U}\right]}{\sum\limits_{i=1}^{m}\left[\overline{x}_{ij}^{L}; \overline{x}_{ij}^{U}\right]} = \left[\frac{\overline{x}_{ij}{}^{L}}{\overline{x}_{ij}^{*U}}; \frac{\overline{x}_{ij}{}^{U}}{\overline{x}_{ij}^{*L}}\right]. \tag{5}$$

where $\overline{x}_{ij}^{*U}$ and $\overline{x}_{ij}^{*L}$ represents the upper and lower limits of the sum by the inverse value criterion for $\left[x_{ij}{}^{L}; x_{ij}^{U}\right]$, respectively.

Step 4. Weighting the normalized rough matrix using Equation (6):

$$\begin{gathered} W_n = \left[w_{ij}^{L}; w_{ij}^{U}\right]_{m \times n} \\ w_{ij} = w_j^{L} \times n_{ij,}^{L} \quad i = O, 1, 2, \ldots, m \\ w_{ij} = w_j^{U} \times n_{ij,}^{U} \quad i = O, 1, 2, \ldots, m. \end{gathered} \tag{6}$$

Step 5. Determining the matrix S_i by summing all the values per rows from the previous weighted matrix, Equation (7).

$$\begin{gathered} S_i = \left[s_{ij}^{L}; s_{ij}^{U}\right] \\ s_{ij}^{L} = \sum w_{ij}^{L}; s_{ij}^{U} = \sum w_{ij}^{U}; \quad j = O, 1, 2, \ldots, n. \end{gathered} \tag{7}$$

Step 6. Calculation of the degree of usefulness applying Equation (8):

$$K_i = \left[\frac{s_{ij}{}^{L}}{s_o^{U}}; \frac{s_{ij}{}^{U}}{s_o^{L}}\right]. \tag{8}$$

where S_o indicates the value of the best alternative.

Step 7: Ranking the alternatives in decreasing order. The highest value is the best solution, while the smallest value represents the worst solution.

To adequately solve decision-making problems with vague or imprecise information according to Mardani et al. [61], the fuzzy set theory [62–64] and aggregation operator theory [65,66] have become powerful tools. As opposed to fuzzy sets theory, which requires a subjective approach in determining partial functions and fuzzy set boundaries, rough set theory determines set boundaries based on real values and depends on the degree of certainty of the decision maker. Since rough set theory deals solely with internal knowledge, i.e., operational data, there is no need to rely on assumption models. In other words, when applying rough sets, only the structure of the given data is used instead of various additional/external parameters. The logic of rough set theory is based solely on data that speak for themselves. When dealing with rough sets, the measurement of uncertainty is based on the vagueness already contained in the data. In this way, the objective indicators contained in the data can be determined. In addition, rough set theory is suitable for application on sets characterized by irrelevant data where the use of statistical methods does not seem appropriate [67].

The benefits of the novel rough ARAS approach in relation to other approaches developed so far can be seen throughout several options. The first one refers to a relatively small number of steps that are needed for a complete calculation according to this approach; then taking into account the optimal alternative as a possibility for more precise decision-making; and reducing uncertainty and subjectivity in a decision-making process using rough instead of crisp numbers.

The implementation of this approach was demonstrated on the case study of transportation companies in Libya, Serbia and Bosnia and Herzegovina.

4. Case Study

The developed model for performance evaluation in transportation companies in developing countries consists of a total of 23 steps divided into five phases: initial data collection, the application of the SWARA (step-wise weight assessment ratio analysis) method [68] for the selection of performance indicators, data collection related to performance measurement in developing countries, the development and implementation of the novel rough ARAS approach and sensitivity analysis. The proposed model is shown in Figure 1 where steps and sub-steps of all phases are presented.

Recognizing the need for forming such a model and completing performance measurement started at the end of 2017, when a part of the team for research and development of the model was formed. The first ideas, a review of the relevant literature, and discussion on indicators for performance measurement with managers of transportation companies, were carried out exactly in that period. In the period of January–March 2018, the first phase of data collection was completed, while the second phase was completed at the beginning of April. The third phase related to data re-collection was carried out in the period of April–June. Along with the idea of the development of this model, it also started the development of the novel rough ARAS algorithm, which is one of key contributions of this paper, and lasted until May 2018. Taking into account the aforementioned and later the application of the developed rough ARAS approach in June it can be concluded that the fourth phase lasted longer than five months. The last phase was carried out in July. When the timing of all the phases of the proposed model is taken into account, it can be concluded that the complete research took about nine months.

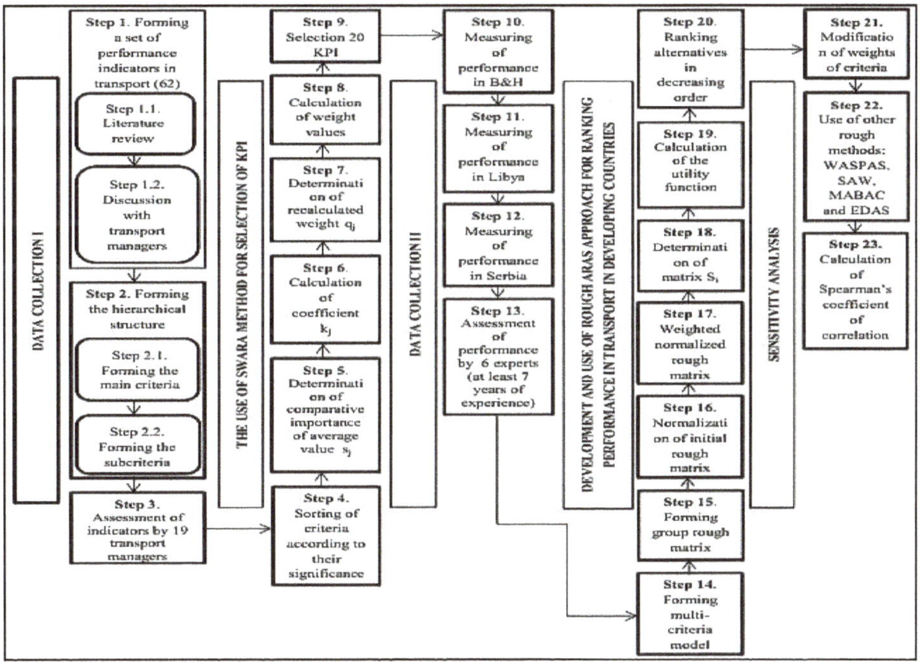

Figure 1. The proposed model.

The first phase is data collection and consists of three steps. The first step is forming a set of 62 performance indicators in the field of transportation based on a detailed literature review, as well as opinions and discussions with managers from various transportation companies. The second step is forming a hierarchical structure made up of the main criteria and sub-criteria, which is given in detail in Figure 2 and explained in Table 1. The third step of the first phase is the assessment of 62 performance indicators by managers from 19 different transportation companies.

Determining the significance of criteria according to different authors [69,70] is one of the most important stages in the decision-making process, so the second phase represents the use of the SWARA method in order to perform the selection of key performance indicators. It consists of six steps, with the first five steps representing the integral steps of the SWARA method. At the very beginning, it is necessary to sort the criteria according to their importance, i.e., it is necessary that experts rank the criteria according to their significance. Then, it is necessary to determine the comparative importance of the average value. The third, fourth and fifth steps are related to the calculation and determination of coefficient and weight values. The sixth step of the second phase is a selection of 20 key performance indicators.

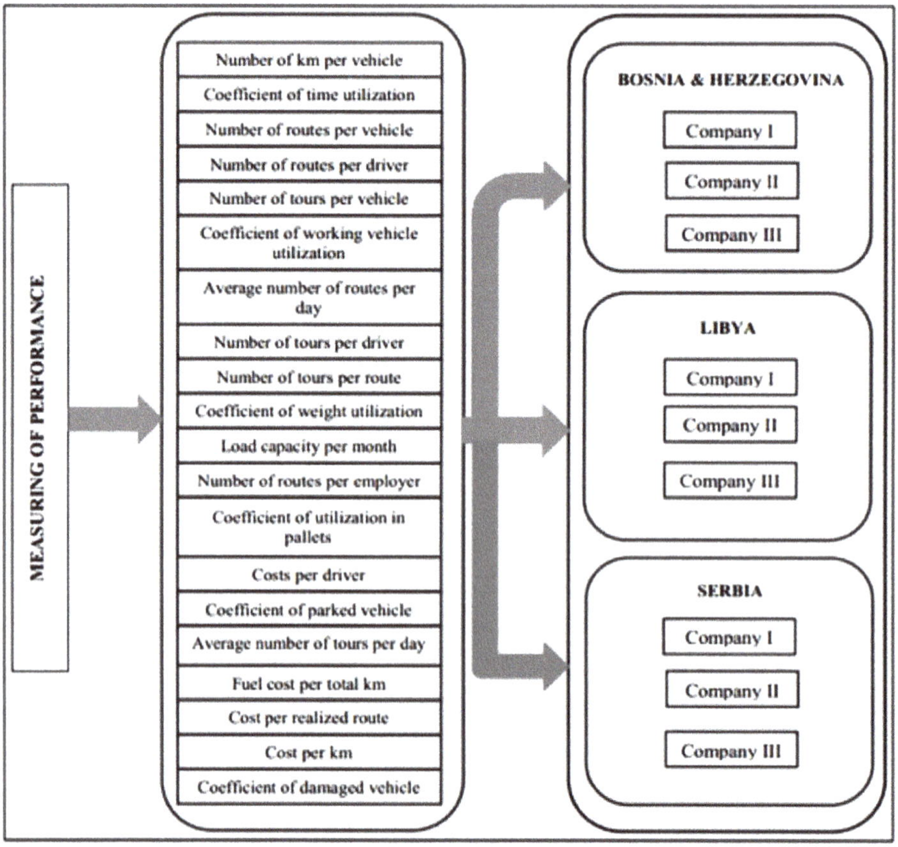

Figure 2. A hierarchical structure of performance measurement in transportation companies in three different countries.

Table 1. Explanation of indicators used in this research.

Indicator	Explanation
Coefficient of time utilization	Coefficient of time utilization (total working time of vehicles (h)/total number of available hours per month, whereby 1 day = 24 h) × 100 (%)
Number of km per vehicle	Number of km per vehicle (total number of kilometers/total number of vehicles)
Number of routes per vehicle	Number of routes per vehicle (total number of routes/total number of vehicles)
Number of routes per driver	Number of routes per driver (total number of routes/total number of drivers)
Number of tours per vehicle	Number of tours per vehicle (total number of tours/total number of vehicles)
Coefficient of working vehicle utilization	Coefficient of working vehicle utilization (total number of working vehicles per month/number of vehicles per day x number of working days) × 100 (%)
Average number of routes per day	Average number of routes per day (total number of routes/number of working days)
Number of tours per driver	Number of tours per driver (total number of tours/total number of drivers)
Number of tours per route	Number of tours per route (total number of tours/total number of routes)
Coefficient of weight utilization	Coefficient of weight utilization (total weight of transported goods/total available weight load capacity of vehicle per month) × 100 (%)
Load capacity per month	Load capacity per month (total available weight load capacity of vehicle per day × average number of tours per type of vehicle per day × number of working days in a month)
Number of routes per employer	Number of routes per employer (total number of routes/total number of employees)
Coefficient of utilization in pallets	Coefficient of utilization in pallets (total number of transported pallets/number of available pallet spaces) × 100 (%)
Costs per driver	Costs per driver (total transport costs/total number of drivers)
Coefficient of parked vehicle	Coefficient of parked vehicle (total number of vehicles in parking space per month/number of vehicles per day × number of working days) × 100 (%)
Average number of tours per day	Average number of tours per day (total number of tours/number of working days)
Fuel cost per total km	Fuel cost per total km (total fuel cost/total number of crossed kilometers)
Cost per realized route	Cost per realized route (total transport costs/total number of routes)
Cost per km	Cost per km (total transport costs/total number of crossed kilometers)
Coefficient of damaged vehicle	Coefficient of damaged vehicle (total number of damaged vehicles per month/number of vehicles per day × number of working days) × 100 (%)

The third phase is also a collection of data for the next part of the work related to the measurement of 20 identified performance indicators in developing countries, in this case Bosnia and Herzegovina, Libya and Serbia. The managers of nine transportation companies located in the aforementioned countries filled in the tables by inputting their own values for each performance indicator. After that, the performance indicators were evaluated by six experts with at least seven years of experience in the field of transportation, and located in the territory of Bosnia and Herzegovina and Serbia.

The fourth phase relates to the development and application of the novel rough ARAS approach for evaluating measured performance indicators in transportation companies in developing countries. It consists of seven steps that are explained in detail in the section related to work methods.

The fifth and also the last phase of the proposed model refers to a sensitivity analysis. The sensitivity analysis was carried out in three steps. The first step is the change in the weight of criteria, and the second step is the application of different methods of multi-criteria decision-making, which is given in detail in the fifth section of the paper. Several methods are applied to the same multi-criteria model in order to better represent the difference in the ranks of transportation companies with a change in approach. The last step is the determination of Spearman's coefficient of correlation, on the basis of which the stability of the obtained results, i.e., the ranking of alternatives, can be determined. The SCC is calculated for both phases of the sensitivity analysis, i.e., for the rank of alternatives according to formed scenarios in which the weights of criteria are changed and in comparison to other approaches.

Figure 2 shows a hierarchical structure with 20 key performance indicators identified by evaluations by experts from 19 different transportation companies. At the first level of the hierarchy, there is a goal that is related to performance measurement in transportation companies in developing countries. The second level of the hierarchy consists of a total of 20 performance indicators that are identified based on the first part of this study, which is shown in [71]. At the third level, there are values of measured performance indicators in transportation companies in developing countries, i.e., in Bosnia and Herzegovina, Libya and Serbia.

As already noted in the section explaining the proposed model, a detailed explanation of the used indicators is given in Table 1 to enable transport managers to input their own values of indicators in an easier and more understandable way.

The values of performance indicators, on the basis of which the alternatives were evaluated, were obtained using the SWARA method in the research [71]. In the initial study, as already mentioned, a total of 62 indicators were evaluated based on 19 decision-makers. The weight coefficients of the criteria in this paper have been multiplied by two in relation to the above research, for the purpose of easier calculation.

In the first step of the proposed approach, a multi-criteria model should be formed, which in this case consists of 20 criteria, nine alternatives and six decision-makers. The alternatives represent three transportation companies from Bosnia and Herzegovina, Libya and Serbia, respectively. In order to be able to form a group rough matrix, it is first necessary to convert individual matrices of all decision-makers (Table 2) into group rough ones.

Table 2. Evaluation of alternatives by six decision-makers.

	A_1						A_2						A_3					
	E_1	E_2	E_3	E_4	E_5	E_6	E_1	E_2	E_3	E_4	E_5	E_6	E_1	E_2	E_3	E_4	E_5	E_6
C_1	9	9	9	7	7	7	7	9	9	7	7	7	7	9	9	7	7	7
C_2	5	7	7	5	5	7	1	3	1	1	1	1	5	7	7	5	5	7
C_3	7	9	7	9	5	7	7	9	7	9	7	7	7	9	5	9	5	7
C_4	7	7	7	7	5	5	7	7	9	9	5	5	7	7	7	7	5	5
C_5	7	9	9	7	7	7	5	7	7	5	5	3	7	9	9	7	7	7
C_6	7	7	9	9	7	7	7	5	7	7	7	7	7	5	7	7	7	7
C_7	5	5	3	3	3	3	5	5	5	3	5	5	5	3	1	3	3	3
C_8	7	7	9	7	7	5	5	5	7	7	5	5	5	5	7	7	5	5
C_9	9	5	7	7	9	7	7	3	5	5	5	3	9	5	7	7	7	5
C_{10}	5	7	7	5	5	5	3	5	3	3	3	5	3	5	3	3	3	5
C_{11}	9	9	9	7	9	9	7	5	3	3	3	5	9	9	9	7	7	7
C_{12}	5	5	5	5	3	5	7	5	5	5	3	5	5	5	3	3	3	5
C_{13}	7	7	7	9	7	9	9	9	9	9	9	9	5	5	3	5	7	7
C_{14}	3	1	1	3	3	5	3	1	1	3	3	5	3	1	1	3	3	5
C_{15}	1	5	3	5	3	3	9	7	7	9	7	9	9	7	7	9	7	9
C_{16}	7	3	3	3	3	3	9	9	9	9	7	9	7	3	3	3	3	3
C_{17}	9	9	9	9	9	9	3	1	1	3	1	3	9	7	9	9	9	9
C_{18}	9	7	7	9	7	7	7	5	5	7	3	5	7	5	5	7	5	7
C_{19}	1	1	1	1	1	1	1	1	3	1	1	1	1	3	3	1	1	1
C_{20}	7	3	5	7	3	7	3	1	1	1	1	1	9	5	7	9	5	9

Table 2. *Cont.*

	A4						A5						A6					
C_1	3	5	5	3	3	3	3	5	7	3	5	5	3	5	5	3	3	3
C_2	3	5	3	3	3	5	3	3	3	3	3	5	3	5	3	3	3	5
C_3	3	3	1	3	1	1	3	3	1	3	1	1	3	3	1	3	1	1
C_4	1	3	1	3	1	1	1	3	1	3	1	1	1	3	1	3	1	1
C_5	3	5	3	3	1	1	3	5	3	3	1	1	3	5	3	3	1	1
C_6	9	9	9	9	9	7	7	7	7	7	7	7	9	9	9	9	9	7
C_7	1	1	1	1	1	1	3	3	3	3	3	3	3	1	1	1	1	1
C_8	1	1	1	1	1	1	3	3	3	3	3	3	1	1	1	1	1	1
C_9	9	5	7	7	7	5	7	5	5	5	5	3	7	5	5	7	5	3
C_{10}	5	7	7	5	5	5	5	7	7	5	5	5	5	9	7	5	5	5
C_{11}	7	5	3	3	3	5	7	5	3	3	3	5	7	5	3	3	3	5
C_{12}	1	1	1	1	1	1	1	1	1	1	1	1	1	1	1	1	1	1
C_{13}	5	3	3	5	5	7	5	5	3	5	7	7	5	3	3	5	5	7
C_{14}	5	3	5	5	5	5	5	3	5	5	5	5	5	3	5	5	5	5
C_{15}	1	1	1	1	1	1	1	1	1	1	1	1	1	1	1	1	1	1
C_{16}	5	1	3	1	1	1	7	3	3	3	3	3	5	1	3	1	1	1
C_{17}	1	1	1	1	1	1	1	1	1	1	1	1	1	1	1	1	1	1
C_{18}	9	7	7	9	7	7	9	9	9	9	7	7	9	7	7	9	7	7
C_{19}	3	3	3	3	3	1	5	7	5	7	7	5	3	3	3	3	3	1
C_{20}	1	1	1	1	1	1	1	1	1	1	1	1	1	1	1	1	1	1
	A7						A8						A9					
C_1	3	5	1	3	1	3	5	7	7	3	5	5	3	5	5	3	3	3
C_2	9	9	9	9	7	9	7	7	7	7	5	9	1	3	1	1	1	1
C_3	5	5	5	7	5	5	7	9	7	9	7	7	9	9	9	9	9	9
C_4	3	5	5	5	5	1	7	7	7	9	5	7	9	9	9	9	9	9
C_5	5	7	7	5	5	3	7	9	9	7	7	7	9	9	9	9	9	9
C_6	1	1	1	3	3	1	5	3	5	5	5	3	1	1	1	3	3	1
C_7	7	9	7	5	7	7	1	1	1	1	1	1	9	9	9	9	9	9
C_8	3	3	5	3	3	5	7	9	9	7	7	7	9	9	9	9	9	9
C_9	9	9	9	9	9	7	7	3	5	5	5	3	3	1	1	1	1	1
C_{10}	5	7	7	5	5	5	9	9	9	7	7	7	9	9	9	7	7	7
C_{11}	1	1	1	1	1	1	3	3	1	1	1	1	1	1	1	1	1	1
C_{12}	5	5	3	3	3	5	9	9	7	9	7	9	9	9	7	9	9	9
C_{13}	7	7	7	7	7	9	7	9	7	9	9	9	9	9	9	9	9	9
C_{14}	7	5	7	7	7	7	9	7	9	9	7	9	1	1	1	1	3	1
C_{15}	1	1	1	1	1	1	1	3	1	1	3	3	1	1	1	1	1	1
C_{16}	7	7	5	5	5	5	5	1	3	1	1	1	9	9	9	7	7	9
C_{17}	5	9	7	5	5	5	3	3	3	3	3	3	1	1	1	1	1	1
C_{18}	7	3	5	5	3	5	7	5	5	5	5	7	1	1	1	1	1	3
C_{19}	3	5	3	5	3	1	1	1	3	1	1	1	3	3	3	3	3	1
C_{20}	3	3	3	3	3	3	1	1	1	3	3	1	1	1	1	1	1	1

The evaluation of the alternatives by six experts shown in Table 2 was carried out on the basis of the linguistic scale defined in [5], taking into account the type of criteria.

The transformation of individual matrices is completed using the Equations (A1)–(A6) in the following way: For the first criterion $\tilde{c}_1 = \{9,9,9,7,7,7\}$ it will be:

$$\underline{Lim}(9) = \frac{1}{6}(9+9+9+7+7+7) = 8.00, \overline{Lim}(9) = 9.00$$

$$\underline{Lim}(7) = 7.00, \overline{Lim}(7) = \frac{1}{6}(9+9+9+7+7+7) = 8.00$$

$$RN(c_1^1) = RN(c_1^2) = RN(c_1^3) = [8.00, 9.00]; RN(c_1^4) = RN(c_1^5) = RN(c_1^6)[7.00, 8.00]$$

$$c_1^L = \frac{c_1^1 + c_1^2 + c_1^3 + c_1^4 + c_1^5 + c_1^6}{n} = \frac{8.00 + 8.00 + 8.00 + 7.00 + 7.00 + 7.00}{6} = 7.50$$

$$c_1^U = \frac{c_1^1 + c_1^2 + c_1^3 + c_1^4 + c_1^5 + c_1^6}{n} = \frac{9.00 + 9.00 + 9.00 + 8.00 + 8.00 + 8.00}{6} = 8.50$$

For the second criterion according to the first alternative $\tilde{c}_2 = \{5,7,7,5,5,7\}$

$$\begin{aligned}
\underline{Lim}(5) &= 5.00, \overline{Lim}(5) = \frac{1}{6}(5+7+7+5+5+7) = 6.00 \\
\underline{Lim}(7) &= \frac{1}{6}(5+7+7+5+5+7) = 6.00, \overline{Lim}(7) = 7.00 \\
RN(c_2^1) = RN(c_2^4) &= RN(c_2^5) = [5.00, 6.00]; RN(c_2^2) = RN(c_2^3) = RN(c_2^6)[6.00, 7.00] \\
c_2^L &= \frac{5.00+6.00+6.00+5.00+5.00+6.00}{6} = 5.50 \\
c_2^U &= \frac{6.00+7.00+7.00+6.00+6.00+7.00}{6} = 6.50
\end{aligned}$$

For the eighth criterion $\tilde{c}_8 = \{7,7,9,7,7,5\}$:

$$\begin{aligned}
\underline{Lim}(7) &= \frac{1}{5}(7+7+7+7+5) = 6.60, \overline{Lim}(7) = \frac{1}{5}(7+7+9+7+7) = 7.40 \\
\underline{Lim}(9) &= \frac{1}{6}(7+7+9+7+7+5) = 7.00, \overline{Lim}(9) = 9.00 \\
\underline{Lim}(5) &= 5.00, \overline{Lim}(5) = \frac{1}{6}(7+7+9+7+7+5) = 7.00 \\
RN(c_8^1) = RN(c_8^2) &= RN(c_8^4) = RN(c_8^5) = [6.60, 7.40]; RN(c_8^3) = [7.00, 9.00]; RN(c_8^6) = [5.00, 7.00] \\
c_8^L &= \frac{6.60+6.60+7.00+6.60+6.60+5.00}{6} = 6.40 \\
c_8^U &= \frac{7.40+7.40+9.00+7.40+7.40+7.00}{6} = 7.60
\end{aligned}$$

For the twentieth criterion according to the first alternative $\tilde{c}_{20} = \{7,3,5,7,7,3\}$

$$\begin{aligned}
\underline{Lim}(3) &= 3.00, \overline{Lim}(3) = \frac{1}{6}(7+3+5+7+7+3) = 5.33 \\
\underline{Lim}(5) &= \frac{1}{3}(3+5+3) = 3.67, \overline{Lim}(5) = \frac{1}{4}(7+5+7+7) = 6.50 \\
\underline{Lim}(7) &= \frac{1}{6}(7+3+5+7+7+3) = 5.33, \overline{Lim}(7) = 7.00 \\
RN(c_{20}^1) = RN(c_{20}^4) &= RN(c_{20}^5) = [5.33, 7.00]; RN(c_{20}^2) = RN(c_{20}^6) = [3.00, 5.33]; RN(c_{20}^3) = [3.67, 6.50] \\
c_{20}^L &= \frac{5.33+3.00+3.67+5.33+5.33+3.00}{6} = 4.28 \\
c_{20}^U &= \frac{7.00+5.33+6.50+7.00+7.00+5.33}{6} = 6.36
\end{aligned}$$

In an identical way, the other values shown in Table 3 are also calculated.

Table 3 shows a part of the calculation that refers to obtaining an aggregated initial group rough matrix for the first alternative. In the same way, other values for other alternatives are calculated, so the initial aggregated rough matrix is presented in Table 4.

Table 4 shows the initial rough matrix from the second step of the developed rough ARAS approach. The last column marked with gray indicates the optimal values of the alternatives obtained by taking maximum or minimum values depending on the type of criteria. The total number of criteria belonging to a beneficial group is 14, while the remaining six criteria are of cost type, and they are C_{14}, C_{15}, C_{17}, C_{18}, C_{19} and C_{20}.

In the third step of the rough ARAS approach, it is necessary to apply a three-phase procedure for normalizing the initial rough matrix. In this step, it is also important which criteria are benefit ones, and which are the cost ones, and, accordingly, the normalization is carried out in the following way.

Applying Equation (2):

$$\text{(a) } n_{11} = \frac{[7.50; 8.50]}{[49.39; 60.61]}$$

i.e., (3):

$$n_{11} = \frac{[7.50; 8.50]}{[60.61; 49.39]} = [0.12; 0.17]$$

The normalized value of the first alternative for the first criterion was obtained. The normalization procedure for all the other criteria that belong to the benefit ones is the same. The second and third phases of normalization include the application of Equations (4) and (5): for cost criteria, Equation (4) should be first applied in order to obtain inverse values for $\left[x_{ij}{}^{L}; x_{ij}^{U}\right]$.

For the first alternative according to the fourteenth criterion, the inverse value (Table 5) is:

$$\text{(b) } \overline{x_{114}} = \left[\frac{1}{3.47}; \frac{1}{1.88}\right] = [0.29, 0.53]$$

Applying Equation (5), a normalized value is obtained (Table 6):

$$\text{(c) } n_{114} = \left[\frac{0.29}{4.45}; \frac{0.53}{2.97}\right] = [0.06, 0.18]$$

Table 5 shows inverse values from the initial rough matrix, i.e., values obtained by applying the second phase of normalization, which is the third step of the developed approach. Table 6 gives an overview of a complete normalized matrix after the application of the above three phases.

Table 3. Calculation of aggregated values for the initial group rough matrix for the first alternative.

	Low Limit						Upper Limit							
	E_1	E_2	E_3	E_4	E_5	E_6	E_1	E_2	E_3	E_4	E_5	E_6	A_1	
C_1	8.00	8.00	8.00	7.00	7.00	7.00	9.00	9.00	9.00	8.00	8.00	8.00	7.50	8.50
C_2	5.00	6.00	6.00	5.00	5.00	6.00	6.00	7.00	7.00	6.00	6.00	7.00	5.50	6.50
C_3	6.50	7.33	6.50	7.33	5.00	6.50	7.80	9.00	7.80	9.00	7.33	7.80	6.53	8.12
C_4	6.33	6.33	6.33	6.33	5.00	5.00	7.00	7.00	7.00	7.00	6.33	6.33	5.89	6.78
C_5	7.00	7.67	7.67	7.00	7.00	7.00	7.67	9.00	9.00	7.67	7.67	7.67	7.22	8.11
C_6	7.00	7.00	7.67	7.67	7.00	7.00	7.67	7.67	9.00	9.00	7.67	7.67	7.22	8.11
C_7	3.67	3.67	3.00	3.00	3.00	3.00	5.00	5.00	3.67	3.67	3.67	3.67	3.22	4.11
C_8	6.60	6.60	7.00	6.60	6.60	5.00	7.40	7.40	9.00	7.40	7.40	7.00	6.40	7.60
C_9	7.33	5.00	6.50	6.50	7.33	6.50	9.00	7.33	7.80	7.80	9.00	7.80	6.53	8.12
C_{10}	5.00	5.67	5.67	5.00	5.00	5.00	5.67	7.00	7.00	5.67	5.67	5.67	5.22	6.11
C_{11}	8.67	8.67	8.67	7.00	8.67	8.67	9.00	9.00	9.00	8.67	9.00	9.00	8.39	8.95
C_{12}	4.67	4.67	4.67	4.67	3.00	4.67	5.00	5.00	5.00	5.00	4.67	5.00	4.39	4.95
C_{13}	7.00	7.00	7.00	7.67	7.00	7.67	7.67	7.67	7.67	9.00	7.67	9.00	7.22	8.11
C_{14}	2.20	1.00	1.00	2.20	2.20	2.67	3.50	2.67	2.67	3.50	3.50	5.00	1.88	3.47
C_{15}	1.00	3.33	2.50	3.33	2.50	2.50	3.33	5.00	3.80	5.00	3.80	3.80	2.53	4.12
C_{16}	3.67	3.00	3.00	3.00	3.00	3.00	7.00	3.67	3.67	3.67	3.67	3.67	3.11	4.23
C_{17}	9.00	9.00	9.00	9.00	9.00	9.00	9.00	9.00	9.00	9.00	9.00	9.00	9.00	9.00
C_{18}	7.67	7.00	7.00	7.67	7.00	7.00	9.00	7.67	7.67	9.00	7.67	7.67	7.22	8.11
C_{19}	1.00	1.00	1.00	1.00	1.00	1.00	1.00	1.00	1.00	1.00	1.00	1.00	1.00	1.00
C_{20}	5.33	3.00	3.67	5.33	3.00	5.33	7.00	5.33	6.50	7.00	5.33	7.00	4.28	6.36

Table 4. The initial rough matrix.

	A_1	A_2	A_3	A_4	A_5	A_6	A_7	A_8	A_9	A_o
C_1	[7.5, 8.5]	[7.22, 8.11]	[7.22, 8.11]	[3.22, 4.11]	[3.88, 5.47]	[3.22, 4.11]	[1.88, 3.47]	[4.53, 6.12]	[3.22, 4.11]	[7.5, 8.5]
C_2	[5.5, 6.5]	[1.06, 1.61]	[5.5, 6.5]	[3.22, 4.11]	[3.06, 3.61]	[3.22, 4.11]	[8.39, 8.95]	[6.4, 7.6]	[1.06, 1.61]	[8.39, 8.95]
C_3	[6.53, 8.12]	[7.22, 8.11]	[6.8]	[1.5, 2.5]	[1.5, 2.5]	[1.5, 2.5]	[5.06, 5.61]	[7.22, 8.11]	[9.9]	[9.9]
C_4	[5.89, 6.78]	[6.8]	[5.89, 6.78]	[1.22, 2.11]	[1.22, 2.11]	[1.22, 2.11]	[3.17, 4.77]	[6.4, 7.6]	[9.9]	[9.9]
C_5	[7.22, 8.11]	[4.53, 6.12]	[7.22, 8.11]	[1.88, 3.47]	[1.88, 3.47]	[1.88, 3.47]	[4.53, 6.12]	[7.22, 8.11]	[9.9]	[9.9]
C_6	[7.22, 8.11]	[6.39, 6.95]	[6.39, 6.95]	[8.39, 8.95]	[7.7]	[8.39, 8.95]	[1.22, 2.11]	[3.89, 4.78]	[1.22, 2.11]	[8.39, 8.95]
C_7	[3.22, 4.11]	[4.39, 4.95]	[2.4, 3.6]	[1.1]	[3.3]	[1.06, 1.61]	[6.4, 7.6]	[1.1]	[9.9]	[9.9]
C_8	[6.4, 7.6]	[5.22, 6.11]	[5.22, 6.11]	[1.1]	[3.3]	[1.1]	[3.22, 4.11]	[7.22, 8.11]	[9.9]	[9.9]
C_9	[6.53, 8.12]	[3.88, 5.47]	[5.88, 7.47]	[5.88, 7.47]	[4.4, 5.6]	[4.53, 6.12]	[8.39, 8.95]	[3.88, 5.47]	[1.06, 1.61]	[8.39, 8.95]
C_{10}	[5.22, 6.11]	[3.22, 4.11]	[3.22, 4.11]	[5.22, 6.11]	[5.22, 6.11]	[5.23, 6.83]	[5.22, 6.11]	[7.5, 8.5]	[7.5, 8.5]	[7.5, 8.5]
C_{11}	[8.39, 8.95]	[3.49, 5.22]	[7.5, 8.5]	[3.49, 5.22]	[3.49, 5.22]	[3.49, 5.22]	[1.1]	[1.22, 2.11]	[1.1]	[8.39, 8.95]
C_{12}	[4.39, 4.95]	[4.4, 5.6]	[3.5, 4.5]	[1.1]	[1.1]	[1.1]	[3.5, 4.5]	[7.89, 8.78]	[8.39, 8.95]	[8.39, 8.95]
C_{13}	[7.22, 8.11]	[9.9]	[4.53, 6.12]	[3.88, 5.47]	[4.53, 6.12]	[3.88, 5.47]	[7.06, 7.61]	[7.89, 8.78]	[9.9]	[9.9]
C_{14}	[1.88, 3.47]	[1.88, 3.47]	[1.88, 3.47]	[4.39, 4.95]	[4.39, 4.95]	[4.39, 4.95]	[6.39, 6.95]	[7.89, 8.78]	[1.06, 1.61]	[1.06, 1.61]
C_{15}	[2.53, 4.12]	[7.5, 8.5]	[7.5, 8.5]	[1.1]	[1.1]	[1.1]	[1.1]	[1.5, 2.5]	[1.1]	[1.1]
C_{16}	[3.11, 4.23]	[8.39, 8.95]	[3.11, 4.23]	[1.23, 2.83]	[3.11, 4.23]	[1.23, 2.83]	[5.22, 6.11]	[1.23, 2.83]	[7.89, 8.78]	[8.39, 8.95]
C_{17}	[9.9]	[1.5, 2.5]	[8.39, 8.95]	[1.1]	[1.1]	[1.1]	[5.23, 6.83]	[3.3]	[1.1]	[1.1]
C_{18}	[7.22, 8.11]	[4.53, 6.12]	[5.5, 6.5]	[7.22, 8.11]	[7.89, 8.78]	[7.22, 8.11]	[3.88, 5.47]	[5.22, 6.11]	[1.06, 1.61]	[1.06, 1.61]
C_{19}	[1.1]	[1.06, 1.61]	[1.22, 2.11]	[2.39, 2.95]	[5.5, 6.5]	[2.39, 2.95]	[2.53, 4.12]	[1.06, 1.61]	[2.39, 2.95]	[1.06, 1.61]
C_{20}	[4.28, 6.36]	[1.06, 1.61]	[6.28, 8.36]	[1.1]	[1.1]	[1.1]	[3.3]	[1.22, 2.11]	[1.1]	[1.1]

Table 5. The inverse values of cost criteria obtained by applying Equation (4).

	C_{14}		C_{15}		C_{17}		C_{18}		C_{19}	
A_1	0.29	0.53	0.24	0.40	0.11	0.11	0.12	0.14	1.00	1.00
A_2	0.29	0.53	0.12	0.13	0.40	0.67	0.16	0.22	0.62	0.94
A_3	0.29	0.53	0.12	0.13	0.11	0.12	0.15	0.18	0.47	0.82
A_4	0.20	0.23	1.00	1.00	1.00	1.00	0.12	0.14	0.34	0.42
A_5	0.20	0.23	1.00	1.00	1.00	1.00	0.11	0.13	0.15	0.18
A_6	0.20	0.23	1.00	1.00	1.00	1.00	0.12	0.14	0.34	0.42
A_7	0.14	0.16	1.00	1.00	0.15	0.19	0.18	0.26	0.24	0.40
A_8	0.11	0.13	0.40	0.67	0.33	0.33	0.16	0.19	0.62	0.94
A_9	0.62	0.94	1.00	1.00	1.00	1.00	0.62	0.94	0.34	0.42
A_o	0.62	0.94	1.00	1.00	1.00	1.00	0.62	0.94	0.62	0.94
SUM	2.97	4.45	6.88	7.33	6.10	6.42	2.39	3.28	4.75	6.48

Table 6. The normalized matrix.

	C1		C2		C3		C8		C9		C10		C19		C20	
A1	0.12	0.17	0.10	0.14	0.10	0.15	0.12	0.15	0.10	0.15	0.08	0.11	0.15	0.21	0.02	0.03
A2	0.12	0.16	0.02	0.04	0.11	0.15	0.09	0.12	0.06	0.10	0.05	0.07	0.10	0.20	0.08	0.14
A3	0.12	0.16	0.10	0.14	0.09	0.15	0.09	0.12	0.09	0.14	0.05	0.07	0.07	0.17	0.02	0.02
A4	0.05	0.08	0.06	0.09	0.02	0.05	0.02	0.02	0.09	0.14	0.08	0.11	0.05	0.09	0.13	0.15
A5	0.06	0.11	0.06	0.08	0.02	0.05	0.05	0.06	0.07	0.11	0.08	0.11	0.02	0.04	0.13	0.15
A6	0.05	0.08	0.06	0.09	0.02	0.05	0.02	0.02	0.07	0.12	0.08	0.12	0.05	0.09	0.13	0.15
A7	0.03	0.07	0.16	0.20	0.08	0.10	0.06	0.08	0.13	0.17	0.08	0.11	0.04	0.08	0.04	0.05
A8	0.07	0.12	0.12	0.17	0.11	0.15	0.13	0.16	0.06	0.10	0.12	0.15	0.10	0.20	0.06	0.12
A9	0.05	0.08	0.02	0.04	0.14	0.17	0.16	0.18	0.02	0.03	0.12	0.15	0.05	0.09	0.13	0.15
Ao	0.12	0.17	0.16	0.20	0.14	0.17	0.16	0.18	0.13	0.17	0.12	0.15	0.10	0.20	0.13	0.15

The fourth step is the aggregation of the normalized rough matrix by multiplying all the values of the normalized matrix with the weighted values of criteria by applying Equation (6). The weighted normalized matrix is shown in Table 7.

Table 7. The weighted normalized rough matrix.

	C_1		C_2		C_8		C_9		C_{10}		C_{19}		C_{20}	
A_1	0.009	0.012	0.007	0.010	0.006	0.008	0.005	0.008	0.004	0.006	0.006	0.008	0.001	0.001
A_2	0.008	0.011	0.001	0.002	0.005	0.007	0.003	0.006	0.003	0.004	0.004	0.008	0.003	0.005
A_3	0.008	0.011	0.007	0.010	0.005	0.007	0.005	0.008	0.003	0.004	0.003	0.007	0.001	0.001
A_4	0.004	0.006	0.004	0.006	0.001	0.001	0.005	0.008	0.004	0.006	0.002	0.003	0.005	0.006
A_5	0.004	0.008	0.004	0.005	0.003	0.003	0.004	0.006	0.004	0.006	0.001	0.001	0.005	0.006
A_6	0.004	0.006	0.004	0.006	0.001	0.001	0.004	0.006	0.004	0.007	0.002	0.003	0.005	0.006
A_7	0.002	0.005	0.011	0.014	0.003	0.004	0.007	0.009	0.004	0.006	0.001	0.003	0.002	0.002
A_8	0.005	0.009	0.008	0.012	0.007	0.009	0.003	0.006	0.006	0.008	0.004	0.008	0.002	0.005
A_9	0.004	0.006	0.001	0.002	0.009	0.010	0.001	0.002	0.006	0.008	0.002	0.003	0.005	0.006
A_o	0.009	0.012	0.011	0.014	0.009	0.010	0.007	0.009	0.006	0.008	0.004	0.008	0.005	0.006

In the fifth step, it is necessary to determine the matrix S_i by summing all the values per rows of the previous weighted matrix (7). After that, applying Equation (8), the degree of usefulness K_i is calculated and in the last seventh step, the ranking of the alternatives according to decreasing order is performed. Operations with rough numbers are performed using Equations (15)–(20). The results of the last three steps of the rough ARAS approach are shown in Table 8.

Table 8. Results and ranking of alternatives using the rough additive ratio assessment (ARAS) approach.

	Si	Ki	Rank
A_1	[0.09, 0.13]	[0.48, 0.89]	2
A_2	[0.08, 0.13]	[0.43, 0.85]	3
A_3	[0.08, 0.12]	[0.41, 0.81]	5
A_4	[0.06, 0.09]	[0.31, 0.59]	8
A_5	[0.06, 0.09]	[0.33, 0.6]	7
A_6	[0.06, 0.09]	[0.31, 0.59]	9
A_7	[0.07, 0.1]	[0.38, 0.71]	6
A_8	[0.08, 0.12]	[0.43, 0.81]	4
A_9	[0.11, 0.15]	[0.58, 1.01]	1

Alternative 9 represents the best solution, which means that the transportation company from Serbia is in first place with the best performance indicators compared to other alternatives. In second and third places are transportation companies from Bosnia and Herzegovina. The worst-ranked companies are from Libya, which is in some way an understandable and expected outcome taking into account recent events in that country.

5. Sensitivity Analysis and Discussion

The sensitivity analysis performed in this paper includes three phases, which determine the sensitivity and validity of the results obtained. The first phase consists of the formation of 10 different scenarios in which the values of weight coefficients are simulated and the rank obtained using Spearman's correlation coefficient is checked. In the second phase, different methods of multi-criteria decision-making are applied in a rough form, and in the third phase, a statistical correlation checking is performed for all applied methods.

Figure 3 shows the ranks of alternatives throughout 10 scenarios formed. The first four scenarios are formed in such a way that five criteria per scenario increase by 0.100 or even 145% and more. In the first scenario, the first five criteria change the value, while the others remain unchanged in sequence to the fourth scenario. The fifth scenario implies a reduction in the value of the first 10 criteria by 100% of their own value. The sixth scenario is formed in the same way as the fifth, with the increased values of the other 10 criteria. In the seventh scenario, the values of C_1–C_{10} criteria are reduced by 100% of their own value, and the C_{11}–C_{20} criteria are increased by the same percentage. In the eighth scenario, the first five criteria are eliminated, so that the alternatives are ranked on the basis of 15 indicators, while in the ninth scenario, the five worst criteria are eliminated. The last, tenth scenario implies that all criteria are equally important with a share of 0.100.

Figure 3 shows the ranks of alternatives through the formed scenarios that have previously been explained in detail. The second, fifth, sixth and tenth scenarios encompass identical ranks as in the initial scenario. In the third scenario, the only difference in rank is the third and eighth alternatives that change places, while in the fourth, the difference is in the first two alternatives that also change positions. Scenarios 7 and 8 have an identical correlation with the initial scenario in which the first two alternatives change places and alternatives 6 and 8, so the correlation is a bit smaller compared to the scenarios previously explained. The first scenario presents greater differences in ranks than the initial scenario, which means that the increase in significance of the first five criteria significantly influences the rank. The biggest difference in the scenarios formed is in the ninth scenario when five of the worst criteria are eliminated, which means that these criteria still play an important role in measuring performance in transportation companies.

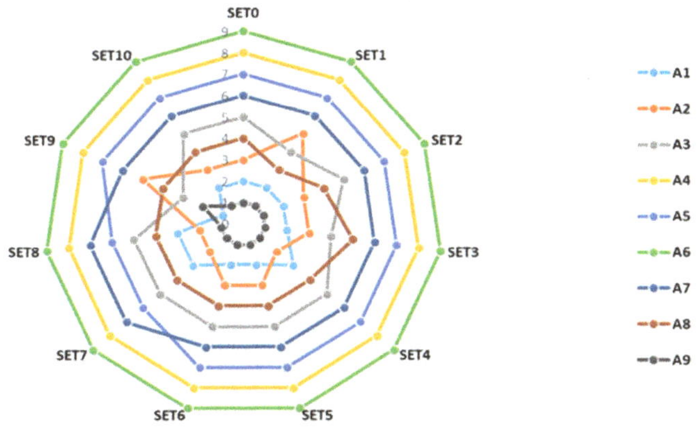

Figure 3. Rank of alternatives through various scenarios.

In Figure 4, Spearman's coefficient of rank correlation is shown in all scenarios compared to the initial one.

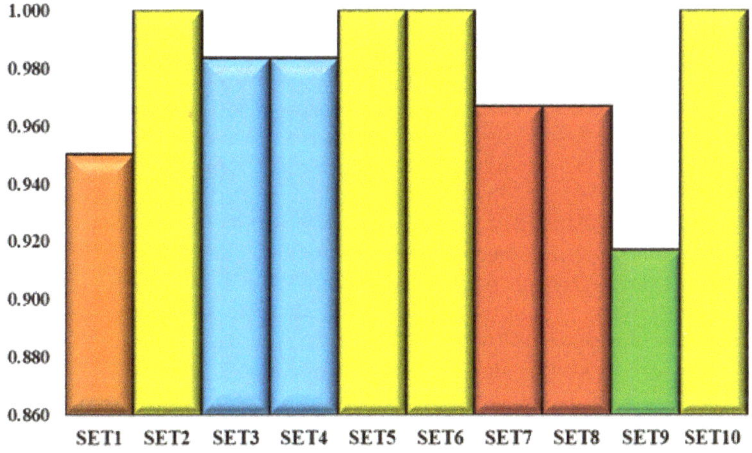

Figure 4. The Spearman's correlation coefficient (SCC) through 10 scenarios.

Figure 4 shows the SCC through all the scenarios for which it can be seen that the model is sensitive to changes of the weights of criteria and that each indicator plays an important role in measuring performance in transportation companies. Spearman's coefficient of correlation ranges in the scope of 0.917–1.00, which represents a high degree of correlation, and the results obtained using the developed rough ARAS model are considered stable. The average SCC value of all 10 scenarios formed in relation to the initial rank is 0.977.

Figure 5 shows the second phase of the sensitivity analysis, which includes the application of different approaches: rough WASPAS (weighted aggregated sum product assessment) [67] rough SAW (simple additive weighting) [5], rough MABAC (multi-attributive border approximation area comparison) [72] and rough EDAS (evaluation based on distance from average solution) [73] to the same multi-criteria model. Extension of the MCDM methods with rough numbers has become common in the last few years. The reason for using a rough SAW, rough MABAC and rough EDAS method is that they in a very short time found application in different areas. These methods are well accepted by the wider academic community, which is confirmed by the number of their citations. Rough WASPAS is a method developed a few months ago and is often used in previous forms (crisp, fuzzy, grey) in the sensitivity analysis [74].

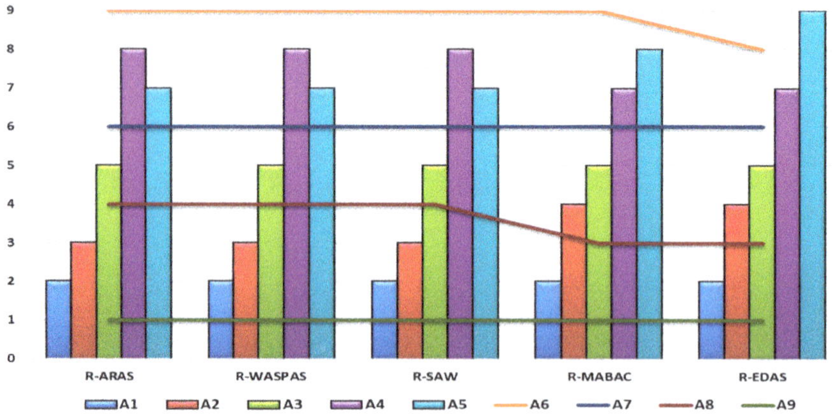

Figure 5. Ranking alternatives according to various rough multi-criteria decision-making models (MCDM) methods.

In Figure 5, it can be seen that the best transportation company does not change the rank in any approach, i.e., in all scenarios it is the best-ranked company. In addition, the following alternatives remain in their original positions by applying all approaches: the first, third and seventh alternatives are in the second, fifth and sixth place, respectively. The second alternative using rough ARAS, rough WASPAS and rough SAW takes the third position, while using rough MABAC and rough EDAS is in fourth place. By applying these two methods, the fourth alternative is best-ranked and takes the seventh position, while applying other approaches it is in the eighth position. The rank of the fifth alternative ranges from the seventh (rough ARAS, rough WASPAS and rough SAW) to ninth place (rough EDAS), while using rough MABAC it is in eighth place. The sixth alternative is worst-ranked in all approaches except rough EDAS when it takes the last position. The eighth alternative occupies the fourth position (rough ARAS, rough WASPAS and rough SAW), while applying rough MABAC and rough EDAS it is in a high third position.

Table 9 shows the third phase in the sensitivity analysis, which includes checking the correlation of ranks using different approaches.

Table 9. Statistical comparison of ranks for different methods.

Methods	R-ARAS	R-WASPAS	R-SAW	R-MABAC	R-EDAS	Average
R-ARAS	1.000	1.000	1.000	0.967	0.933	0.980
R-WASPAS	-	1.000	1.000	0.967	0.933	0.975
R-SAW	-	-	1.000	0.967	0.933	0.967
R-MABAC	-	-	-	1.000	0.983	0.992
R-EDAS	-	-	-	-	1.000	1.000
		Overall average				0.983

Table 9 shows the SCC for all five applied approaches according to which it can be concluded that the ranks in all approaches are in very high correlation. The above is confirmed by the average value of SCC for all approaches which is 0.983. Spearman's correlation coefficient ranges in the scope from 0.933–1.00. The developed rough ARAS has a full correlation of ranks with rough WASPAS and rough SAW. It has a slightly smaller correlation (0.967) with rough MABAC, while the smallest correlation with rough EDAS (0.933). Since rough WASPAS and rough SAW have a full correlation with rough ARAS, it is obvious that they have the same correlation with other approaches as rough ARAS has. Rough MABAC and rough EDAS have a correlation coefficient of 0.983, which is a consequence of the change of rank in two alternatives.

An explanation of the obtained results can be seen throughout the current state of the transport system in the observed countries. Libya is plagued by a poor transport system, which leads to enormous problems of traffic congestion and pollution. The lack of an integrated transport system has made logistics operations in the country primitive. City-to-city freight, sometimes over long distances of more than 1500 kilometers, is only carried out by people-owned vehicles or small shipping companies. This is also the case for logistics within cities, where they are carried out by inadequate fleets and trucks using the inner city roads. For instance, local product distributions are carried out by the same producing companies, which tend to own freight vehicles for distribution within cities, as well as larger trucks for longer haulage. The same applies to some food manufacturers, as well as to some companies specializing in supplying various goods and distributing them to retailers.

Road transportation is the most developed mode of transportation in Bosnia and Herzegovina, although by the middle of 2003, when it obtained the first 11 km of a modern highway, B&H was the only country in the region of south-east Europe that did not have a single kilometer of modern highway. With the construction of new road networks, road freight transportation has gained significance. Companies with their own fleets, which deal with both domestic and international freight transportation, have been established. Most frequently, cargo transportation is performed by furgon trucks, tanks and trailer trucks, but a large number of means of transportation exceeds the permitted age limit. Although road networks have already been constructed and reconstructed, there is still a need for improvement and the construction of a new road infrastructure; because of the aforementioned factors, there is often a delay in road traffic, and as a consequence, the delivery of goods to the recipient regarding transportation at a national level. Regarding international freight transportation, these problems are much less common.

Road transportation in Serbia is at a bit higher in its level of development than in BiH and represents a dynamic and dominant mode of transportation. Although transportation in Serbia is more developed compared to that in B&H, it still largely lags behind the region due to the poor condition of the existing road network, as well as a slow construction of new transport corridors. In the last few years, there has been a tendency for the growth of freight transportation and an increase in the number of road vehicles. However, this growth represents a heavy burden for already bad road infrastructure. The current road transportation is characterized by an ageing vehicle fleet, which affects safety in traffic, and poor road infrastructure causes damage to road vehicles.

It is important to note that in the last few years, the number of transportation companies offering logistics outsourcing in Bosnia and Herzegovina and Serbia has increased, and consequently, the quality of a complete transport system and, to a certain extent, logistics has also increased.

6. Conclusions

In this research, a model was created for performance measurement in transportation companies based on 20 performance indicators, which were identified in the previous study based on the assessment of a large number of decision-makers. The key contribution of this paper is the improvement and enrichment of the methodology for treating uncertainties in the field of group multi-criteria decision-making through the development of a novel rough ARAS approach. The developed approach allows bridging the gap that currently exists in the methodology for measuring and monitoring performance indicators in the logistics subsystem of transportation. The proposed model in this paper provides an opportunity to improve the efficiency of the operations of transportation companies in developing countries, and encourages the development of competitiveness as an important factor in a success of every company. The novel rough ARAS approach is applied to evaluate the performance of nine transportation companies located in the territory of Bosnia and Herzegovina, Libya and Serbia, i.e., in developing countries. The results obtained are checked throughout a sensitivity analysis forming different scenarios, simulating the weight values of criteria and applying different approaches developed over the last three years. The stability of the model is verified throughout a statistical correlation coefficient, which shows a high correlation of ranks in all scenarios. The results show that transportation companies from Serbia and Bosnia and Herzegovina are significantly ahead of the transportation companies from Libya. The measurements presented in this paper can serve as a guide to transportation companies in developing countries, since they provide an insight into the calculated values of the 20 most important performance indicators and, thus, unambiguously show them which indicators to pay attention to in order to rationalize their business. In addition, the measured values of the most significant performance should further stimulate the development of competitiveness both among developing countries and among countries that have developed logistics and good-quality transportation. We ranked nine companies from three different countries, but that does not mean that we optimized this. The model ranked the performance in transportation and the best alternative having the best performance in comparison to another; but this does not mean that the best alternative has optimal performance. Taking all this into account and the methodology developed in this paper, there is a possibility for a post-analysis in all transportation companies evaluated here, where the best-ranked ones can serve others as a benchmark, which can be one of the directions for future research. Also, one of the guidelines for future research in a practical aspect should be constant monitoring and measurement of the worst performance and their improvement. Moreover, improvement from academic and practical aspects can develop the MCDM method which can be used only for the measurement of perfomance in transportation. A similar approach to solve decision problems in a supply chain is developed in [75].

Author Contributions: Each author has participated and contributed sufficiently to take public responsibility for appropriate portions of the content.

Funding: This research received no external funding.

Conflicts of Interest: The authors declare no conflict of interest.

Appendix A

Rough Set Theory and Operations with Rough Numbers

In rough set theory, any vague idea can be represented as a couple of exact concepts based on the lower and upper approximations.

Suppose U is the universe which contains all the objects, Y is an arbitrary object of U, R is a set of t classes $\{G_1, G_2, \ldots, G_t\}$ that cover all the objects in U, $R = \{G_1, G_2, \ldots, G_t\}$. If these classes are ordered as $\{G_1 < G_2 < \ldots < G_t\}$, then $\forall Y \in U, G_q \in R, 1 \leq q \leq t$, the lower approximation $(\underline{Apr}(G_q))$, upper approximation $(\overline{Apr}(G_q))$ and boundary region $(\overline{Bnd}(G_q))$ of class G_q are, defined as:

$$\underline{Apr}(G_q) = \{Y \in U/R(Y) \leq G_q\} \tag{A1}$$

$$\overline{Apr}(G_q) = \{Y \in U/R(Y) \geq G_q\} \tag{A2}$$

$$Bnd(G_q) = \cup\{Y \in U/R(Y) \neq G_q\} = \{Y \in U/R(Y) \geq G_q\} \cup \{Y \in U/R(Y) < G_q\} \tag{A3}$$

Then G_q can be shown as a rough number $(RN(G_q))$, which is determined by its corresponding lower limit $(\underline{Lim}(G_q))$ and upper limit $(\overline{Lim}(G_q))$ where:

$$\underline{Lim}(G_q) = \frac{1}{M_L}\sum R(Y)\Big|Y \in \underline{Apr}(G_q) \tag{A4}$$

$$\overline{Lim}(G_q) = \frac{1}{M_U}\sum R(Y)\big|Y \in \overline{Apr}(G_q) \tag{A5}$$

$$RN(G_q) = \left[\underline{Lim}(G_q), \overline{Lim}(G_q)\right] \tag{A6}$$

where M_L, M_U are the numbers of objects that contained in $\underline{Apr}(G_q)$ and $\overline{Apr}(G_q)$, respectively.

The difference between them is expressed as a rough boundary interval $(IRBnd(G_q))$:

$$IRBnd(G_q) = \overline{Lim}(G_q) - \underline{Lim}(G_q) \tag{A7}$$

The operations for two rough numbers $RN(\alpha) = \left[\underline{Lim}(\alpha), \overline{Lim}(\alpha)\right]$ and $RN(\beta) = \left[\underline{Lim}(\beta), \overline{Lim}(\beta)\right]$ are:

Addition (+) of two rough numbers $RN(\alpha)$ and $RN(\beta)$:

$$RN(\alpha) + RN(\beta) = \left[\underline{Lim}(\alpha) + \underline{Lim}(\beta), \overline{Lim}(\alpha) + \overline{Lim}(\beta)\right] \tag{A8}$$

Subtraction $(-)$ of two rough numbers $RN(\alpha)$ and $RN(\beta)$:

$$RN(\alpha) - RN(\beta) = \left[\underline{Lim}(\alpha) - \overline{Lim}(\beta), \overline{Lim}(\alpha) - \underline{Lim}(\beta)\right] \tag{A9}$$

Multiplication $(\times)$ of two rough numbers $RN(\alpha)$ and $RN(\beta)$:

$$RN(\alpha) \times RN(\beta) = \left[\underline{Lim}(\alpha) \times \underline{Lim}(\beta), \overline{Lim}(\alpha) \times \overline{Lim}(\beta)\right] \tag{A10}$$

Division $(/)$ of two rough numbers $RN(\alpha)$ and $RN(\beta)$:

$$RN(\alpha) / RN(\beta) = \left[\underline{Lim}(\alpha) / \overline{Lim}(\beta), \overline{Lim}(\alpha) / \underline{Lim}(\beta)\right] \tag{A11}$$

Scalar multiplication of rough number $RN(\alpha)$, where μ is a non-zero constant:

$$\mu \times RN(\alpha) = \left[\mu \times \underline{Lim}(\alpha), \mu \times \overline{Lim}(\alpha)\right] \tag{A12}$$

References

1. Grabara, J.; Kolcun, M.; Kot, S. The role of information systems in transport logistics. *Int. J. Educ. Res.* **2014**, *2*, 1–8.
2. Borzacchiello, M.T.; Torrieri, V.; Nijkamp, P. An operational information systems architecture for assessing sustainable transportation planning: Principles and design. *Eval. Progr. Plan.* **2009**, *32*, 381–389. [CrossRef] [PubMed]
3. Ghiani, G.; Laporte, G.; Musmanno, R. *Introduction to Logistics Systems Planning and Control*; John Wiley & Sons: West Sussex, UK, 2004.
4. Hashemkhani Zolfani, S.; Zavadskas, E.K.; Turskis, Z. Design of products with both international and local perspectives based on yin-yang balance theory and SWARA method. *Ecomic Res.-Ekomsk Istrživnj* **2013**, *26*, 153–166. [CrossRef]
5. Stević, Ž.; Pamučar, D.; Zavadskas, E.K.; Ćirović, G.; Prentkovskis, O. The Selection of Wagons for the Internal Transport of a Logistics Company: A Novel Approach Based on Rough BWM and Rough SAW Methods. *Symmetry* **2017**, *9*, 264. [CrossRef]
6. Guasch, J.L.; Kogan, J. Inventories and logistic costs in developing countries: Levels and determinants—A red flag for competitiveness and growth. *Revista de la Competencia y de la Propiedad Int.* **2006**, 5.
7. Rantasila, K.; Ojala, L. Measurement of national-level logistics costs and performance. In *Transport Forum Discussion Paper*; International Transport Forum: Paris, France, 2012. [CrossRef]
8. Enoma, A.; Allen, S. Developing key performance indicators for airport safety and security. *Facilities* **2007**, *25*, 296–315. [CrossRef]
9. Kosanke, L.; Schultz, M. Key Performance Indicators for Performance-Based Airport Management from the perspective of airport operations. In Proceedings of the Fifth Air Transport and Operations Symposium (ATOS), Delft, Netherlands, 7–11 September 2015.

10. Gillen, D.; Lall, A. Developing measures of airport productivity and performance: An application of data envelopment analysis. *Transp. Res. E Logist. Transp. Rev.* **1997**, *33*, 261–273. [CrossRef]
11. Lindberg, C.-F.; Tan, S.; Yan, J.Y.; Starfelt, F. Key performance indicators improve industrial performance. *Energy Procedia* **2015**, *75*, 1785–1790. [CrossRef]
12. Lapide, L. What about measuring supply chain performance. *Achiev. Supply Chain Excell. Technol.* **2000**, *2*, 287–297.
13. Chae, B. Developing key performance indicators for supply chain: An industry perspective. *Supply Chain Manag. Int. J.* **2009**, *14*, 422–428. [CrossRef]
14. Parmenter, D. *Key Performance Indicators: Developing, Implementing, and Using Winning KPIs*; John Wiley & Sons: Hoboken, NJ, USA, 2015.
15. Jumadi, H.; Zailani, S. Integrating green innovations in logistics services towards logistics services sustainability: A conceptual paper. *Environ. Res. J.* **2010**, *4*, 261–271. [CrossRef]
16. Solakivi, T.; Ojala, L.; Laari, S.; Lorentz, H.; Toyli, J.; Malmsten, J.; Viherlehto, N. *Finland State of Logistics 2014*; University of Turku: Turku, Finland, 2015.
17. *Drivers of Logistics Performance: A Case Study of Turkey*; International Transport Forum (ITF); OECD: Paris, France, 2015.
18. Ekici, Ş.Ö.; Kabak, Ö.; Ülengin, F. Linking to compete: Logistics and global competitiveness interaction. *Transp. Policy* **2016**, *48*, 117–128. [CrossRef]
19. Hoekman, B.; Nicita, A. Trade policy, trade costs, and developing country trade. *World Dev.* **2011**, *39*, 2069–2079. [CrossRef]
20. Kim, I.; Min, H. Measuring supply chain efficiency from a green perspective. *Manag. Res. Rev.* **2011**, *34*, 1169–1189. [CrossRef]
21. Martí, L.; Puertas, R.; García, L. The importance of the logistics performance index in international trade. *Appl. Econ.* **2014**, *46*, 2982–2992. [CrossRef]
22. Çemberci, M.; Civelek, M.E.; Canbolat, N. The moderator effect of global competitiveness index on dimensions of logistics performance index. *Procedia Soc. Behav. Sci.* **2015**, *195*, 1514–1524. [CrossRef]
23. Civelek, M.E.; Uca, N.; Cemberci, M. The mediator effect of logistics performance index on the relation between global competitiveness index and gross domestic product. *Eur. Sci. J.* **2015**, *11*, 368–375. [CrossRef]
24. Uca, N.; Ince, H.; Sumen, H. The mediator effect of logistics performance index on the relation between corruption perception index and foreign trade volume. *Eur. Sci. J.* **2016**, *12*, 37–45. [CrossRef]
25. Gopal, P.; Thakkar, J. A review on supply chain performance measures and metrics: 2000–2011. *Int. J. Prod. Perform. Manag.* **2012**, *61*, 518–547. [CrossRef]
26. Rajesh, R.; Pugazhendhi, S.; Ganesh, K.; Ducq, Y.; Koh, S.L. Generic balanced scorecard framework for third party logistics service provider. *Int. J. Prod. Econ.* **2012**, *140*, 269–282. [CrossRef]
27. Gutierrez, D.M.; Scavarda, L.F.; Fiorencio, L.; Martins, R.A. Evolution of the performance measurement system in the logistics department of a broadcasting company: An action research. *Int. J. Prod. Econ.* **2015**, *160*, 1–12. [CrossRef]
28. Wang, L.; Zhang, H.; Zeng, Y.R. Fuzzy analytic hierarchy process (FAHP) and balanced scorecard approach for evaluating performance of third-party logistics (TPL) enterprises in Chinese context. *Afr. J. Bus. Manag.* **2012**, *6*, 521–529. [CrossRef]
29. Keebler, J.S.; Plank, R.E. Logistics performance measurement in the supply chain: A benchmark. *Benchmarking Int. J.* **2009**, *16*, 785–798. [CrossRef]
30. Chia, A.; Goh, M.; Hum, S. Performance measurement in supply chain entities: Balanced scorecard perspective. *Benchmarking Int. J.* **2009**, *16*, 605–620. [CrossRef]
31. Jothimani, D.; Sarmah, S.P. Supply chain performance measurement for third party logistics. *Benchmarking Int. J.* **2014**, *21*, 944–963. [CrossRef]
32. Poveda-Bautista, R.; Baptista, D.C.; García-Melón, M. Setting competitiveness indicators using BSC and ANP. *Int. J. Prod. Res.* **2012**, *50*, 4738–4752. [CrossRef]
33. Norin, A. Airport Logistics: Modeling and Optimizing the Turn-around Process. Ph.D. Thesis, Linköping University, Norrköping, Sweden, 2008.
34. Humphreys, I.; Francis, G. Performance measurement: A review of airports. *Int. J. Transp. Manag.* **2002**, *1*, 79–85. [CrossRef]

35. Morrell, P.; Lu, C.H.-Y. Aircraft noise social cost and charge mechanisms—A case study of amsterdam airport schiphol. *Transp. Res. Part D Transp. Environ.* **2000**, *5*, 305–320. [CrossRef]
36. Ignaccolo, M. Environmental capacity: Noise pollution at catania-fontanarossa international airport. *J. Air Transp. Manag.* **2000**, *6*, 191–199. [CrossRef]
37. Konsta, K.; Plomaritou, E. Key performance indicators (kpis) and shipping companies performance evaluation: The case of greek tanker shipping companies. *Int. J. Bus. Manag.* **2012**, *7*. [CrossRef]
38. Nathnail, E.; Gogas, M.; Adamos, G. Urban freight terminals: A sustainability cross-case analysis. *Transp. Res. Procedia* **2016**, *16*, 394–402. [CrossRef]
39. Bentaleb, F.; Mabrouki, C.; Semma, A. Key performance indicators evaluation and performance measurement in dry port-seaport system: A multi criteria approach. *J. ETA Marit. Sci.* **2015**, *3*, 97–116. [CrossRef]
40. Stoilova, S.; Kunchev, L. Study of criteria for evaluation of transportation with intermodal transport. In Proceedings of the 16th International Scientific Conference Engineering for Rural Development, Jelgava, Latvia, 24–26 May 2017.
41. Carlucci, D. Evaluating and selecting key performance indicators: An ANP-based model. *Meas. Bus. Excell.* **2010**, *14*, 66–76. [CrossRef]
42. Alvandi, M.; Fazli, S.; Yazdani, L.; Aghaee, M. An integrated mcdm method in ranking bsc perspectives and key performance indicators (kpis). *Manag. Sci. Lett.* **2012**, *2*, 995–1004. [CrossRef]
43. Mladenovic, G.; Vajdic, N.; Wündsch, B.; Temeljotov-Salaj, A. Use of key performance indicators for ppp transport projects to meet stakeholders' performance objectives. *Built Environ. Proj. Asset Manag.* **2013**, *3*, 228–249. [CrossRef]
44. Podgórski, D. Measuring operational performance of osh management system-a demonstration of ahp-based selection of leading key performance indicators. *Saf. Sci.* **2015**, *73*, 146–166. [CrossRef]
45. Stević, Ž. Izbor i merenje ključnih indikatora performansi u skladišnom sistemu. In Proceedings of the Internacionalni Naučni Skup SM 2015 Strategijski Menadžment i Sistemi Podrške Odlučivanju u Strategijskom Menadžmentu, Subotica-Palić, Serbia, 21 May 2015; pp. 931–938.
46. Adler, N.; Liebert, V.; Yazhemsky, E. Benchmarking airports from a managerial perspective. *Omega* **2013**, *41*, 442–458. [CrossRef]
47. Chung, T.W.; Ahn, W.C.; Jeon, S.M.; Van Thai, V. A benchmarking of operational efficiency in Asia Pacific international cargo airports. *Asian J. Shipp. Logist.* **2015**, *31*, 85–108. [CrossRef]
48. Kılkış, Ş.; Kılkış, Ş. Benchmarking airports based on a sustainability ranking index. *J. Clean. Prod.* **2016**, *130*, 248–259. [CrossRef]
49. Maclean, M.; Harvey, C.; Clegg, S.R. Conceptualizing historical organization studies. *Acad. Manag. Rev.* **2016**, *41*, 609–632. [CrossRef]
50. Schmidberger, S.; Bals, L.; Hartmann, E.; Jahns, C. Ground handling services at european hub airports: Development of a performance measurement system for benchmarking. *Int. J. Prod. Econ.* **2009**, *117*, 104–116. [CrossRef]
51. Akyuz, G.A.; Erkan, T.E. Supply chain performance measurement: A literature review. *Int. J. Prod. Res.* **2010**, *48*, 5137–5155. [CrossRef]
52. Yew Wong, C.; Karia, N. Explaining the competitive advantage of logistics service providers: A resource-based view approach. *Int. J. Prod. Econ.* **2010**, *128*, 51–67. [CrossRef]
53. Qureshi, M.N.; Kumar, D.; Kumar, P. An integrated model to identify and classify the key criteria and their role in the assessment of 3PL services providers. *Asia Pac. J. Mark. Logist.* **2008**, *20*, 227–249. [CrossRef]
54. Shaik, M.N.; Abdul-Kader, W. Performance measurement of reverse logistics enterprise: A comprehensive and integrated approach. *Meas. Bus. Excell.* **2012**, *16*, 23–34. [CrossRef]
55. Zavadskas, E.K.; Stević, Ž.; Tanackov, I.; Prentkovskis, O. A novel multicriteria approach–Rough Step-Wise Weight Assessment Ratio Analysis method (R-SWARA) and its application in logistics. *Stud. Inf. Control* **2018**, *27*, 97–106. [CrossRef]
56. Zavadskas, E.K.; Turskis, Z. A new additive ratio assessment (aras) method in multicriteria decision-making. *Technol. Econ. Dev. Econ.* **2010**, *16*, 159–172. [CrossRef]
57. Turskis, Z.; Zavadskas, E.K. A new fuzzy additive ratio assessment method (aras-f). Case study: The analysis of fuzzy multiple criteria in order to select the logistic centers location. *Transport* **2010**, *25*, 423–432. [CrossRef]
58. Turskis, Z.; Zavadskas, E.K. A novel method for multiple criteria analysis: Grey Additive Ratio Assessment (ARAS-G) method. *Informatica* **2010**, *21*, 597–610.

59. Pawlak, Z. *Rough Sets: Theoretical Aspects of Reasoning about Data*; Springer: Berlin, Germany, 1991.
60. Pawlak, Z. Anatomy of conflicts. *Bull. Eur. Assoc. Theor. Comput. Sci.* **1993**, *50*, 234–247.
61. Mardani, A.; Nilashi, M.; Zavadskas, E.K.; Awang, S.R.; Zare, H.; Jamal, N.M. Decision making methods based on fuzzy aggregation operators: Three decades review from 1986 to 2017. *Int. J. Inf. Technol. Decis. Mak.* **2018**, *17*, 391–466. [CrossRef]
62. Nowaková, J.; Prílepok, M.; Snášel, V. Medical image retrieval using vector quantization and fuzzy s-tree. *J. Med. Syst.* **2016**, *41*. [CrossRef]
63. Kumar, A.; Kumar, D.; Jarial, S.K. A hybrid clustering method based on improved artificial bee colony and fuzzy C-Means algorithm. *Int. J. Artif. Intell.* **2017**, *15*, 24–44.
64. Lukovac, V.; Popović, M. Fuzzy Delphi approach to defining a cycle for assessing the performance of military drivers. *Decis. Mak. Appl. Manag. Eng.* **2018**, *1*, 67–81. [CrossRef]
65. Medina, J.; Ojeda-Aciego, M. Multi-adjoint t-concept lattices. *Inf. Sci.* **2010**, *180*, 712–725. [CrossRef]
66. Pozna, C.; Minculete, N.; Precup, R.-E.; Kóczy, L.T.; Ballagi, Á. Signatures: Definitions, operators and applications to fuzzy modelling. *Fuzzy Sets Syst.* **2012**, *201*, 86–104. [CrossRef]
67. Stojić, G.; Stević, Ž.; Antuchevičienė, J.; Pamučar, D.; Vasiljević, M. A Novel Rough WASPAS Approach for Supplier Selection in a Company Manufacturing PVC Carpentry Products. *Information* **2018**, *9*, 121. [CrossRef]
68. Keršuliene, V.; Zavadskas, E.K.; Turskis, Z. Selection of rational dispute resolution method by applying new step-wise weight assessment ratio analysis (swara). *J. Bus. Econ. Manag.* **2010**, *11*, 243–258. [CrossRef]
69. Petrović, G.S.; Madić, M.; Antucheviciene, J. An approach for robust decision making rule generation: Solving transport and logistics decision making problems. *Expert Syst. Appl.* **2018**, *106*, 263–276. [CrossRef]
70. Karavidić, Z.; Projović, D. A multi-criteria decision-making (MCDM) model in the security forces operations based on rough sets. *Decis. Mak. Appl. Manag. Eng.* **2018**, *1*, 97–120. [CrossRef]
71. Radović, D.; Stević, Ž. Evaluation and selection of KPI in transport using SWARA method. *Transp. Logist.* **2018**, *44*, 60–68.
72. Roy, J.; Chatterjee, K.; Bandhopadhyay, A.; Kar, S. Evaluation and selection of Medical Tourism sites: A rough AHP based MABAC approach. *arXiv* **2016**, arXiv:1606.08962. [CrossRef]
73. Stević, Ž.; Pamučar, D.; Vasiljević, M.; Stojić, G.; Korica, S. Novel integrated multi-criteria model for supplier selection: Case study construction company. *Symmetry* **2017**, *9*, 279. [CrossRef]
74. Vesković, S.; Stević, Ž.; Stojić, G.; Vasiljević, M.; Milinković, S. Evaluation of the railway management model by using a new integrated model DELPHI-SWARA-MABAC. *Decis. Mak. Appl. Manag. Eng.* **2018**. [CrossRef]
75. Dey, B.; Bairagi, B.; Sarkar, B.; Sanyal, S.K. Multi objective performance analysis: A novel multi-criteria decision making approach for a supply chain. *Comput. Ind. Eng.* **2016**, *94*, 105–124. [CrossRef]

symmetry

MDPI

Article

A New Evaluation for Solving the Fully Fuzzy Data Envelopment Analysis with Z-Numbers

Ali Namakin, Seyyed Esmaeil Najafi *, Mohammad Fallah and Mehrdad Javadi

Department of Industrial Engineering, Science and Research Branch, Islamic Azad University, Tehran 46818-53617, Iran; a_namakin@azad.ac.ir (A.N.); Mohammad.fallah43@yahoo.com (M.F.); mjavadi@azad.ac.ir (M.J.)

* Correspondence: seyedesmailnajafi@gmail.com or najafi1515@yahoo.com

Received: 19 August 2018; Accepted: 29 August 2018; Published: 5 September 2018

Abstract: There are numerous models for solving the efficiency evaluation in data envelopment analysis (DEA) with fuzzy input and output data. However, because of the limitation of those strategies, they cannot be implemented for solving fully fuzzy DEA (FFDEA). Furthermore, in real-world problems with imprecise data, fuzziness is not sufficient to consider, and the reliability of the information is also very vital. To overcome these flaws, this paper presented a new method for solving the fully fuzzy DEA model where all parameters are Z-numbers. The new approach is primarily based on crisp linear programming and has a simple structure. Moreover, it is proved that the only existing method to solve FFDEA with Z-numbers is not valid. An example is also presented to illustrate the efficiency of our proposed method and provide an explanation for the content of the paper.

Keywords: data envelopment analysis; Z-numbers; full fuzzy environment; fuzzy efficiency

1. Introduction

Data envelopment analysis (DEA) is a linear programming method for measuring the relative efficiencies of homogeneous decision-making units (DMUs) without knowing production functions (i.e., just by utilizing input and output information) [1,2]. The DEA technique has just been effectively connected in various cases such as broadcasting companies, banking institutions, R&D organizations, health care services, manufacturing, telecommunications, and supply chain management. Classical DEA models, such as CCR and BCC models [1,2], need crisp inputs and outputs, which are not typically accessible in real-world applications. However, the observed values of the input and output information in real-world issues are imprecise or vague [3–12] and imprecise evaluations could also be the end result of the unquantifiable, incomplete and non-available facts.

It is beneficial to consider the information of experts about the parameters as fuzzy data. The idea of the fuzzy set became established in [13]. After this, many researchers have applied this theory to different problems; see [14–22] and references therein. There are also many studies reported utilizing fuzzy set theory in DEA; see [23–33] and references therein.

Sengupta was the first to analyze DEA models in a fuzzy environment [23]. Triantis and Girod [24] combine DEA and fuzzy parametric programming to handle random measurement errors in input and output data. Kao and Liu [24] accompanied the simple concept of transforming a fuzzy DEA model to a crisp DEA model and presented an approach to measure the efficiencies of the DMUs. According to fuzzy arithmetic operations and fuzzy comparisons among fuzzy numbers, Dia [26] planned a model of fuzzy DEA (FDEA). Garcia et al. [27] applied the possibility of DEA version for failure mode and effects analysis (FMEA) and proposed a fuzzy DEA method to figure out ranking indices. Wang et al. [28] developed fuzzy DEA models with fuzzy inputs and outputs through fuzzy arithmetic. To obtain the efficiencies of DMUs as fuzzy numbers, they converted the proposed fuzzy DEA models into three linear programming (LP) models. Wang and Chin [29] utilized a fuzzy expected value approach and suggested the optimistic and the pessimistic efficiencies to solve FDEA. Emrouznejad et al. [30] presented a taxonomy of the FDEA methods and provided a classification scheme with six categories. Puri and Yadav [31] applied the α-cut approach and provided a cross-efficiency technique for a fuzzy DEA model with undesirable fuzzy outputs. Wanke et al. [32] developed a new FDEA model and, using the bootstrap-truncated regressions, evaluated the efficiency of the Mozambican Banks. Hatami-Marbini et al. [33] proposed a comprehensive cross-efficiency fuzzy DEA approach for supplier evaluation. Readers can also refer to [34–43] for reviews of more recent research on fuzzy DEA approaches.

However, because of the limitations of the above methods, they cannot be implemented for solving fully fuzzy DEA (FFDEA), where all the inputs and outputs, as well as the decision variables, are fuzzy numbers. To the best of our knowledge, there is not much research on FFDEA. Hatami-Marbini et al. [44] utilized a fully fuzzy linear programming problem presented by an FFDEA model. Kazemi and Alimi [45] proposed an FFDEA model based on the ranking function. The feature of this proposed model is that it considers three situations for the problem and solves them simultaneously. Puri and Yadav [46] developed a new FFDEA model with undesirable factors and applied this model on multi-component FFDEA. Khaleghi et al. [47], based on simplex techniques and multi-objective optimization, obtain the fuzzy efficiency of FFDEA model.

Although the fuzzy set theory has been introduced as a powerful tool to quantify vague data, and several authors have suggested various fuzzy methods in DEA, there is a key inadequacy in past methodologies. A critical problem is that in classical fuzzy sets, the degree of sureness of information is not taken into account. When dealing with real information, fuzziness is not enough to take into account and the reliability of the information is vital. Recently, Sotoudeh-Anvari et al. [48] suggested a new FFDEA in a fuzzy situation with Z-numbers. The Z-number is a novel fuzzy notion which has more potential to articulate the vague circumstances of real applications. A Z-number has two components used to express a value of an arbitrary variable X, $Z = (A, B)$, where A is an assessment of a value of X and B is an assessment of certainty of A [49]. Sotoudeh-Anvari et al.'s model [48] is based on Kang et al.'s [50] and Allahviranloo et al.'s [51] methods and has low computational intricacy. However, the method has some flaws and is not valid.

In this paper, it is shown that the mentioned model is not true. To solve this drawback, we improved the model and propose a new algorithm. The rest of this work is ordered as follows: Section 2 presents some essential concepts regarding fuzzy set theory and converting a Z-number into a fuzzy number. In Section 3, we study the DEA, fuzzy DEA and fully fuzzified DEA models, briefly. Sotoudeh-Anvari et al.'s model [48] is reviewed in Section 4. Section 5 explains the shortcoming of the existing model [48]. The new FFDEA model with Z-numbers is also proposed in this section. A numerical example is given in Section 6. Finally, the paper is concluded.

2. Preliminaries

We start with some fundamental notations and starter results that we seek advice from later. For details, we refer to [1,13,18,21,48–50].

2.1. Definitions

Definition 1 [18]. *A fuzzy subset $\widetilde{A}$ of a set X is defined by its membership function $\mu_{\widetilde{A}} : X \rightarrow [0,1]$, where the value of $\mu_{\widetilde{A}}(x)$ at x shows the grade of membership of x in $\widetilde{A}$.*

Definition 2 [18]. *A triangular fuzzy number (TFNs) $\widetilde{A}$ can be defined by (a,b,c), where $c \geq b \geq a$. The membership function $\mu_{\widetilde{A}}(x)$ is given by (1):*

$$\mu_{\widetilde{A}}(x) = \begin{cases} \frac{x-a}{b-a} & a \leq x \leq b \\ \frac{x-c}{b-c} & b \leq x \leq c \\ 0 & \text{otherwise} \end{cases} \tag{1}$$

Definition 3 [21]. *Let $\widetilde{A} = (a,b,c)$ be a triangular fuzzy number. Then $\widetilde{A}$ is called a non-negative fuzzy number if and only if $a \geq 0$.*

Definition 4 [21]. *Let $\widetilde{A} = (a,b,c)$ be a triangular fuzzy number. Then $\widetilde{A}$ is called an unrestricted fuzzy number if $a,b,c \in R$.*

Definition 5 [21]. *Consider $\widetilde{A} = (a,b,c)$ and $\widetilde{B} = (d,e,f)$ as two triangular fuzzy numbers, then we have:*

(i) $\widetilde{A} \oplus \widetilde{B} = (a,b,c) \oplus (d,e,f) = (a+d, b+e, c+f)$,
(ii) $\widetilde{A} - \widetilde{B} = (a,b,c) - (d,e,f) = (a-f, b-e, c-d)$,
(iii) $\widetilde{A} \otimes \widetilde{B} = (\min(\gamma), be, \max(\gamma))$ *where,* $\gamma = \{ad, af, cd, cf\}$.

Definition 6 [21]. *Consider $\widetilde{A} = (a,b,c)$ and $\widetilde{B} = (d,e,f)$ as two triangular fuzzy numbers. Then these numbers are equal if and only if $a = d, b = e$ and $c = f$.*

Definition 7 [21]. *Consider $\widetilde{A} = (a,b,c)$ as a triangular fuzzy number. Then the ranking function of $\widetilde{A}$ is defined as follows:*

$$R(\widetilde{A}) = \frac{1}{4}(a + 2b + c)$$

Definition 8 [21]. *Suppose $\widetilde{A}$ and $\widetilde{B}$ be two triangular fuzzy numbers, then*

(i) $\widetilde{A} \leq \widetilde{B}$ *if and only if* $R(\widetilde{A}) \leq R(\widetilde{B})$.
(ii) $\widetilde{A} < \widetilde{B}$ *if and only if* $R(\widetilde{A}) < R(\widetilde{B})$.

Definition 9 [48]. *A triangular fuzzy number can also be defined as $\widetilde{A} = (M, \alpha, \beta)$ which is referred to as a left right (L-R) fuzzy number. M Is the central value, α is the left width (spread) and β is the right width (spread). The membership function also has the following form:*

$$\mu_{\widetilde{A}}(x) = \begin{cases} \frac{x-M+\alpha}{\alpha} & M-\alpha \leq x \leq M \\ \frac{M-x+\beta}{\beta} & M \leq x \leq M+\beta \\ 0 & otherwise \end{cases} \tag{2}$$

Remark 1. *By Definitions 2 and 6, we can see that a triangular fuzzy number $\widetilde{A} = (a,b,c)$ can be represented as $\widetilde{A} = (M, \alpha, \beta)$, where $a = M - \alpha$, $b = M$ and $c = M + \beta$.*

So, based on Definitions 3–5, we have:

Definition 10. *A L-R fuzzy number $\widetilde{A} = (M, \alpha, \beta)$ is called a non-negative (positive) fuzzy number if and only if $M - \alpha \geq 0 \quad (M - \alpha > 0)$.*

Definition 11. *Let* $\widetilde{A} = (M^A, \alpha^A, \beta^A)$ *and* $\widetilde{B} = (M^B, \alpha^B, \beta^B)$ *be two triangular fuzzy numbers and* λ *is a non-fuzzy number. Then we have:*

$$\widetilde{A} + \widetilde{B} = (M^A, \alpha^A, \beta^A) + (M^B, \alpha^B, \beta^B) = (M^A + M^B, \alpha^A + \alpha^B, \beta^A + \beta^B) \tag{3}$$

$$\widetilde{A} - \widetilde{B} = (M^A, \alpha^A, \beta^A) - (M^B, \alpha^B, \beta^B) = (M^A - M^B, \alpha^A + \beta^B, \beta^A + \alpha^B) \tag{4}$$

$$\lambda\widetilde{A} = (\lambda M^A, \lambda\alpha^A, \lambda\beta^A) \qquad \lambda > 0 \tag{5}$$

And for non-negative fuzzy numbers $\widetilde{A}, \widetilde{B}$ *the multiplication is defined as follows:*

$$\widetilde{A} \otimes \widetilde{B} = (M^A M^B, M^A\alpha^B + M^B\alpha^A - \alpha^A\alpha^B, M^A\beta^B + M^B\beta^A + \beta^A\beta^B)$$

Definition 12. *Let* $\widetilde{A} = (M^A, \alpha^A, \beta^A)$ *be any triangular fuzzy number, then the ranking function of* $\widetilde{A}$ *is as follows:*

$$R(\widetilde{A}) = M^A + \frac{\beta^A - \alpha^A}{4}.$$

Definition 13 [49]**.** *A Z-number is an ordered pair of fuzzy numbers indicated as* $Z = (\widetilde{A}, \widetilde{B})$*. The first component,* $\widetilde{A}$*, is a fuzzy restriction and the second component,* $\widetilde{B}$*, is a level of reliability of the first component. For ease, A and B are supposed to be triangular fuzzy numbers.*

2.2. Converting a Z-Number into a Fuzzy Number

In order to make more computations, the Z-number should be transformed into a usual fuzzy number. Kang et al. [50] presented an efficient and very easy to implement approach, called Kang et al.'s method, for turning a Z-number into a classical fuzzy number based on the fuzzy expectation. Kang et al.'s method is described as follows:

Suppose a Z-number is $Z = (\widetilde{A}, \widetilde{B})$.

Step 1: Convert the second part $(\widetilde{B})$ into a crisp number. Defuzzification transforms a fuzzy number into a crisp value. The most commonly used defuzzification technique is the centroid defuzzification approach. This calculation is completed by using Equation (6).

$$\alpha = \frac{\int x\mu_{\widetilde{B}}(x)dx}{\int \mu_{\widetilde{B}}(x)dx} \tag{6}$$

We noted that when $\widetilde{B} = (b_1, b_2, b_3)$, Equation (6) becomes as follows:

$$\alpha = \frac{b_1 + b_2 + b_3}{3} \tag{7}$$

Step 2: The weighted Z-number can be defined as:

$$\widetilde{Z}^\alpha = \{(x, \mu_{\widetilde{A}^\alpha}) | \mu_{\widetilde{A}^\alpha} = \alpha\mu_{\widetilde{A}}(x), x \in [0,1]\} \tag{8}$$

Step 3: By multiplying $\sqrt{\alpha}$, convert the weighted Z-number into the following classical fuzzy number:

$$\widetilde{Z}' = \sqrt{\alpha} \times \widetilde{A}^\alpha = (\sqrt{\alpha} \times a, \sqrt{\alpha} \times b, \sqrt{\alpha} \times c, \sqrt{\alpha} \times d) \tag{9}$$

In this way, the Z-number is transformed into a conventional fuzzy number.

3. DEA, FDEA and FFDEA

The efficiency of a DMU is established as the ratio of sum weighted output to sum weighted input, subjected to happen between one and zero.

Let p-th DMU (DMUp) be under consideration, then the CCR model for the relative efficiency is as follows [1]:

$$\begin{aligned} \theta_p{}^* = \quad & \max \frac{\sum_{r=1}^{s} u_r y_{rp}}{\sum_{i=1}^{m} v_i x_{ip}} \\ s.t. \quad & \\ & \frac{\sum_{r=1}^{s} u_r y_{rj}}{\sum_{i=1}^{m} v_i x_{ij}} \leq 1, \quad \forall j \\ & u_r, v_i \geq 0 \qquad \forall r, i. \end{aligned} \tag{10}$$

In this model, each DMU (suppose that we have n DMUs) uses m inputs x_{ij} $(i = 1,2,\ldots,m)$, to obtain s outputs y_{rj} $(r = 1,2,\ldots,s)$. Here $u_r (r = 1,2,\ldots,s)$ and $v_i (i = 1,2,\ldots,m)$ are the weights of the ith input and rth output. This fractional program is calculated for every DMU to find out its best input and output weights. To simplify the computation, the nonlinear program shown as (10) can be converted to a linear program (LP) and the model is called the CCR model:

$$\begin{aligned} & \theta_p{}^* = \max \sum_{r=1}^{s} u_r y_{rp} \\ s.t: \quad & \\ & \sum_{i=1}^{m} v_i x_{ip} = 1 \\ & \sum_{r=1}^{s} u_r y_{rj} - \sum_{i=1}^{m} v_i x_{ij} \leq 0, \qquad \forall j \\ & u_r, v_i \geq 0 \qquad \forall r, i. \end{aligned} \tag{11}$$

We solve Equation (11) n-times to work out the efficiency of n DMUs. If $\theta_p{}^* = 1$, we say that the DMUp is efficient, otherwise it is inefficient.

Fuzzy DEA (FDEA) is a strong method for evaluating the efficiency of DMUs with imprecise information. The fuzzy CCR model is defined as follows:

$$\begin{aligned} \theta_p{}^* = \quad & \max \sum_{r=1}^{s} u_r \tilde{y}_{rp} \\ s.t: \quad & \\ & \sum_{i=1}^{m} v_i \tilde{x}_{ip} = 1 \\ & \sum_{r=1}^{s} u_r \tilde{y}_{rj} - \sum_{i=1}^{m} v_i \tilde{x}_{ij} \leq 0, \qquad \forall j \\ & u_r, v_i \geq 0 \qquad \forall r, i. \end{aligned} \tag{12}$$

where, $\tilde{x}_{ij}(i = 1,2,\ldots,m)$ and $\tilde{y}_{rj}(r = 1,2,\ldots,s)$ are fuzzy inputs and fuzzy outputs for the jth DMU (DMUj).

If all input and output data and all parameters are characterized by fuzzy numbers, we call this problem a fully fuzzy DEA (FFDEA) with the following model:

$$\begin{aligned} \theta_p{}^* = \; & \max \sum_{r=1}^{s} \widetilde{u}_r \widetilde{y}_{rp} \\ s.t: \; & \\ & \sum_{i=1}^{m} \widetilde{v}_i \widetilde{x}_{ip} = 1 \\ & \sum_{r=1}^{s} \widetilde{u}_r \widetilde{y}_{rj} - \sum_{i=1}^{m} \widetilde{v}_i \widetilde{x}_{ij} \leq 0, \qquad \forall j \\ & \widetilde{u}_r, \widetilde{v}_i \geq 0 \qquad \forall r, i \end{aligned} \tag{13}$$

where $\widetilde{x}_{ij} = (x_{ij}^M, x_{ij}^\alpha, x_{ij}^\beta)$, $\widetilde{y}_{ij} = (y_{ij}^M, y_{ij}^\alpha, y_{ij}^\beta)$ and their weights given by $\widetilde{u}_r = (u_r^M, u_r^\alpha, u_r^\beta)$,$\widetilde{v}_r = (v_r^M, v_r^\alpha, v_r^\beta)$. Using the Charnes and Cooper transformation, we have the following model:

$$\begin{aligned} & \max \quad \theta_p = \sum_{r=1}^{s} (u_r^M, u_r^\alpha, u_r^\beta) \otimes (y_{rp}^M, y_{rp}^\alpha, y_{rp}^\beta) \\ & s.t.(v_r^M, v_r^\alpha, v_r^\beta) \otimes (x_{ip}^M, y_{ip}^\alpha, y_{ip}^\beta) = (1,0,0) \\ & \sum_{r=1}^{s} (u_r^M, u_r^\alpha, u_r^\beta) \otimes (y_{rj}^M, y_{rj}^\alpha, y_{rj}^\beta) - \sum_{i=1}^{m} (u_r^M, u_r^\alpha, u_r^\beta) \otimes (x_{ij}^M, y_{ij}^\alpha, y_{ij}^\beta) \leq (0,0,0) \quad \forall j \\ & (u_r^M, u_r^\alpha, u_r^\beta), (v_r^M, v_r^\alpha, v_r^\beta) \geq 0 \qquad \forall r, j \end{aligned} \tag{14}$$

4. Sotoudeh-Anvari et al.'s Algorithm

Sotoudeh-Anvari et al. [48] proposed a regular technique to extend the DEA to the fully fuzzy environment with Z-numbers. Their model is as follows:

Method 1.

Step 1: Consider the inputs and outputs of each DMU as well as their weights by using Z-numbers. Using the Charnes and Cooper transformation we have:

$$\begin{aligned} \theta_p{}^{\approx} = \; & \max \sum_{r=1}^{s} \widetilde{\widetilde{u}}_r \widetilde{\widetilde{y}}_{rp} \\ s.t: \; & \\ & \sum_{i=1}^{m} \widetilde{\widetilde{v}}_i \widetilde{\widetilde{x}}_{ip} = 1 \\ & \sum_{r=1}^{s} \widetilde{\widetilde{u}}_r \widetilde{\widetilde{y}}_{rj} - \sum_{i=1}^{m} \widetilde{\widetilde{v}}_i \widetilde{\widetilde{x}}_{ij} \leq 0, \qquad \forall j \\ & \widetilde{\widetilde{u}}_r, \widetilde{\widetilde{v}}_i \geq 0 \qquad \forall r, i \end{aligned} \tag{15}$$

where $\approx$ point out the Z-numbers.

Step 2: Using the Kang et al. [50] model, convert Z-numbers into usual fuzzy numbers. Then the inputs and outputs of each DMU convert into $\widetilde{x}_{ij} = (x_{ij}^{M}, x_{ij}^{\alpha}, x_{ij}^{\beta})$ and $\widetilde{y}_{ij} = (y_{ij}^{M}, y_{ij}^{\alpha}, y_{ij}^{\beta})$. Furthermore, their weights will be $\widetilde{u}_r = (u_r^{M}, u_r^{\alpha}, u_r^{\beta})$, $\widetilde{v}_r = (v_r^{M}, v_r^{\alpha}, v_r^{\beta})$ and we have:

$$
\begin{aligned}
&\max \quad \widetilde{\theta}_p = \sum_{r=1}^{s} (u_r^{M}, u_r^{\alpha}, u_r^{\beta}) \otimes (y_{rp}^{M}, y_{rp}^{\alpha}, y_{rp}^{\beta}) \\
&s.t. \\
&\sum_{i=1}^{m} (v_i^{M}, v_i^{\alpha}, v_i^{\beta}) \otimes (x_{ip}^{M}, y_{ip}^{\alpha}, y_{ip}^{\beta}) \approx (1,0,0) \\
&\sum_{r=1}^{s} (u_r^{M}, u_r^{\alpha}, u_r^{\beta}) \otimes (y_{rj}^{M}, y_{rj}^{\alpha}, y_{rj}^{\beta}) - \sum_{i=1}^{m} (u_r^{M}, u_r^{\alpha}, u_r^{\beta}) \otimes (x_{ij}^{M}, y_{ij}^{\alpha}, y_{ij}^{\beta}) \leq (0,0,0) \quad \forall j \\
&(u_r^{M}, u_r^{\alpha}, u_r^{\beta}), (v_i^{M}, v_i^{\alpha}, v_i^{\beta}) \geq 0 \qquad \forall r, i.
\end{aligned} \tag{16}
$$

Step 3: The fuzzy DEA Equation (16) can be transformed into the following DEA model:

$$
\begin{aligned}
&\max \quad \widetilde{\theta}_p = \sum_{r=1}^{s} (u_r^{M}(y_{rp}^{M} + y_{rp}^{\beta} - y_{rp}^{\alpha}), u_r^{\beta} y_{rp}^{M}, u_r^{\alpha} y_{rp}^{M}) \\
&s.t. \\
&\sum_{i=1}^{m} (v_i^{M}(x_{ip}^{M} + x_{ip}^{\beta} - x_{ip}^{\alpha}), v_i^{\beta} x_{ip}^{M}, v_i^{\alpha} x_{ip}^{M}) = 1, \\
&\sum_{r=1}^{s} (u_r^{M}(y_{rj}^{M} + y_{rj}^{\beta} - y_{rj}^{\alpha}), u_r^{\beta} y_{rj}^{M}, u_r^{\alpha} y_{rj}^{M}) - \sum_{i=1}^{m} (v_i^{M}(x_{ij}^{M} + x_{ij}^{\beta} - x_{ij}^{\alpha}), v_i^{\beta} x_{ij}^{M}, v_i^{\alpha} x_{ij}^{M}) \leq (0,0,0) \quad \forall j \\
&(u_r^{M}, u_r^{\alpha}, u_r^{\beta}), (v_i^{M}, v_i^{\alpha}, v_i^{\beta}) \geq 0 \qquad \forall r, i.
\end{aligned} \tag{17}
$$

Step 4: Convert the fuzzy DEA Equation (17) into the following LP model:

$$
\begin{aligned}
&\max \quad \theta_P = R(\widetilde{\theta}_P) = \sum_{r=1}^{s} \left[u_r^{M} \left(y_{rp}^{M} + (\tfrac{1}{4}) y_{rp}^{\beta} - (\tfrac{1}{4}) y_{rp}^{\alpha} \right) + u_r^{\beta} \left(\left(\tfrac{1}{4}\right) y_{rp}^{M} \right) - u_r^{\alpha} \left(\left(\tfrac{1}{4}\right) y_{rp}^{M} \right) \right] \\
&s.t. \\
&\sum_{i=1}^{m} \left[v_i^{M} \left(x_{ip}^{M} + \left(\tfrac{1}{4}\right) x_{ip}^{\beta} - \left(\tfrac{1}{4}\right) x_{ip}^{\alpha} \right) + v_i^{\beta} \left(\left(\tfrac{1}{4}\right) x_{ip}^{M} \right) - v_i^{\alpha} \left(\left(\tfrac{1}{4}\right) x_{ip}^{M} \right) \right] = 1 \\
&\sum_{r=1}^{s} \left[u_r^{M} \left(y_{rj}^{M} + \left(\tfrac{1}{4}\right) y_{rj}^{\beta} - \left(\tfrac{1}{4}\right) y_{rj}^{\alpha} \right) \right] + u_r^{\beta} \left(\left(\tfrac{1}{4}\right) y_{rj}^{M} \right) - u_r^{\alpha} \left(\left(\tfrac{1}{4}\right) y_{rj}^{M} \right) \leq \\
&\sum_{r=1}^{s} \left[v_i^{M} \left(x_{ij}^{M} + (\tfrac{1}{4}) x_{ij}^{\beta} - (\tfrac{1}{4}) x_{ij}^{\alpha} \right) + v_i^{\beta} \left(\left(\tfrac{1}{4}\right) x_{ij}^{M} \right) - v_i^{\alpha} \left(\left(\tfrac{1}{4}\right) x_{ij}^{M} \right) \right], \quad \forall j \\
&\qquad u_r^{M} - u_r^{\alpha} \geq 0, \qquad \forall r \\
&\qquad u_r^{M} - \left(\tfrac{1}{4}\right) u_r^{\alpha} + \left(\tfrac{1}{4}\right) u_r^{\beta} \geq 0 \qquad \forall r \\
&\qquad v_i^{M} - u_i^{\alpha} \geq 0, \qquad \forall i \\
&\qquad v_i^{M} - \left(\tfrac{1}{4}\right) v_i^{\alpha} + \left(\tfrac{1}{4}\right) v_i^{\beta} \geq 0 \qquad \forall i \\
&\qquad u_r^{\alpha} \geq 0, u_r^{\beta} \geq 0, \qquad \forall r \\
&\qquad v_i^{\alpha} \geq 0, v_i^{\beta} \geq 0, \qquad \forall i.
\end{aligned} \tag{18}
$$

Step 5: Run Equation (18) and obtain the optimal solutions of $u_r^{M^*}, u_r^{\alpha^*}, u_r^{\beta^*}, v_i^{M^*}, v_i^{\alpha^*}$ and $v_i^{\beta^*}$.

5. Main Results

In this section, we explain the shortcomings of Sotoudeh-Anvari et al.'s Algorithm [48] and present the new Algorithm.

5.1. The Shortcoming of the Existing Algorithm

From Definition 11, if $\widetilde{A} = (M^A, \alpha^A, \beta^A)$ and $\widetilde{B} = (M^B, \alpha^B, \beta^B)$ are two non-negative triangular fuzzy numbers then:

$$\widetilde{A} \otimes \widetilde{B} = (M^A M^B, M^A \alpha^B + M^B \alpha^A - \alpha^A \alpha^B, M^A \beta^B + M^B \beta^A + \beta^A \beta^B).$$

Nonetheless, it is evident from Step 3 of Method 1 that Sotoudeh-Anvari et al. [38], have utilized the wrong product:

$$\widetilde{A} \otimes \widetilde{B} = (M^A(M^B + \beta^B - \alpha^B), \beta^A M^B, \alpha^A M^B),$$

to transform the fuzzy DEA Equation (16) into the fuzzy DEA Equation (17). Henceforth, Method 1, proposed by Sotoudeh-Anvari et al. [48], is not substantial in its present frame.

5.2. Improvement Model for FFDEA with Z-Numbers

In this section, to remove the mentioned shortcoming, we proposed an improved model for fully fuzzy DEA with Z-numbers.

Method 2.

Step 1: Consider the DEA model that the inputs and outputs of each DMU as well as their weights are Z-numbers.

Step 2: Using Kang et al.'s [50] model, convert Z-numbers into usual fuzzy numbers and obtain a fully fuzzy DEA model with triangular fuzzy numbers.

Step 3: Using Definition 11, the fully fuzzy DEA model of Step 2 can be transformed into the following model:

$$\begin{aligned}
&\max \quad \widetilde{\theta}_p = \sum_{r=1}^{s} (u_r^M y_{rp}^M, u_r^M y_{rp}^\alpha + y_{rp}^M u_r^\alpha - u_r^\alpha y_{rp}^\alpha, u_r^M y_{rp}^\beta + y_{rp}^M u_r^\beta + u_r^\beta y_{rp}^\beta) \\
&s.t. \\
&\sum_{i=1}^{m} (v_i^M x_{ip}^M, v_i^M x_{ip}^\alpha + x_{ip}^M v_i^\alpha - v_i^\alpha x_{ip}^\alpha, v_i^M x_{ip}^\beta + x_{ip}^M v_i^\beta + v_i^\beta x_{ip}^\beta) \approx 1, \\
&\sum_{r=1}^{s} (u_r^M y_{rj}^M, u_r^M y_{rj}^\alpha + y_{rj}^M u_r^\alpha - u_r^\alpha y_{rj}^\alpha, u_r^M y_{rj}^\beta + y_{rj}^M u_r^\beta + u_r^\beta y_{rj}^\beta) \leq \\
&\qquad \sum_{i=1}^{m} (v_i^M x_{ij}^M, v_i^M x_{ij}^\alpha + x_{ij}^M v_i^\alpha - v_i^\alpha x_{ij}^\alpha, v_i^M x_{ij}^\beta + x_{ij}^M v_i^\beta + v_i^\beta x_{ij}^\beta) \qquad \forall j \\
&(u_r^M, u_r^\alpha, u_r^\beta), (v_i^M, v_i^\alpha, v_i^\beta) \geq 0 \qquad \forall r, i
\end{aligned} \tag{19}$$

Step 4: Based on Definitions 7 and 8, convert the fuzzy DEA Equation (19) into the following model:

$$\begin{aligned}
&\max \quad \theta_p = R(\widetilde{\theta}_p) = \sum_{r=1}^{s} R((u_r^M y_{rp}^M, u_r^M y_{rp}^\alpha + y_{rp}^M u_r^\alpha - u_r^\alpha y_{rp}^\alpha, u_r^M y_{rp}^\beta + y_{rp}^M u_r^\beta + u_r^\beta y_{rp}^\beta)) \\
&\sum_{i=1}^{m} R((v_i^M x_{ip}^M, v_i^M x_{ip}^\alpha + x_{ip}^M v_i^\alpha - v_i^\alpha x_{ip}^\alpha, v_i^M x_{ip}^\beta + x_{ip}^M v_i^\beta + v_i^\beta x_{ip}^\beta)) = R(1,1,1), \\
&\sum_{r=1}^{s} R((u_r^M y_{rj}^M, u_r^M y_{rj}^\alpha + y_{rj}^M u_r^\alpha - u_r^\alpha y_{rj}^\alpha, u_r^M y_{rj}^\beta + y_{rj}^M u_r^\beta + u_r^\beta y_{rj}^\beta)) \leq \\
&\qquad \sum_{i=1}^{m} R((v_i^M x_{ij}^M, v_i^M x_{ij}^\alpha + x_{ij}^M v_i^\alpha - v_i^\alpha x_{ij}^\alpha, v_i^M x_{ij}^\beta + x_{ij}^M v_i^\beta + v_i^\beta x_{ij}^\beta)) \qquad \forall j \\
&R(u_r^M, u_r^\alpha, u_r^\beta) \geq (0,0,0), R(v_i^M, v_i^\alpha, v_i^\beta) \geq (0,0,0), \qquad \forall r, i \\
&u_r^M - u_r^\alpha \geq 0, \quad \forall r, \\
&v_i^M - v_i^\alpha \geq 0, \quad \forall i, \\
&u_r^\alpha \geq 0, u_r^\beta \geq 0, \quad \forall r \\
&v_i^\alpha \geq 0, v_i^\beta \geq 0, \quad \forall i.
\end{aligned} \tag{20}$$

Step 5: Based on Definition 12, convert Equation (20) into the following model:

$$
\begin{aligned}
&\max \quad \theta_P = R(\widetilde{\theta}_P) = \sum_{r=1}^{s}\left[u_r^M y_{rp}^M + \tfrac{1}{4}[u_r^M y_{rp}^\beta + y_{rp}^M u_r^\beta + u_r^\beta y_{rp}^\beta] - \tfrac{1}{4}[u_r^M y_{rp}^\alpha + y_{rp}^M u_r^\alpha - u_r^\alpha y_{rp}^\alpha]\right] \\
&s.t. \\
&\sum_{i=1}^{m}\left[v_i^M x_{ip}^M + \tfrac{1}{4}[v_i^M x_{ip}^\beta + x_{ip}^M v_i^\beta + v_i^\beta x_{ip}^\beta] - \tfrac{1}{4}[v_i^M x_{ip}^\alpha + x_{ip}^M v_i^\alpha - v_i^\alpha x_{ip}^\alpha]\right] = 1, \\
&\sum_{r=1}^{s}\left[u_r^M y_{rj}^M + \tfrac{1}{4}[u_r^M y_{rj}^\beta + y_{rj}^M u_r^\beta + u_r^\beta y_{rj}^\beta] - \tfrac{1}{4}[u_r^M y_{rj}^\alpha + y_{rj}^M u_r^\alpha - u_r^\alpha y_{rj}^\alpha]\right] \leq \\
&\sum_{i=1}^{m}\left[v_i^M x_{ij}^M + \tfrac{1}{4}[v_i^M x_{ij}^\beta + x_{ij}^M v_i^\beta + v_i^\beta x_{ij}^\beta] - \tfrac{1}{4}[v_i^M x_{ij}^\alpha + x_{ij}^M v_i^\alpha - v_i^\alpha x_{ij}^\alpha]\right], \qquad \forall j \\
&u_r^M - u_r^\alpha \geq 0, \qquad \forall r \\
&u_r^M - \left(\tfrac{1}{4}\right)u_r^\alpha + \left(\tfrac{1}{4}\right)u_r^\beta \geq 0 \qquad \forall r \\
&v_i^M - u_i^\alpha \geq 0, \qquad \forall i \\
&v_i^M - \left(\tfrac{1}{4}\right)v_i^\alpha + \left(\tfrac{1}{4}\right)v_i^\beta \geq 0 \qquad \forall i \\
&u_r^\alpha \geq 0, u_r^\beta \geq 0, \qquad \forall r \\
&v_i^\alpha \geq 0, v_i^\beta \geq 0, \qquad \forall i.
\end{aligned}
\tag{21}
$$

Step 6: Obtain the optimal solutions of $u_r^{M^*}, u_r^{\alpha^*}, u_r^{\beta^*}, v_i^{M^*}, v_i^{\alpha^*}$ and $v_i^{\beta^*}$.

6. Numerical Example

Sotoudeh-Anvari et al. [48], tackled the existing problem taken from Guo and Tanaka [52] to represent their proposed approach. However, as discussed in Section 5, there are some shortcomings in Method 1. Therefore, the results of Sotoudeh-Anvari et al. [48] are likewise not correct. In this section, the correct consequences of this problem are solved by Method 2.

Problem 1. *Consider five DMUs with two inputs and two outputs where all the input and output data are designed as Z-numbers (see Table 1).*

Table 1. Five decision-making units (DMUs) with two Z-number inputs and two Z-number outputs. M = medium. MH = medium high. H = high. VH = very high.

DMU	Inputs 1	Inputs 2	Outputs 1	Outputs 2
A	((4.51, 5.16, 5.80), MH)	((2.125, 2.348, 2.572), VH)	((2.870, 3.110, 3.349), H)	((4.545, 4.904, 5.263), H)
B	((3.46, 3.46, 3.46), H)	((1.674, 1.794, 1.913), H)	((2.460, 2.460, 2.460), VH)	((4.263, 4.521, 4.780), MH)
C	((4.92, 5.48, 6.04), VH)	((2.631, 3.11, 3.588), H)	((3.229, 3.827, 4.425), H)	((4.809, 5.704, 6.599), VH)
D	((4.80, 5.80, 6.79), M)	((2.460, 2.572, 2.684), VH)	((2.971, 3.229, 3.746), MH)	((7.779, 8.062, 8.345), M)
E	((7.05, 7.77, 8.49), H)	((4.647, 5.293, 5.938), MH)	((6.223, 7.213, 8.203), M)	((7.775, 8.851, 9.928), H)

In Table 1, the linguistic variables need to be transformed into triangular fuzzy numbers which are listed in Table 2.

Table 2. Linguistic variables for measuring of the reliability of Z-numbers.

Linguistic Term	Abbreviation	Corresponding TFNs
Very Low	VL	(0.1, 0.2, 0.3)
Low	L	(0.2, 0.3, 0.4)
Medium Low	ML	(0.3, 0.4, 0.5)
Medium	M	(0.4, 0.5, 0.6)
Medium High	MH	(0.5, 0.6, 0.7)
High	H	(0.6, 0.7, 0.8)
Very High	VH	(0.7, 0.8, 0.9)

By using Kang et al.'s [50] model, the Z-number values of Table 1 are converted into fuzzy numbers, which are listed in Table 3.

Table 3. Data transformation from Z-numbers into fuzzy numbers.

DMU	Inputs 1	Inputs 2	Outputs 1	Outputs 2
A	(4, 0.5, 0.5)	(2.1, 0.2, 0.2)	(2.6, 0.2, 0.2)	(4.1, 0.3, 0.3)
B	(2.9, 0, 0)	(1.5, 0.1, 0.1)	(2.2, 0, 0)	(3.5, 0.2, 0.2)
C	(4.9, 0.5, 0.5)	(2.6, 0.4, 0.4)	(3.2, 0.5, 0.5)	(5.1, 0.8, 0.8)
D	(4.1, 0.7, 0.7)	(2.3, 0.1, 0.1)	(2.5, 0.2, 0.4)	(5.7, 0.2, 0.2)
E	(6.1, 0.2, 1)	(4.1, 0.5, 0.5)	(5.1, 0.7, 0.7)	(7.4, 0.9, 0.9)

Now, we use Method 2 to solve the performance assessment problem. For example, Method 2 for DMU_A can be used as follows:

Step 1: Obtain a fully fuzzy DEA model with triangular fuzzy numbers:

$$
\begin{aligned}
&\max \quad \widetilde{\theta}_A \approx (2.6,0.2,0.2) \otimes (u_1^M, u_1^\alpha, u_1^\beta) + (4.1,0.3,0.3) \otimes (u_2^M, u_2^\alpha, u_2^\beta) \\
&s.t. \\
&(4,0.5,0.5) \otimes (v_1^M, v_1^\alpha, v_1^\beta) + (2.1,0.2,0.2) \otimes (v_2^M, v_2^\alpha, v_2^\beta) \approx 1, \\
&\left[(2.6,0.2,0.2) \otimes (u_1^M, u_1^\alpha, u_1^\beta) + (4.1,0.3,0.3) \otimes (u_2^M, u_2^\alpha, u_2^\beta)\right] - \\
&\qquad \left[(4,0.5,0.5) \otimes (v_1^M, v_1^\alpha, v_1^\beta) + (2.1,0.2,0.2) \otimes (v_2^M, v_2^\alpha, v_2^\beta)\right] \leq (0,0,0), \\
&\left[(2.2,0,0) \otimes (u_1^M, u_1^\alpha, u_1^\beta) + (3.5,0.2,0.2) \otimes (u_2^M, u_2^\alpha, u_2^\beta)\right] - \\
&\qquad \left[(2.9,0,0) \otimes (v_1^M, v_1^\alpha, v_1^\beta) + (1.5,0.1,0.1) \otimes (v_2^M, v_2^\alpha, v_2^\beta)\right] \leq (0,0,0), \\
&\left[(3.2,0.5,0.5) \otimes (u_1^M, u_1^\alpha, u_1^\beta) + (5.1,0.8,0.8) \otimes (u_2^M, u_2^\alpha, u_2^\beta)\right] - \\
&\qquad \left[(4.9,0.5,0.5) \otimes (v_1^M, v_1^\alpha, v_1^\beta) + (2.6,0.4,0.4) \otimes (v_2^M, v_2^\alpha, v_2^\beta)\right] \leq (0,0,0), \\
&\left[(2.5,0.2,0.4) \otimes (u_1^M, u_1^\alpha, u_1^\beta) + (5.7,0.9,0.9) \otimes (u_2^M, u_2^\alpha, u_2^\beta)\right] - \\
&\qquad \left[(4.1,0.7,0.7) \otimes (v_1^M, v_1^\alpha, v_1^\beta) + (2.3,0.1,0.1) \otimes (v_2^M, v_2^\alpha, v_2^\beta)\right] \leq (0,0,0), \\
&\left[(5.1,0.7,0.7) \otimes (u_1^M, u_1^\alpha, u_1^\beta) + (7.4,0.9,0.9) \otimes (u_2^M, u_2^\alpha, u_2^\beta)\right] - \\
&\qquad \left[(6.1,0.2,1) \otimes (v_1^M, v_1^\alpha, v_1^\beta) + (4.1,0.5,0.5) \otimes (v_2^M, v_2^\alpha, v_2^\beta)\right] \leq (0,0,0), \\
&(u_r^M, u_r^\alpha, u_r^\beta), (v_i^M, v_i^\alpha, v_i^\beta) \geq 0 \qquad \forall r, i
\end{aligned}
\tag{22}
$$

Step 2: Using the Definition 11, the fully fuzzy DEA model of Step 2 can be transformed into the following model:

$$
\begin{aligned}
&\max \quad \widetilde{\theta}_A \approx (2.6u_1^M, 0.2u_1^M + 2.4u_1^\alpha, 0.2u_1^M + 2.8u_1^\beta) + (4.1u_2^M, 0.3u_2^M + 3.8u_2^\alpha, 0.3u_2^M + 4.4u_2^\beta) \\
&s.t. \\
&(4v_1^M, 0.5v_1^M + 3.5v_1^\alpha, 0.5v_1^M + 4.5v_1^\beta) + (2.1v_2^M, 0.2v_2^M + 1.9v_2^\alpha, 0.2v_2^M + 2.3v_2^\beta) \approx (1,1,1), \\
&(2.6u_1^M, 0.2u_1^M + 2.4u_1^\alpha, 0.2u_1^M + 2.8u_1^\beta) + (4.1u_2^M, 0.3u_2^M + 3.8u_2^\alpha, 0.3u_2^M + 4.4u_2^\beta) \\
&\qquad - (4v_1^M, 0.5v_1^M + 3.5v_1^\alpha, 0.5v_1^M + 4.5v_1^\beta) - (2.1v_2^M, 0.2v_2^M + 1.9v_2^\alpha, 0.2v_2^M + 2.3v_2^\beta) \leq (0,0,0), \\
&(2.2u_1^M, 2.2u_1^\alpha, 2.2u_1^\beta) + (3.5u_2^M, 0.2u_2^M + 3.3u_2^\alpha, 0.2u_2^M + 3.7u_2^\beta) \\
&\qquad - (2.9v_1^M, 2.9v_1^\alpha, 2.9v_1^\beta) - (1.5v_2^M, 0.1v_2^M + 1.4v_2^\alpha, 0.1v_2^M + 1.6v_2^\beta) \leq (0,0,0), \\
&(3.2u_1^M, 0.5u_1^M + 2.7u_1^\alpha, 0.5u_1^M + 3.7u_1^\beta) + (5.1u_2^M, 0.8u_2^M + 4.3u_2^\alpha, 0.8u_2^M + 5.9u_2^\beta) \\
&\qquad - (4.9v_1^M, 0.5v_1^M + 4.4v_1^\alpha, 0.5v_1^M + 5.4v_1^\beta) - (2.6v_2^M, 0.4v_2^M + 2.2v_2^\alpha, 0.4v_2^M + 3v_2^\beta) \leq (0,0,0), \\
&(2.5u_1^M, 0.2u_1^M + 2.3u_1^\alpha, 0.4u_1^M + 2.9u_1^\beta) + (5.7u_2^M, 0.2u_2^M + 5.5u_2^\alpha, 0.2u_2^M + 5.9u_2^\beta) \\
&\qquad - (4.1v_1^M, 0.7v_1^M + 3.4v_1^\alpha, 0.7v_1^M + 4.8v_1^\beta) - (2.3v_2^M, 0.1v_2^M + 2.2v_2^\alpha, 0.1v_2^M + 2.4v_2^\beta) \leq (0,0,0), \\
&(5.1u_1^M, 0.7u_1^M + 4.4u_1^\alpha, 0.7u_1^M + 5.8u_1^\beta) + (7.4u_2^M, 0.9u_2^M + 6.5u_2^\alpha, 0.9u_2^M + 8.3u_2^\beta) \\
&\qquad - (6.1v_1^M, 0.2v_1^M + 5.9v_1^\alpha, v_1^M + 7.1v_1^\beta) - (4.1v_2^M, 0.5v_2^M + 3.6v_2^\alpha, 0.5v_2^M + 4.6v_2^\beta) \leq (0,0,0), \\
&(u_r^M, u_r^\alpha, u_r^\beta), (v_i^M, v_i^\alpha, v_i^\beta) \geq (0,0,0) \qquad \forall r, i
\end{aligned}
\tag{23}
$$

Step 3: Based on Definitions 7 and 8, convert the above fuzzy DEA model to the following model:

$$
\begin{aligned}
&\max \quad \theta_A = R(\widetilde{\theta}_A) = R\Big((2.6u_1^M, 0.2u_1^M + 2.4u_1^\alpha, 0.2u_1^M + 2.8u_1^\beta) + (4.1u_2^M, 0.3u_2^M + 3.8u_2^\alpha, 0.3u_2^M + 4.4u_2^\beta)\Big) \\
&s.t. \\
&R\Big((4v_1^M, 0.5v_1^M + 3.5v_1^\alpha, 0.5v_1^M + 4.5v_1^\beta) + (2.1v_2^M, 0.2v_2^M + 1.9v_2^\alpha, 0.2v_2^M + 2.3v_2^\beta)\Big) = R(1,1,1), \\
&R\left(\begin{array}{l} (2.6u_1^M, 0.2u_1^M + 2.4u_1^\alpha, 0.2u_1^M + 2.8u_1^\beta) + (4.1u_2^M, 0.3u_2^M + 3.8u_2^\alpha, 0.3u_2^M + 4.4u_2^\beta) \\ \quad - (4v_1^M, 0.5v_1^M + 3.5v_1^\alpha, 0.5v_1^M + 4.5v_1^\beta) - (2.1v_2^M, 0.2v_2^M + 1.9v_2^\alpha, 0.2v_2^M + 2.3v_2^\beta) \end{array}\right) \leq R(0,0,0), \\
&R\left(\begin{array}{l} (2.2u_1^M, 2.2u_1^\alpha, 2.2u_1^\beta) + (3.5u_2^M, 0.2u_2^M + 3.3u_2^\alpha, 0.2u_2^M + 3.7u_2^\beta) \\ \quad - (2.9v_1^M, 2.9v_1^\alpha, 2.9v_1^\beta) - (1.5v_2^M, 0.1v_2^M + 1.4v_2^\alpha, 0.1v_2^M + 1.6v_2^\beta) \end{array}\right) \leq R(0,0,0), \\
&R\left(\begin{array}{l} (3.2u_1^M, 0.5u_1^M + 2.7u_1^\alpha, 0.5u_1^M + 3.7u_1^\beta) + (5.1u_2^M, 0.8u_2^M + 4.3u_2^\alpha, 0.8u_2^M + 5.9u_2^\beta) \\ \quad - (4.9v_1^M, 0.5v_1^M + 4.4v_1^\alpha, 0.5v_1^M + 5.4v_1^\beta) - (2.6v_2^M, 0.4v_2^M + 2.2v_2^\alpha, 0.4v_2^M + 3v_2^\beta) \end{array}\right) \leq R(0,0,0), \\
&R\left(\begin{array}{l} (2.5u_1^M, 0.2u_1^M + 2.3u_1^\alpha, 0.4u_1^M + 2.9u_1^\beta) + (5.7u_2^M, 0.2u_2^M + 5.5u_2^\alpha, 0.2u_2^M + 5.9u_2^\beta) \\ \quad - (4.1v_1^M, 0.7v_1^M + 3.4v_1^\alpha, 0.7v_1^M + 4.8v_1^\beta) - (2.3v_2^M, 0.1v_2^M + 2.2v_2^\alpha, 0.1v_2^M + 2.4v_2^\beta) \end{array}\right) \leq R(0,0,0), \\
&R\left(\begin{array}{l} (5.1u_1^M, 0.7u_1^M + 4.4u_1^\alpha, 0.7u_1^M + 5.8u_1^\beta) + (7.4u_2^M, 0.9u_2^M + 6.5u_2^\alpha, 0.9u_2^M + 8.3u_2^\beta) \\ \quad - (6.1v_1^M, 0.2v_1^M + 5.9v_1^\alpha, v_1^M + 7.1v_1^\beta) - (4.1v_2^M, 0.5v_2^M + 3.6v_2^\alpha, 0.5v_2^M + 4.6v_2^\beta) \end{array}\right) \leq R(0,0,0), \\
&R(u_r^M, u_r^\alpha, u_r^\beta) \geq R(0,0,0), \qquad \forall r \\
&R(v_i^M, v_i^\alpha, v_i^\beta) \geq R(0,0,0) \qquad \forall i \\
&u_r^M - u_r^\alpha \geq 0, \quad \forall r, \\
&v_i^M - v_i^\alpha \geq 0, \quad \forall i, \\
&u_r^\alpha \geq 0, u_r^\beta \geq 0, \quad \forall r \\
&v_i^\alpha \geq 0, v_i^\beta \geq 0, \quad \forall i.
\end{aligned}
\tag{24}
$$

Step 4: Based on Definition 12, convert the above model to the following model:

$$
\begin{aligned}
&\max \ \theta_A = \Big((2.6u_1^M - 0.6u_1^\alpha + 0.7u_1^\beta) + (4.1u_2^M - 0.95u_2^\alpha + 1.1u_2^\beta)\Big) \\
&s.t. \\
&\Big((4v_1^M - 0.875v_1^\alpha + 1.125v_1^\beta) + (2.1v_2^M - 0.475v_2^\alpha + 0.575v_2^\beta)\Big) = 1, \\
&(2.6u_1^M - 0.6u_1^\alpha + 0.7u_1^\beta) + (4.1u_2^M - 0.95u_2^\alpha + 1.1u_2^\beta) - \\
&\Big((4v_1^M - 0.875v_1^\alpha + 1.125v_1^\beta) + (2.1v_2^M - 0.475v_2^\alpha + 0.575v_2^\beta)\Big) \leq 0, \\
&(2.2u_1^M - 0.55u_1^\alpha + 0.55u_1^\beta) + (3.5u_2^M - 0.825u_2^\alpha + 0.925u_2^\beta) - \\
&\Big((2.9v_1^M - 0.725v_1^\alpha + 0.725v_1^\beta) + (1.5v_2^M - 0.35v_2^\alpha + 0.4v_2^\beta)\Big) \leq 0, \\
&(3.2u_1^M - 0.675u_1^\alpha + 0.925u_1^\beta) + (5.1u_2^M - 1.075u_2^\alpha + 1.475u_2^\beta) - \\
&\Big((4.9v_1^M - 1.1v_1^\alpha + 1.35v_1^\beta) + (2.6v_2^M - 0.55v_2^\alpha + 0.75v_2^\beta)\Big) \leq 0, \\
&(2.55u_1^M - 0.575u_1^\alpha + 0.725u_1^\beta) + (5.7u_2^M - 1.2u_2^\alpha + 1.65u_2^\beta) - \\
&\Big((4.1v_1^M - 0.85v_1^\alpha + 1.2v_1^\beta) + (2.3v_2^M - 0.55v_2^\alpha + 0.6v_2^\beta)\Big) \leq 0, \\
&(5.1u_1^M - 1.1u_1^\alpha + 1.45u_1^\beta) + (7.4u_2^M - 1.625u_2^\alpha + 2.075u_2^\beta) - \\
&\Big((6.3v_1^M - 1.475v_1^\alpha + 1.775v_1^\beta) + (4.1v_2^M - 0.9v_2^\alpha + 1.15v_2^\beta)\Big) \leq 0, \\
&u_1^M - 0.25u_1^\alpha + 0.25u_1^\beta \geq 0, u_2^M - 0.25u_2^\alpha + 0.25u_2^\beta \geq 0, \\
&v_1^M - 0.25v_1^\alpha + 0.25v_1^\beta \geq 0, v_2^M - 0.25v_2^\alpha + 0.25v_2^\beta \geq 0, \\
&u_1^M - u_1^\alpha \geq 0, u_2^M - u_2^\beta \geq 0, \\
&v_1^M - v_1^\alpha \geq 0, v_2^M - v_2^\alpha \geq 0, \\
&u_r^\alpha \geq 0, u_r^\beta \geq 0, \qquad r = 1,2, \\
&v_i^\alpha \geq 0, v_i^\beta \geq 0, \qquad i = 1,2.
\end{aligned}
\tag{25}
$$

After computations with Lingo, we have the following optimal information for DMU_A:

$$u_1^* = (0.08682, 0.08682, 0.36267), u_2^* = (0.107402, 0, 0), v_1^* = (0.15182, 0, 0), v_2^* = (0.132123, 0, 0.20046).$$

And,

$$\widetilde{\theta}^*{}_A = (\widetilde{\theta}_1^M, \widetilde{\theta}_1^\alpha, \widetilde{\theta}_1^\beta) = (0.6660, 0.2579, 1.0651).$$

Similarly, for other DMUs, we report the results in Table 4.

Table 4. Fuzzy efficiencies of the other DMUs.

	DMU_B	DMU_C	DMU_D	DMU_E
u_1^*	$(0, 0, 0)$	$(0, 0, 1.08108)$	$(0, 0, 0)$	$(0, 0, 0.68965)$
u_2^*	$(0, 0, 1.08108)$	$(0, 0, 0)$	$(0.17544, 0, 0)$	$(0, 0, 0)$
v_1^*	$(0, 0, 1.158997)$	$(0, 0, 0.45747)$	$(0.24390, 0, 0)$	$(0.16949, 0, 0)$
v_2^*	$(0, 0, 0.39932)$	$(0.14282, 0, 0.14783)$	$(0, 0, 0)$	$(0, 0, 0)$
$\widetilde{\theta}^*$	$(0, 0, 4)$	$(0, 0, 4)$	$(1, 0.1579, 0.1579)$	$(0, 0, 4)$

By these results, we can see that DMU_A is inefficient, and the others are efficient. Further, we have used Wang et al.'s model [28] for comparing and ranking fuzzy efficiencies.

From Table 5, the DMUs are fully ranked in terms of their fuzzy efficiencies as follows:

$$B = C = E \overset{60.57\%}{\geq} D \overset{56.92\%}{\geq} A.$$

Table 5. The matrix of the degree of preference for fuzzy efficiencies obtained by model [28].

DMUs	*A*	*B*	*C*	*D*	*E*
A	-	0.4308	0.4308	0.3943	0.4308
B	0.5692	-	0.5	0.5557	0.5
C	0.5692	0.5	-	0.5557	0.5
D	0.6257	0.4443	0.4443	-	0.4443
E	0.5692	0.5	0.5	0.5557	-

According to our model, DMU_B, DMU_C and DMU_E are efficient. However, by the model of [48], we can see that DMU_B, DMU_D and DMU_E are efficient. Even if the results were the same, because the model of [48] uses the wrong strategy, it would still not be valid. We take note that this example is utilized to demonstrate the computational procedure of the proposed technique and such a comparison is insignificant. Although the suggested procedure has been employed to a numerical case, the same frames could be used, with some adjustment, to other benchmarking problems.

7. Conclusions and Future Work

Within the past few years, a developing interest has appeared in fully fuzzy DEA (FFDEA) and presently there are numerous strategies to solve it. However, in original fuzzy sets, the certainty of the information is approximately ignored. Z-number is a suitable measure for comprehensive explanation of real-life information and has that extra ability of being able to depict uncertain information. This paper explains the drawbacks of Sotoudeh-Anvari et al.'s method [48] in the fully fuzzy DEA to rank DMUs with imprecise data. We show that the mentioned method did not consider the fuzzy axioms, so it may produce incorrect rankings in some cases. To remove the existing drawbacks, we presented a new fully fuzzified DEA, where all decision parameters and variables are Z-numbers. Based on our results, we can see that Sotoudeh-Anvari et al.'s method [48] should not be used for evaluating the best relative fuzzy efficiencies of DMUs. Furthermore, our model used Wang et al.'s model [28] for comparing and ranking fuzzy efficiencies. It is explained that our model provides the right evaluation of the relative efficiency of a DMU under ambiguous circumstances and gives more reliable results. In spite of the fact that the model, arithmetic operations and results introduced here have exhibited the viability of our approach, it might be additionally applied in other fuzzy DEA problem such as network FFDEA, FFDEA with common weights, and FFDEA with undesirable outputs. For future work, we plan to consider these issues.

Author Contributions: A.N. and S.E.N. contributed to generating the research ideas, designed the research, analyzed the data and developed the paper. M.F. and M.J. provided extensive advice throughout the study regarding the research design, methodology, and revised the manuscript. All the authors have read and approved the final manuscript.

Funding: This research received no external funding.

Acknowledgments: The authors would like to thank all the three reviewers and the editor for their valuable suggestions which have led to an improvement in both the quality and clarity of this paper.

Conflicts of Interest: The authors declare no conflict of interest.

References

1. Charnes, A.; Cooper, W.W.; Rhodes, E. Measuring the efficiency of decision making units. *Eur. J. Oper. Res.* **1978**, *2*, 429–444. [CrossRef]
2. Banker, R.D.; Charnes, A.; Cooper, W.W. Some models for estimating technical and scale inefficiencies in data envelopment analysis. *Manag. Sci.* **1984**, *30*, 1078–1092. [CrossRef]
3. Cooper, W.W.; Park, K.S.; Yu, G. IDEA and AR-IDEA: Models for dealing with imprecise data in DEA. *Manag. Sci.* **1999**, *45*, 597–607. [CrossRef]
4. Despotis, D.K.; Smirlis, Y.G. Data envelopment analysis with imprecise data. *Eur. J. Oper. Res.* **2002**, *140*, 24–36. [CrossRef]
5. Kao, C. Interval efficiency measures in data envelopment analysis with imprecise data. *Eur. J. Oper. Res.* **2006**, *174*, 1087–1099. [CrossRef]
6. Emrouznejad, A.; Parker, B.R.; Tavares, G. Evaluation of research in efficiency and productivity: A survey and analysis of the first 30 years of scholarly literature in DEA. *Socio-Econ. Plan. Sci.* **2008**, *42*, 151–157. [CrossRef]
7. Jahanshahloo, G.R.; Abbasian-Naghneh, S. Data envelopment analysis with imprecise data. *Appl. Math. Sci.* **2011**, *5*, 3089–3106.
8. Liu, J.S.; Lu, L.Y.; Lu, W.M. Research fronts in data envelopment analysis. *Omega* **2016**, *58*, 33–45. [CrossRef]
9. Wei, G.; Wang, J. A comparative study of robust efficiency analysis and data envelopment analysis with imprecise data. *Expert Syst. Appl.* **2017**, *81*, 28–38. [CrossRef]
10. Chen, Y.; Cook, W.D.; Du, J.; Hu, H.; Zhu, J. Bounded and discrete data and Likert scales in data envelopment analysis: Application to regional energy efficiency in China. *Ann. Oper. Res.* **2017**, *255*, 347–366. [CrossRef]
11. Toloo, M.; Keshavarz, E.; Hatami-Marbini, A. Dual-role factors for imprecise data envelopment analysis. *Omega* **2018**, *77*, 15–31. [CrossRef]
12. Zhou, X.; Xu, Z.; Yao, L.; Tu, Y.; Lev, B.; Pedrycz, W. A novel Data Envelopment Analysis model for evaluating industrial production and environmental management system. *J. Clean. Prod.* **2018**, *170*, 773–788. [CrossRef]
13. Zadeh, L.A. Fuzzy sets. *Inf. Control* **1965**, *8*, 338–353. [CrossRef]
14. Bellman, R.E.; Zadeh, L.A. Decision-making in a fuzzy environment. *Manag. Sci.* **1970**, *17*, 141–164. [CrossRef]
15. Tanaka, H.; Asai, K. A formulation of fuzzy linear programming based on comparison of fuzzy number. *Control Cybern.* **1984**, *13*, 185–194.
16. Campos, L.; Verdegay, J.L. Linear programming problems and ranking of fuzzy numbers. *Fuzzy Sets Syst.* **1989**, *32*, 1–11. [CrossRef]
17. Zimmermann, H.J. *Fuzzy Set Theory—And Its Applications*; Springer Science Business Media: Berlin, Germany, 2011.
18. Edalatpanah, S.A.; Shahabi, S. A new two-phase method for the fuzzy primal simplex algorithm. *Int. Rev. Pure Appl. Math.* **2012**, *8*, 157–164.

19. Saberi Najafi, H.; Edalatpanah, S.A.; Dutta, H. A nonlinear model for fully fuzzy linear programming with fully unrestricted variables and parameters. *Alex. Eng. J.* **2016**, *55*, 2589–2595. [CrossRef]
20. Saberi Najafi, S.; Edalatpanah, S.A. A Note on A new method for solving fully fuzzy linear programming problems. *Appl. Math. Model.* **2013**, *37*, 7865–7867. [CrossRef]
21. Rodríguez, R.M.; Martínez, L.; Herrera, F.; Torra, V. A Review of Hesitant Fuzzy Sets: Quantitative and Qualitative Extensions. In *Fuzzy Logic in Its 50th Year*; Springer International Publishing: Cham, Switzerland, 2016; pp. 109–128.
22. Das, S.K.; Edalatpanah, S.A.; Mandal, T. A proposed model for solving fuzzy linear fractional programming problem: Numerical Point of View. *J. Comput. Sci.* **2018**, *25*, 367–375. [CrossRef]
23. Sengupta, J.K. A fuzzy systems approach in data envelopment analysis. *Comput. Math. Appl.* **1992**, *24*, 259–266. [CrossRef]
24. Triantis, K.; Girod, O. A mathematical programming approach for measuring technical efficiency in a fuzzy environment. *J. Prod. Anal.* **1998**, *10*, 85–102. [CrossRef]
25. Kao, C.; Liu, S.T. Fuzzy efficiency measures in data envelopment analysis. *Fuzzy Sets Syst.* **2000**, *113*, 427–437. [CrossRef]
26. Dia, M. A model of fuzzy data envelopment analysis. *Inf. Syst. Oper. Res.* **2004**, *42*, 267–279. [CrossRef]
27. Garcia, P.A.A.; Schirru, R.; Melo, P.F.F.E. A fuzzy data envelopment analysis approach for FMEA. *Prog. Nucl. Energy* **2005**, *46*, 359–373. [CrossRef]
28. Wang, Y.M.; Luo, Y.; Liang, L. Fuzzy data envelopment analysis based upon fuzzy arithmetic with an application to performance assessment of manufacturing enterprises. *Expert Syst. Appl.* **2009**, *36*, 5205–5211. [CrossRef]
29. Wang, Y.M.; Chin, K.S. Fuzzy data envelopment analysis: A fuzzy expected value approach. *Expert Syst. Appl.* **2011**, *38*, 11678–11685. [CrossRef]
30. Emrouznejad, A.; Tavana, M.; Hatami-Marbini, A. The state of the art in fuzzy data envelopment analysis. In *Performance Measurement with Fuzzy Data Envelopment Analysis*; Springer: Berlin/Heidelberg, Germany, 2014; pp. 1–45.
31. Puri, J.; Yadav, S.P. A fuzzy DEA model with undesirable fuzzy outputs and its application to the banking sector in India. *Expert Syst. Appl.* **2014**, *41*, 6419–6432. [CrossRef]
32. Wanke, P.; Barros, C.P.; Emrouznejad, A. Assessing productive efficiency of banks using integrated Fuzzy-DEA and bootstrapping: A case of Mozambican banks. *Eur. J. Oper. Res.* **2016**, *249*, 378–389. [CrossRef]
33. Hatami-Marbini, A.; Agrell, P.J.; Tavana, M.; Khoshnevis, P. A flexible cross-efficiency fuzzy data envelopment analysis model for sustainable sourcing. *J. Clean. Prod.* **2017**, *142*, 2761–2779. [CrossRef]
34. Nastis, S.A.; Bournaris, T.; Karpouzos, D. Fuzzy data envelopment analysis of organic farms. *Oper. Res.* **2017**, 1–14. [CrossRef]
35. Lio, W.; Liu, B. Uncertain data envelopment analysis with imprecisely observed inputs and outputs. *Fuzzy Optim. Decis. Mak.* **2017**, *17*, 357–373. [CrossRef]
36. Hu, C.F.; Liu, F.B. Data envelopment analysis with non-LR type fuzzy data. *Soft Comput.* **2017**, *21*, 5851–5857. [CrossRef]
37. Hu, C.K.; Liu, F.B.; Hu, C.F. Efficiency measures in fuzzy data envelopment analysis with common weights. *J. Ind. Manag. Optim.* **2017**, *13*, 237–249. [CrossRef]
38. Izadikhah, M.; Khoshroo, A. Energy management in crop production using a novel Fuzzy Data Envelopment Analysis model. *RAIRO-Oper. Res.* **2018**, *52*, 595–617. [CrossRef]
39. Amirkhan, M.; Didehkhani, H.; Khalili-Damghani, K.; Hafezalkotob, A. Mixed uncertainties in data envelopment analysis: A fuzzy-robust approach. *Expert Syst. Appl.* **2018**, *103*, 218–237. [CrossRef]
40. Rezaee, M.J.; Jozmaleki, M.; Valipour, M. Integrating dynamic fuzzy C-means, data envelopment analysis and artificial neural network to online prediction performance of companies in stock exchange. *Phys. A Stat. Mech. Appl.* **2018**, *489*, 78–93. [CrossRef]
41. Wang, S.; Yu, H.; Song, M. Assessing the efficiency of environmental regulations of large-scale enterprises based on extended fuzzy data envelopment analysis. *Ind. Manag. Data Syst.* **2018**, *118*, 463–479. [CrossRef]
42. Mu, W.; Kanellopoulos, A.; van Middelaar, C.E.; Stilmant, D.; Bloemhof, J.M. Assessing the impact of uncertainty on benchmarking the eco-efficiency of dairy farming using fuzzy data envelopment analysis. *J. Clean. Prod.* **2018**, *189*, 709–717. [CrossRef]

43. Zhou, X.; Xu, Z.; Chai, J.; Yao, L.; Wang, S.; Lev, B. Efficiency evaluation for banking systems under uncertainty: A multi-period three-stage DEA model. *Omega* **2018**, in press. [CrossRef]
44. Hatami-Marbini, A.; Tavana, M.; Ebrahimi, A. A fully fuzzified data envelopment analysis model. *Int. J. Inf. Decis. Sci.* **2011**, *3*, 252–264.
45. Kazemi, M.; Alimi, A. A fully fuzzy approach to data envelopment analysis. *J. Math. Comput. Sci.* **2014**, *11*, 238–245. [CrossRef]
46. Puri, J.; Yadav, S.P. A fully fuzzy approach to DEA and multi-component DEA for measuring fuzzy technical efficiencies in the presence of undesirable outputs. *Int. J. Syst. Assur. Eng. Manag.* **2015**, *6*, 268–285. [CrossRef]
47. Khaleghi, S.; Noura, A.; Lotfi, F.H. Measuring Efficiency and Ranking Fully Fuzzy DEA. *Indian J. Sci. Technol.* **2015**, *8*. [CrossRef]
48. Sotoudeh-Anvari, A.; Najafi, E.; Sadi-Nezhad, S. A new data envelopment analysis in fully fuzzy environment on the base of the degree of certainty of information. *J. Intell. Fuzzy Syst.* **2016**, *30*, 3131–3142. [CrossRef]
49. Zadeh, L.A. A note on Z-numbers. *Inf. Sci.* **2011**, *181*, 2923–2932. [CrossRef]
50. Kang, B.; Wei, D.; Li, Y.; Deng, Y. A method of converting Z-number to classical fuzzy number. *J. Inf. Comput. Sci.* **2012**, *9*, 703–709.
51. Allahviranloo, T.; Hosseinzadeh Lotfi, F.; Kiasary, M.K.; Kiani, N.A.; Alizadeh, L. Solving fully fuzzy linear programming problem by the ranking function. *Appl. Math. Sci.* **2008**, *2*, 19–32.
52. Guo, P.; Tanaka, H. Fuzzy DEA: A perceptual evaluation method. *Fuzzy Sets Syst.* **2001**, *119*, 149–160. [CrossRef]

symmetry

MDPI

Article

An Extended Step-Wise Weight Assessment Ratio Analysis with Symmetric Interval Type-2 Fuzzy Sets for Determining the Subjective Weights of Criteria in Multi-Criteria Decision-Making Problems

Mehdi Keshavarz-Ghorabaee [1], Maghsoud Amiri [1], Edmundas Kazimieras Zavadskas [2,*], Zenonas Turskis [2] and Jurgita Antucheviciene [2]

1 Department of Industrial Management, Faculty of Management and Accounting, Allameh Tabataba'i University, Tehran 1489684511, Iran; m.keshavarz_gh@yahoo.com (M.K.-G.); amiri@atu.ac.ir (M.A.)
2 Department of Construction Management and Real Estate, Vilnius Gediminas Technical University, Sauletekio al. 11, LT-10223 Vilnius, Lithuania; zenonas.turskis@vgtu.lt (Z.T.); jurgita.antucheviciene@vgtu.lt (J.A.)
* Correspondence: edmundas.zavadskas@vgtu.lt; Tel.: +370-5-274-4910

Received: 19 March 2018; Accepted: 30 March 2018; Published: 31 March 2018

Abstract: Determination of subjective weights, which are based on the opinions and preferences of decision-makers, is one of the most important matters in the process of multi-criteria decision-making (MCDM). Step-wise Weight Assessment Ratio Analysis (SWARA) is an efficient method for obtaining the subjective weights of criteria in the MCDM problems. On the other hand, decision-makers may express their opinions with a degree of uncertainty. Using the symmetric interval type-2 fuzzy sets enables us to not only capture the uncertainty of information flexibly but also to perform computations simply. In this paper, we propose an extended SWARA method with symmetric interval type-2 fuzzy sets to determine the weights of criteria based on the opinions of a group of decision-makers. The weights determined by the proposed approach involve the uncertainty of decision-makers' preferences and the symmetric form of the weights makes them more interpretable. To show the procedure of the proposed approach, it is used to determine the importance of intellectual capital dimensions and components in a company. The results show that the proposed approach is efficient in determining the subjective weights of criteria and capturing the uncertainty of information.

Keywords: multi-criteria decision-making (MCDM); group decision-making; interval type-2 fuzzy set (IT2FS); subjective weights; criteria weights; Step-wise Weight Assessment Ratio Analysis (SWARA)

1. Introduction

In discrete multi-criteria decision-making (MCDM) processes, we are usually faced with a set of alternatives that need to be evaluated with respect to a set of criteria. The weights of criteria are important components in the process of MCDM. There are two types of criteria weights: objective weights and subjective weights [1]. The objective weights are derived from the information of decision-matrices and the subjective weights are determined based on the information provided by decision-makers [2]. In other words, the subjective weights refer to the decision-makers' subjective perceptions and judgments about the relative importance of criteria [3]. Entropy, CRITIC (Criteria Importance Through Inter-criteria Correlation) and Standard Deviation (SD) are some of the popular methods for obtaining objective weights; on the other hand, the analytic hierarchy process (AHP), analytic network process (ANP), SIMOS (a method proposed by Simos [4]) and simple multi-attribute ranking technique (SMART) are the most prevalent methods for determination of subjective weights [5].

Another method for determining subjective criteria weights is Step-wise Weight Assessment Ratio Analysis (SWARA). The focus of this study is on the extending of this method.

The SWARA is an efficient and relatively new method which was proposed by Kersuliene et al. [6] for the determination of subjective weights in multi-criteria decision-making problems. This method has lower computational complexity compared to some other methods like AHP. Many researchers have utilized the SWARA method in different MCDM problems. Dehnavi et al. [7] proposed a novel hybrid approach based on the SWARA method, geographical information system (GIS) and adaptive neuro-fuzzy inference system (ANFIS) to evaluate landslide susceptible areas in Iran. Karabasevic et al. [8] developed a framework for multi-criteria assessment of personnel using the ARAS (Additive Ratio ASsessment) and SWARA methods under uncertainty handled by fuzzy sets. Nakhaei et al. [9] proposed a new approach based on SWARA and SMART for the rapid assessment of the vulnerability of office buildings to a blast. They presented a case study of Swiss Re Tower and used the proposed approach. Shukla et al. [10] integrated the SWARA method with PROMETHEE (Preference Ranking Organization METHod for Enrichment of Evaluations) and applied it to a comprehensive enterprise resource planning (ERP) system selection framework. Işık and Adalı [11] developed a hybrid MCDM approach based on the SWARA and operational competitiveness ratings analysis (OCRA) methods. They used SWARA to determine criteria weights and OCRA to rank the alternatives in a hotel selection problem. Mavi et al. [12] presented an integrated approach based on SWARA and MOORA (Multi-Objective Optimization on the basis of Ratio Analysis) in the fuzzy environment. Using a combination of sustainability and risk factors, the presented approach was used to evaluate third-party reverse logistics provider. Panahi et al. [13] used the SWARA and GIS to define copper prospectivity mapping in a region in Iran. They validated the obtained results by the operating receiver characteristics (ORC) technique. Stanujkic et al. [14] proposed an MCDM approach that allows the decision-makers involved in a negotiation process to express their preferences in a better way. Their proposed approach, called ARCAS (Additive Ratio Compromise ASsessment), is based on the ARAS and SWARA methods. Karabašević et al. [15] developed an expert-based method for the determination of subjective weights of criteria in a multi-criteria decision-making problem by integrating the SWARA and Delphi methods. Stanujkic et al. [16] presented a modified version of the SWARA method in which the most appropriate alternative is selected on the basis of negotiation. They used an empirical example of the personnel selection problem to illustrate their method. Urosevic et al. [17] developed an integrated MCDM approach based on the SWARA and WASPAS (Weighted Aggregated Sum Product Assessment) methods. They applied the developed approach to an example of personnel selection in the tourism sector. Jamali et al. [18] analyzed the competitive strategies of a LARG (Lean, Agile, Resilient and Green) supply chain in Iranian cement industries using SWARA, SWOT (Strengths, Weaknesses, Opportunities and Threats) and SPACE (Strategic Position and Action Evaluation) techniques. Juodagalvienė et al. [19] proposed a hybrid multi-criteria decision-making approach based on the SWARA and EDAS (Evaluation based on Distance from Average Solution) methods for assessment of architectural shapes of the buildings. Tayyar and Durmu [20] made a comparison between three weighting methods including Max100, SWARA and Pairwise Comparison (or AHP). The results of their study showed that the variation of the weights determined by the Pairwise Comparison method is higher than that of SWARA and Max100 methods. Valipour et al. [21] presented a hybrid MCDM approach based on the SWARA and COPRAS (COmplex PRoportional ASsessment) methods and applied it to a case study of risk assessment in a deep foundation excavation project in Iran. Keshavarz Ghorabaee et al. [22] developed a fuzzy hybrid approach to deal with MCDM problems based on the SWARA, CRITIC and EDAS methods and they used it for the assessment of construction equipment considering the sustainability dimensions. Dahooie et al. [23] proposed a framework of competency with five criteria for choosing the best information technology (IT) expert. They used the SWARA method for weighting the criteria and a grey ARAS (ARAS-G) method for evaluation of alternatives. Interested readers are referred to the recent survey paper presented by Mardani et al. [24].

The fuzzy set theory is an efficient tool to handle the inexact and imprecise information given by experts or decision-makers in an MCDM process [25–31]. Type-1 fuzzy sets are the basic or ordinary fuzzy sets which was introduced by Zadeh [32]. As an extension of the type-1 fuzzy sets, Zadeh [33] presented the concept of type-2 fuzzy sets. By increasing the degree of fuzziness, we can handle inexact and imprecise information more efficiently. Hence, we can say that type-2 fuzzy sets (T2FSs) can improve the process of capturing the uncertainty. On the other hand, type-2 fuzzy sets are computationally intensive and this can be considered as a disadvantage of them. To deal with this matter, we can use interval type-2 fuzzy sets (IT2FSs) in which secondary membership functions are interval sets [34,35]. A symmetric interval type-2 fuzzy set (SIT2FS) is a sub-class of IT2FSs and its primary membership function has symmetrical properties [36]. Using the symmetric IT2FSs simplifies the computations and leads to more interpretable results.

Many studies have been made on interval type-2 fuzzy sets and their applications in real-world problems. This type of fuzzy sets has also been applied to several MCDM processes in past years [37,38]. Soner et al. [39] presented a hybrid MCDM methodology based on the AHP and VIKOR (in Serbian: VIseKriterijumska Optimizacija I Kompromisno Resenje) methods with interval type-2 fuzzy sets. They also provided a practical application of the methodology in maritime transportation industry. Keshavarz Ghorabaee et al. [40] proposed a novel multi-objective methodology for supplier evaluation and order allocation in supply chains based on the EDAS method and IT2FSs and applied it to a problem considering environmental criteria. Qin et al. [41] developed a multi-criteria group decision making approach based upon interval type-2 fuzzy sets, the prospect theory and the TODIM (an acronym in Portuguese of interactive and multi-criteria decision making) method. They also introduced a new distance measure for IT2FSs using α-cut technique. Baykasoğlu and Gölcük [42] presented an approach by integrating TOPSIS (Technique for Order Preference by Similarity to Ideal Solution) with DEMATEL (Decision Making Trial and Evaluation Laboratory) to handle MCDM problems under the uncertain environment defined by IT2FSs. They used the presented approach in a SWOT-based strategy selection problem. Deveci et al. [43] proposed an interval type-2 fuzzy TOPSIS method to deal with multi-criteria decision-making problems under uncertainty. To verify their method, a case study of airline route selection in Turkey was also presented by them. Zhong and Yao [44] developed an extended ELECTRE (ELimination Et Choix Traduisant la REalité) method to deal with multi-criteria group decision making with interval type-2 fuzzy sets based on new measures of proximity and entropy of IT2FSs. Kundu et al. [45] introduced a new ranking method for IT2FSs by using a relative preference index and generalized credibility measures. Based on the ranking method, they presented a fuzzy multi-criteria decision-making methodology and applied it to the transportation mode selection problem. Qin et al. [46] developed a new approach to handle multi-criteria group decision-making problems under interval type-2 fuzzy environment. Their method is based on the LINMAP (Linear Programming Models with the Aid of Multidimensional Analysis of Preference) method and they showed the application of it in supplier selection problems. Ju et al. [47] proposed an interval type-2 fuzzy multi-criteria group decision-making approach with unknown weight information. They integrated the grey relational projection (GRP) and AHP methods in the proposed approach. Yao and Wang [48] presented some factors including hesitant factor, fuzzy factor and interval factor to quantify the hesitancy, fuzziness and interval information related to an IT2FS. Then, based on these factors, they introduced a cross-entropy and an MCDM approach. Liu et al. [49] proposed an integrated approach for multi-criteria group decision-making based on the ANP and VIKOR methods. In a sustainable supplier selection problem, they obtained the weights of criteria by using ANP and made the final evaluation of suppliers via VIKOR. Wu et al. [50] developed a new methodology based on the TOPSIS method to handle large scale group decision making problems with social network information in an interval type-2 fuzzy environment. They demonstrated the feasibility of the methodology using an illustrative example. For more information about the applications of IT2FSs in multi-criteria decision-making, interested readers can refer to a comprehensive review presented by Celik et al. [51].

In this paper, the SWARA method is extended with interval type-2 fuzzy sets. Using interval type-2 fuzzy sets helps us to deal more effectively with the uncertainty of experts' or decision-makers' opinions, judgements and preferences. Accordingly, the subjective weights determined by the extended SWARA could be more justified for the decision-making process. To extend the SWARA method, we utilize a sub-class of IT2FSs called symmetric interval type-2 fuzzy sets. The computations of the process can be simplified when we use symmetric IT2FSs. Moreover, the subjective weights resulted in the form of symmetric IT2FSs is more interpretable. Therefore, in the proposed approach, we can quantify the uncertain information given by decision-makers efficiently and present interpretable subjective weights for criteria.

The rest of this paper is organized as follows. In Section 2, firstly, the concepts and definitions related to type-2 fuzzy sets, interval T2FSs and symmetric IT2FSs are presented and then an extended SWARA method with symmetric IT2FSs is detailed. In Section 3, we illustrated the proposed approach by means of an example of determining the importance of intellectual capital dimensions and components in a company. In Section 4, conclusions are presented.

2. Methodology

In this section, we first present the concepts of type-2 fuzzy sets and definitions and arithmetic operations of symmetric interval type-2 fuzzy sets. Then a new approach based on the SWARA method is proposed to determine the subjective weights of criteria under uncertainty.

2.1. Concepts and Definitions

The concept of type-2 fuzzy sets was firstly presented by Zadeh [33] as an extension of type-1 fuzzy sets (ordinary fuzzy sets). Type-2 fuzzy sets help us to increase the fuzziness of a relation and this can increase the ability to deal with inexact information in a logically correct manner [52]. In the following, we present the important and basic concepts and definitions related to T2FSs.

Definition 1. *A type-2 fuzzy set $\tilde{\tilde{\mathcal{U}}}$ has a two dimensional membership function which can be defined as follows [35]:*

$$\tilde{\tilde{\mathcal{U}}} = \int_{x\in X}\int_{\mu\in\mathcal{D}_X} \omega_{\tilde{\tilde{\mathcal{U}}}}(x,\mu)/(x,\mu) \tag{1}$$

In Equation (1), x is the primary variable with domain X and μ is the secondary variable. $\mathcal{D}_X$ denotes the primary membership function and $\omega_{\tilde{\tilde{\mathcal{U}}}}(x,\mu)$ shows the secondary grade, where $\omega_{\tilde{\tilde{\mathcal{U}}}}(x,\mu) \in [0,\ 1]$ and $\mathcal{D}_X \subseteq [0,\ 1]$. In this equation, the union of all admissible values of x and μ is represented by $\int\int$.

Definition 2. *In a type-2 fuzzy set $\tilde{\tilde{\mathcal{U}}}$, if all values of the secondary grade are equal to one ($\omega_{\tilde{\tilde{\mathcal{U}}}}(x,\ \mu) = 1$), then $\tilde{\tilde{\mathcal{U}}}$ is an interval type-2 fuzzy set which can be described as follows [35]:*

$$\tilde{\tilde{\mathcal{U}}} = \int_{x\in X}\int_{\mu\in\mathcal{D}_X} 1/(x,\mu) \tag{2}$$

Definition 3. *Footprint of Uncertainty (FOU) is the union of all primary memberships described by the area between its bounds called Upper Membership Function (UMF) and Lower Membership Function (LMF). The UMF and LMF of an IT2FS $\tilde{\tilde{\mathcal{U}}}$ are ordinary (type-1) fuzzy sets and their membership functions can be represented by $\overline{\mu}_{\tilde{\tilde{\mathcal{U}}}}(x)$ and $\underline{\mu}_{\tilde{\tilde{\mathcal{U}}}}(x)$, respectively [35].*

Definition 4. *If the FOU of an interval type-2 fuzzy set $\tilde{\tilde{\mathcal{U}}}$ is symmetrical about $x = u$, it can be considered as a sub-class called symmetric IT2FSs. A SIT2FS has the following properties [36].*

$$\overline{\mu}_{\tilde{\tilde{U}}}(u+x) = \overline{\mu}_{\tilde{\tilde{U}}}(u-x) \tag{3}$$

$$\underline{\mu}_{\tilde{\tilde{U}}}(u+x) = \underline{\mu}_{\tilde{\tilde{U}}}(u-x) \tag{4}$$

Definition 5. *A symmetric IT2FS $\tilde{\tilde{U}}$ is a triangular SIT2FS if its UMF and LMF are symmetric triangular fuzzy sets. Based on the definition of symmetric triangular fuzzy sets presented by Ma et al. [53], this kind of fuzzy sets can be defined using the following membership functions.*

$$\overline{\mu}_{\tilde{\tilde{U}}}(x) = \begin{cases} (x-u+\overline{\delta})/\overline{\delta}, & u-\overline{\delta} \le x \le u \\ \frac{(u+\overline{\delta}-x)}{\overline{\delta}}, & u \le x \le u+\overline{\delta} \\ 0, & otherwise \end{cases} \tag{5}$$

$$\underline{\mu}_{\tilde{\tilde{U}}}(x) = \begin{cases} (x-u+\underline{\delta})/\underline{\delta}, & u-\underline{\delta} \le x \le u \\ \frac{(u+\underline{\delta}-x)}{\underline{\delta}}, & u \le x \le u+\underline{\delta} \\ 0, & otherwise \end{cases} \tag{6}$$

where u is the center of the set and $\overline{\delta}$ and $\underline{\delta}$ ($\overline{\delta} > \underline{\delta}$) denote the fuzziness values of the UMF and LMF, respectively. If $\overline{\delta} = \underline{\delta}$ then this fuzzy set is transformed into an ordinary symmetric triangular fuzzy set. Here we assume that both the UMF and LMF are normal triangular fuzzy sets. The graphical representation of this fuzzy set is shown in Figure 1. This kind of fuzzy sets can also be defined as a triplet $(u,\ \underline{\delta},\ \overline{\delta})$.

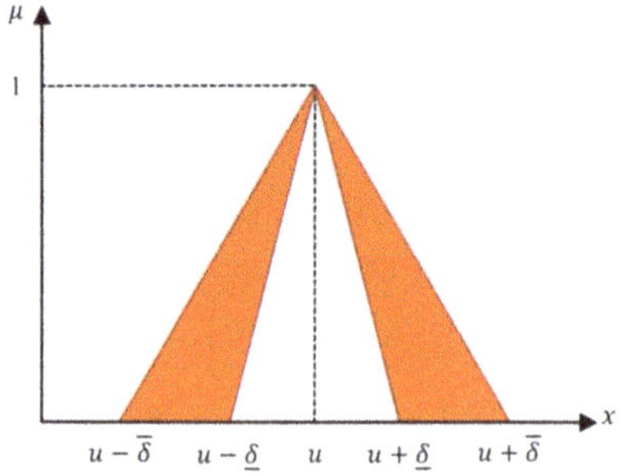

Figure 1. An example of triangular symmetric IT2FS.

Definition 6. *In the membership function of a triangular symmetric IT2FS (Equations (5) and (6)), we can define the fuzziness values ($\overline{\delta}$ and $\underline{\delta}$) by using a coefficient of the center value (u). If $\overline{\delta} = \overline{\rho}u$ and $\underline{\delta} = \underline{\rho}u$, then the range $[\underline{\rho}, \overline{\rho}]$ is defined here as the interval of uncertainty about the center (IUC).*

Definition 7. *Let $\tilde{\tilde{M}}$ and $\tilde{\tilde{N}}$ be two triangular symmetric IT2FSs where $\tilde{\tilde{M}} = (m, \underline{\delta}_m, \overline{\delta}_m)$ and $\tilde{\tilde{N}} = (n, \underline{\delta}_n, \overline{\delta}_n)$ and d is a positive crisp number. Then using the arithmetic operations of IT2FSs, we can define the following operations for these fuzzy sets [54].*

$$\tilde{\tilde{M}} \oplus \tilde{\tilde{N}} = (m+n, \underline{\delta}_m + \underline{\delta}_n,\ \overline{\delta}_m + \overline{\delta}_n) \tag{7}$$

$$\tilde{\tilde{M}} + d = (m+d,\ \underline{\delta}_m,\ \overline{\delta}_m) \tag{8}$$

$$\tilde{\tilde{M}} \cdot d = (m \cdot d, \underline{\delta}_m \cdot d, \overline{\delta}_m \cdot d) \tag{9}$$

Definition 8. *Let us denote by* $\mathcal{T}(\tilde{\tilde{\mathcal{U}}})$ *the defuzzified value of an interval type-2 fuzzy set* $\tilde{\tilde{\mathcal{U}}}$. *If* $\tilde{\tilde{\mathcal{U}}}$ *is a triangular symmetric IT2FSs and* $\tilde{\tilde{\mathcal{U}}} = (u, \underline{\delta}, \overline{\delta})$, *the defuzzified value of* $\tilde{\tilde{\mathcal{U}}}$ *is equal to u* ($\mathcal{T}\left(\tilde{\tilde{\mathcal{U}}}\right) = u$).

2.2. An Extended SWARA with Symmetric IT2FSs

The subjective weights of criteria are usually determined based on the opinion and judgment of the decision-maker(s) or expert(s) in a multi-criteria decision-making process. Because these weights are very important, there have been several methods for determination of them. One of the efficient methods for determination of subjective weights of criteria is the SWARA method [6]. In this section, we propose an extended SWARA method based on the concepts and definitions of symmetric interval type-2 fuzzy sets to determine the subjective criteria weights. The subjective criteria weights can be more sensible if they determined based on the judgments and preferences of a group of decision-makers. The extended SWARA proposed in this study has the ability of determining the weights of criteria for a group decision-making problem. Suppose that we want to determine the subjective weights of M criteria based on the opinions of N decision-makers. The steps of determination of criteria weights using the extended SWARA with SIT2FSs are presented as follows:

Step 1. After formation of a group of decision-makers, each decision-maker sorts the criteria in a descending order based on the expected importance of them, that is, the most important criterion is ranked first. Let us denote by τ_{jk} the rank of jth criterion based on the opinion of kth decision-maker ($j \in \{1, 2, \ldots, M\}$ and $k \in \{1, 2, \ldots, N\}$).

Step 2. In this step, we start with the criterion which has the second rank in the sorted list of Step 1 ($\tau_{jk} = 2$) and get the relative importance of each criterion ($S_{\tau_{jk}}$ and τ_{jk} = 2, 3, ..., M) compared to the previous criterion of the list from each decision-maker. Also, the decision-makers are asked to give the IUC related to each value of $S_{\tau_{jk}}$. Suppose that $[\underline{\rho}_{\tau_{jk}}, \overline{\rho}_{\tau_{jk}}]$ is the IUC of $S_{\tau_{jk}}$ values. Then, based on Definitions 5 and 6, the symmetric triangular interval type-2 fuzzy set related to decision-makers' expression is represented as follows:

$$\tilde{\tilde{S}}_{\tau_{jk}} = \left(S_{\tau_{jk}}, \underline{\delta s}_{\tau_{jk}}, \overline{\delta s}_{\tau_{jk}}\right) \tag{10}$$

where $\underline{\delta s}_{\tau_{jk}} = \underline{\rho}_{\tau_{jk}} S_{\tau_{jk}}$ and $\overline{\delta s}_{\tau_{jk}} = \overline{\rho}_{\tau_{jk}} S_{\tau_{jk}}$.

Step 3. In this step, we calculate the values of $\tilde{\tilde{K}}_{\tau_{jk}} = \left(K_{\tau_{jk}}, \underline{\delta k}_{\tau_{jk}}, \overline{\delta k}_{\tau_{jk}}\right)$ for each decision-maker and $\tau_{jk} > 1$ as follows:

$$\tilde{\tilde{K}}_{\tau_{jk}} = 1 + \tilde{\tilde{S}}_{\tau_{jk}} \tag{11}$$

Step 4. Calculation of the relative weighting factors $\tilde{\tilde{Q}}_{\tau_{jk}}$ is made in this step by the following equation.

$$\tilde{\tilde{Q}}_{\tau_{jk}} = \begin{cases} \left(1, \underline{\delta q}_1, \overline{\delta q}_1\right) & if\, \tau_{jk} = 1 \\ \left(\dfrac{\mathcal{T}\left(\tilde{\tilde{Q}}_{\tau_{jk}-1}\right)}{\mathcal{T}\left(\tilde{\tilde{K}}_{\tau_{jk}}\right)^2}\right) \cdot \tilde{\tilde{K}}_{\tau_{jk}} & if\, \tau_{jk} > 1 \end{cases} \tag{12}$$

where $\underline{\delta q}_1 = \frac{1}{M-1} \sum_{\tau_{jk}=2}^{M} \underline{\delta k}_{\tau_{jk}}$ and $\overline{\delta q}_1 = \frac{1}{M-1} \sum_{\tau_{jk}=2}^{M} \overline{\delta k}_{\tau_{jk}}$. By using these formulas, we involve a degree of uncertainty in the first rank criterion. Although the average interval of the other criteria is used in this study, this interval can be set by the decision- makers.

Step 5. Using the following equation, we can determine the subjective weights for kth decision-maker in the form of symmetric interval type-2 fuzzy sets:

$$\tilde{\tilde{w}}_{jk} = \left(1/\mathcal{T}\left(\underset{\tau_{jk}=1}{\overset{M}{\oplus}} \tilde{\tilde{Q}}_{\tau_{jk}}\right)\right) \cdot \tilde{\tilde{Q}}_{\tau_{jk}} \tag{13}$$

Step 6. Finally, the aggregated subjective weight of each criterion is calculated as follows:

$$\tilde{\tilde{w}}_j = \frac{1}{N} \underset{k=1}{\overset{N}{\oplus}} \tilde{\tilde{w}}_{jk} \tag{14}$$

In a hierarchical structure, a criterion may have some sub-criteria. To determine the subjective weights of sub-criteria, firstly, we should fulfill the presented steps to obtain the subjective weights of each criterion; secondly, the subjective weights of sub-criteria of each criterion should be determined by the same steps and finally, the global subjective weights of sub-criteria ($\tilde{\tilde{w}}_j^G$) are calculated by multiplication of the subjective weights of sub-criteria and their upper level criteria. In this study, we use the defuzzified values of the subjective weights of criteria in the upper level of hierarchical structure to keep the symmetrical property of the weights.

If $\tilde{\tilde{w}}_r$ shows the subjective weight of rth criterion and $\tilde{\tilde{w}}_{rl}$ denote the subjective weight of lth sub-criterion of rth criterion. Then the global subjective weight of lth sub-criterion ($\tilde{\tilde{w}}_{rl}^G$) is calculated as follows:

$$\tilde{\tilde{w}}_{rl}^G = \mathcal{T}\left(\tilde{\tilde{w}}_r\right) \cdot \tilde{\tilde{w}}_{rl} \tag{15}$$

To make the proposed approach clear, we present the flowchart of the procedure in Figure 2.

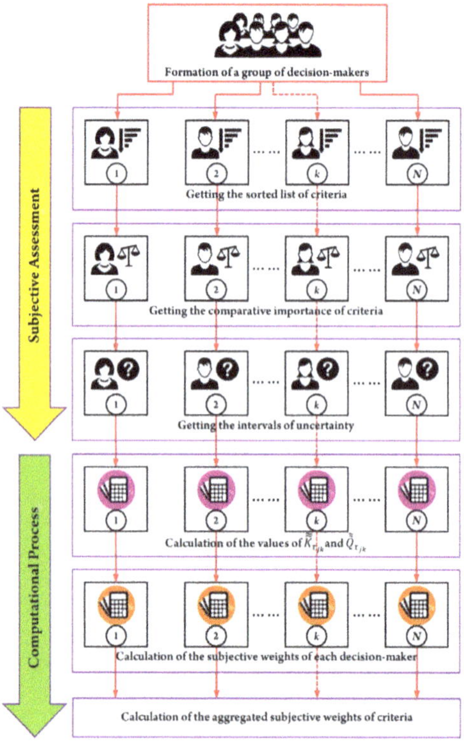

Figure 2. The procedure of the extended SWARA with symmetric IT2FSs.

3. Illustrative Example

In this section, the extended SWARA with SIT2FSs is used to determine the importance of intellectual capital dimensions and components in a company. The hierarchical structure of the multi-criteria decision-making problem is defined in Figure 3 based on the study of Bozbura and Beskese [55]. According to Figure 3, we have three main criteria which include some sub-criteria. Suppose that three experts (decision-makers) from the company are appointed to perform the process of the criteria and sub-criteria assessment. Based on the hierarchical structure, firstly, the proposed approach is used to determine the importance or weights of the main criteria, then the weights of the sub-criteria of each criterion are calculated. Therefore, the procedure of the proposed approach should be performed four times (Once for the main criteria and three for the sub-criteria of them).

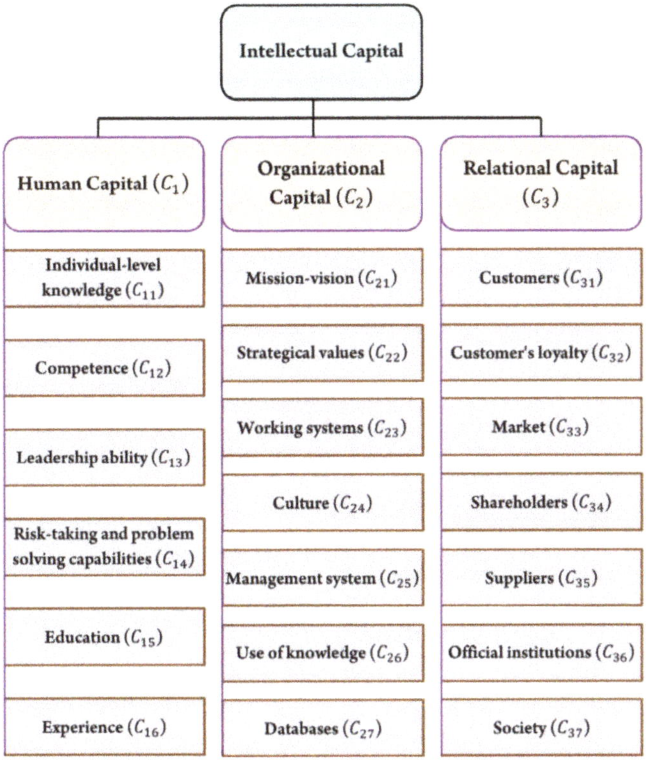

Figure 3. The hierarchical structure of intellectual capital dimensions.

The information given by the first (DM_1), second (DM_2) and third (DM_3) decision-makers about the main criteria are shown in Table 1 and the information related to sub-criteria are presented in Tables 2–4. As can be seen in these tables, according to Step 2 of the proposed approach, the relative importance of each criterion and sub-criterion compared to the previous item of the sorted list ($S_{\tau_{jk}}$) is transformed into a triangular SIT2FS based on the IUC provided by each decision-makers and Equation (10). The values of $\tilde{\tilde{K}}_{\tau_{jk}}$ and $\tilde{\tilde{Q}}_{\tau_{jk}}$, which are calculated using Equations (11) and (12) in Steps 3 and 4, are also presented in Tables 1–4.

Table 1. The information related to the main criteria.

	Sorted	$S_{\o jk}$	IUC	$\tilde{\tilde{S}}_{\o jk}$	$\tilde{\tilde{K}}_{\o jk}$	$\tilde{\tilde{Q}}_{\o jk}$
DM_1	C_1	—	—	—	—	(1, 0.086, 0.098)
	C_2	0.1	[1.1, 1.2]	(0.1, 0.11, 0.12)	(1.1, 0.11, 0.12)	(0.909, 0.091, 0.099)
	C_3	0.05	[1.25, 1.5]	(0.05, 0.063, 0.075)	(1.05, 0.063, 0.075)	(0.866, 0.052, 0.062)
DM_2	C_2	—	—	—	—	(1, 0.094, 0.11)
	C_1	0.05	[1.15, 1.2]	(0.05, 0.058, 0.06)	(1.05, 0.058, 0.06)	(0.952, 0.052, 0.054)
	C_3	0.1	[1.3, 1.6]	(0.1, 0.13, 0.16)	(1.1, 0.13, 0.16)	(0.866, 0.102, 0.126)
DM_3	C_2	—	—	—	—	(1, 0.306, 0.36)
	C_1	0.5	[1.1, 1.3]	(0.5, 0.55, 0.65)	(1.5, 0.55, 0.65)	(0.667, 0.244, 0.289)
	C_3	0.05	[1.25, 1.4]	(0.05, 0.063, 0.07)	(1.05, 0.063, 0.07)	(0.635, 0.038, 0.042)

As an example, here we calculate the value of $\tilde{\tilde{S}}_2$ for the second rank main criteria of the first decision-maker. According to Table 1, the value of S_2 is equal to 0.1 for DM_1 and the interval of uncertainty about the center is [1.1, 1.2] for S_2. Based on Step 2 and Definition 6, we can calculate $\underline{\delta s}_2$ and $\overline{\delta s}_2$ as follows: $\underline{\delta s}_2 = \underline{\rho}_2\, S_2 = 1.1 \times 0.1 = 0.11$ and $\overline{\delta s}_2 = \overline{\rho}_2 S_2 = 1.2 \times 0.1 = 0.12$. Then we have $\tilde{\tilde{S}}_2 = (0.1, 0.11, 0.12)$. The other values of $S_{\tau_{jk}}$ are calculated in the same way.

Table 2. The information related to the sub-criteria of the first criterion (C_1).

	Sorted	$S_{\o jk}$	IUC	$\tilde{\tilde{S}}_{\o jk}$	$\tilde{\tilde{K}}_{\o jk}$	$\tilde{\tilde{Q}}_{\o jk}$
DM_1	C_{11}	—	—	—	—	(1, 0.177, 0.194)
	C_{16}	0.2	[1.05, 1.15]	(0.2, 0.21, 0.23)	(1.2, 0.21, 0.23)	(0.833, 0.146, 0.16)
	C_{15}	0.25	[1.1, 1.2]	(0.25, 0.275, 0.3)	(1.25, 0.275, 0.3)	(0.667, 0.147, 0.16)
	C_{14}	0.15	[1.05, 1.1]	(0.15, 0.158, 0.165)	(1.15, 0.158, 0.165)	(0.58, 0.079, 0.083)
	C_{13}	0.1	[1.25, 1.5]	(0.1, 0.125, 0.15)	(1.1, 0.125, 0.15)	(0.527, 0.06, 0.072)
	C_{12}	0.1	[1.15, 1.25]	(0.1,0.115,0.125)	(1.1, 0.115, 0.125)	(0.479, 0.05, 0.054)
DM_2	C_{16}	—	—	—	—	(1, 0.162, 0.181)
	C_{11}	0.25	[1.15, 1.25]	(0.25, 0.288, 0.313)	(1.25, 0.288, 0.313)	(0.8, 0.184, 0.2)
	C_{14}	0.05	[1.15, 1.25]	(0.05, 0.058, 0.063)	(1.05, 0.058, 0.063)	(0.762, 0.042, 0.045)
	C_{15}	0.1	[1.05, 1.1]	(0.1, 0.105, 0.11)	(1.1, 0.105, 0.11)	(0.693, 0.066, 0.069)
	C_{12}	0.2	[1.25,1.5]	(0.2, 0.25, 0.3)	(1.2, 0.25, 0.3)	(0.577, 0.12, 0.144)
	C_{13}	0.1	[1.1, 1.2]	(0.1, 0.11, 0.12)	(1.1, 0.11, 0.12)	(0.525, 0.052, 0.057)
DM_3	C_{11}	—	—	—	—	(1, 0.191, 0.209)
	C_{15}	0.05	[1.15, 1.25]	(0.05, 0.058, 0.063)	(1.05, 0.058, 0.063)	(0.952, 0.052, 0.057)
	C_{16}	0.1	[1.05, 1.1]	(0.1, 0.105, 0.11)	(1.1, 0.105, 0.11)	(0.866, 0.083, 0.087)
	C_{14}	0.3	[1.3, 1.4]	(0.3, 0.39, 0.42)	(1.3, 0.39, 0.42)	(0.666, 0.2, 0.215)
	C_{12}	0.25	[1.1, 1.2]	(0.25, 0.275, 0.3)	(1.25, 0.275, 0.3)	(0.533, 0.117, 0.128)
	C_{13}	0.1	[1.25, 1.5]	(0.1, 0.125, 0.15)	(1.1, 0.125, 0.15)	(0.484, 0.055, 0.066)

Table 3. The information related to the sub-criteria of the second criterion (C_2).

	Sorted	$S_{\o jk}$	IUC	$\tilde{\tilde{S}}_{\o jk}$	$\tilde{\tilde{K}}_{\o jk}$	$\tilde{\tilde{Q}}_{\o jk}$
DM_1	C_{25}	—	—	—	—	(1, 0.235, 0.255)
	C_{23}	0.3	[1.2, 1.3]	(0.3, 0.36, 0.39)	(1.3, 0.36, 0.39)	(0.769, 0.213, 0.231)
	C_{26}	0.15	[1.1, 1.15]	(0.15, 0.165, 0.173)	(1.15, 0.165, 0.173)	(0.669, 0.096, 0.1)
	C_{27}	0.35	[1.05, 1.15]	(0.35, 0.368, 0.402)	(1.35, 0.368, 0.402)	(0.495, 0.135, 0.148)
	C_{24}	0.2	[1.1, 1.25]	(0.2, 0.22, 0.25)	(1.2, 0.22, 0.25)	(0.413, 0.076, 0.086)
	C_{21}	0.15	[1.25, 1.3]	(0.15, 0.188, 0.195)	(1.15, 0.188, 0.195)	(0.359, 0.059, 0.061)
	C_{22}	0.1	[1.1, 1.2]	(0.1, 0.11, 0.12)	(1.1, 0.11, 0.12)	(0.326, 0.033, 0.036)

Table 3. *Cont.*

	Sorted	$S_{\o_{jk}}$	IUC	$\tilde{\tilde{S}}_{\o_{jk}}$	$\tilde{\tilde{K}}_{\o_{jk}}$	$\tilde{\tilde{Q}}_{\o_{jk}}$
DM_2	C_{23}	—	—	—	—	(1, 0.238, 0.252)
	C_{25}	0.4	[1.1, 1.15]	(0.4, 0.44, 0.46)	(1.4, 0.44, 0.46)	(0.714, 0.224, 0.235)
	C_{27}	0.2	[1.2, 1.25]	(0.2, 0.24, 0.25)	(1.2, 0.24, 0.25)	(0.595, 0.119, 0.124)
	C_{26}	0.2	[1.3, 1.35]	(0.2, 0.26, 0.27)	(1.2, 0.26, 0.27)	(0.496, 0.107, 0.112)
	C_{21}	0.1	[1.15, 1.25]	(0.1, 0.115, 0.125)	(1.1, 0.115, 0.125)	(0.451, 0.047, 0.051)
	C_{22}	0.15	[1.1, 1.25]	(0.15, 0.165, 0.188)	(1.15, 0.165, 0.188)	(0.392, 0.056, 0.064)
	C_{24}	0.2	[1.05, 1.1]	(0.2, 0.21, 0.22)	(1.2, 0.21, 0.22)	(0.327, 0.057, 0.06)
DM_3	C_{25}	—	—	—	—	(1, 0.156, 0.168)
	C_{26}	0.1	[1.05, 1.1]	(0.1, 0.105, 0.11)	(1.1, 0.105, 0.11)	(0.909, 0.087, 0.091)
	C_{23}	0.1	[1.15, 1.25]	(0.1, 0.115,0.125)	(1.1, 0.115, 0.125)	(0.826, 0.086, 0.094)
	C_{21}	0.15	[1.15, 1.3]	(0.15, 0.173,0.195)	(1.15, 0.173, 0.195)	(0.719, 0.108, 0.122)
	C_{27}	0.25	[1.1, 1.2]	(0.25, 0.275,0.3)	(1.25, 0.275, 0.3)	(0.575, 0.126, 0.138)
	C_{24}	0.2	[1.05, 1.1]	(0.2, 0.21,0.22)	(1.2, 0.21, 0.22)	(0.479, 0.084, 0.088)
	C_{22}	0.05	[1.15, 1.2]	(0.05, 0.058,0.06)	(1.05, 0.058, 0.06)	(0.456, 0.025, 0.026)

Table 4. The information related to the sub-criteria of the third criterion (C_3).

	Sorted	$S_{\o_{jk}}$	IUC	$\tilde{\tilde{S}}_{\o_{jk}}$	$\tilde{\tilde{K}}_{\o_{jk}}$	$\tilde{\tilde{Q}}_{\o_{jk}}$
DM_1	C_{33}	—	—	—	—	(1, 0.145, 0.153)
	C_{34}	0.2	[1.25, 1.3]	(0.2, 0.25, 0.26)	(1.2, 0.25, 0.26)	(0.833, 0.174, 0.181)
	C_{31}	0.25	[1.1, 1.15]	(0.25, 0.275, 0.288)	(1.25, 0.275, 0.288)	(0.667, 0.147, 0.153)
	C_{32}	0.05	[1.05, 1.1]	(0.05, 0.053, 0.055)	(1.05, 0.053, 0.055)	(0.635, 0.032, 0.033)
	C_{35}	0.05	[1.2, 1.3]	(0.05, 0.06, 0.065)	(1.05, 0.06, 0.065)	(0.605, 0.035, 0.037)
	C_{36}	0.1	[1.25, 1.35]	(0.1, 0.125, 0.135)	(1.1, 0.125, 0.135)	(0.55, 0.062, 0.067)
	C_{37}	0.1	[1.1, 1.15]	(0.1, 0.11, 0.115)	(1.1, 0.11, 0.115)	(0.5, 0.05, 0.052)
DM_2	C_{34}	—	—	—	—	(1, 0.183, 0.2)
	C_{33}	0.3	[1.15, 1.25]	(0.3, 0.345, 0.375)	(1.3, 0.345, 0.375)	(0.769, 0.204, 0.222)
	C_{32}	0.1	[1.05, 1.15]	(0.1, 0.105, 0.115)	(1.1, 0.105, 0.115)	(0.699, 0.067, 0.073)
	C_{31}	0.25	[1.25, 1.4]	(0.25, 0.313, 0.35)	(1.25, 0.313, 0.35)	(0.559, 0.14, 0.157)
	C_{35}	0.05	[1.1, 1.2]	(0.05, 0.055, 0.06)	(1.05, 0.055, 0.06)	(0.533, 0.028, 0.03)
	C_{37}	0.1	[1.05, 1.1]	(0.1, 0.105, 0.11)	(1.1, 0.105, 0.11)	(0.484, 0.046, 0.048)
	C_{36}	0.15	[1.15, 1.25]	(0.15, 0.173, 0.188)	(1.15, 0.173, 0.188)	(0.421, 0.063, 0.069)
DM_3	C_{33}	—	—	—	—	(1, 0.205, 0.225)
	C_{34}	0.25	[1.15, 1.25]	(0.25, 0.288, 0.313)	(1.25, 0.288, 0.313)	(0.8, 0.184, 0.2)
	C_{31}	0.1	[1.1, 1.2]	(0.1, 0.11, 0.12)	(1.1, 0.11, 0.12)	(0.727, 0.073, 0.079)
	C_{35}	0.1	[1.25, 1.5]	(0.1, 0.125, 0.15)	(1.1, 0.125, 0.15)	(0.661, 0.075, 0.09)
	C_{32}	0.15	[1.1, 1.2]	(0.15, 0.165, 0.18)	(1.15, 0.165, 0.18)	(0.575, 0.082, 0.09)
	C_{37}	0.2	[1.15, 1.2]	(0.2, 0.23, 0.24)	(1.2, 0.23, 0.24)	(0.479, 0.092, 0.096)
	C_{36}	0.3	[1.05, 1.15]	(0.3, 0.315, 0.345)	(1.3, 0.315, 0.345)	(0.369, 0.089, 0.098)

According to the values of $\tilde{\tilde{K}}_{\tau_{jk}}$ and $\tilde{\tilde{Q}}_{\tau_{jk}}$ and Equation (13) of Step 5, the subjective weights of the main criteria and their sub-criteria can be calculated for each decision-maker. The results of this step are presented in Table 5.

Table 5. The subjective weights of the criteria and sub-criteria for each decision-maker ($\widetilde{\widetilde{w}}_{jk}$).

	DM_1	DM_2	DM_3
C_1	(0.36, 0.031, 0.035)	(0.338, 0.019, 0.019)	(0.29, 0.106, 0.126)
C_{11}	(0.245, 0.043, 0.047)	(0.184, 0.042, 0.046)	(0.222, 0.042, 0.046)
C_{12}	(0.117, 0.012, 0.013)	(0.132, 0.028, 0.033)	(0.118, 0.026, 0.028)
C_{13}	(0.129, 0.015, 0.018)	(0.12, 0.012, 0.013)	(0.108, 0.012, 0.015)
C_{14}	(0.142, 0.019, 0.02)	(0.175, 0.01, 0.01)	(0.148, 0.044, 0.048)
C_{15}	(0.163, 0.036, 0.039)	(0.159, 0.015, 0.016)	(0.212, 0.012, 0.013)
C_{16}	(0.204, 0.036, 0.039)	(0.23, 0.037, 0.042)	(0.192, 0.018, 0.019)
C_2	(0.328, 0.033, 0.036)	(0.355, 0.033, 0.039)	(0.434, 0.133, 0.156)
C_{21}	(0.089, 0.015, 0.015)	(0.113, 0.012, 0.013)	(0.145, 0.022, 0.025)
C_{22}	(0.081, 0.008, 0.009)	(0.099, 0.014, 0.016)	(0.092, 0.005, 0.005)
C_{23}	(0.191, 0.053, 0.057)	(0.252, 0.06, 0.063)	(0.166, 0.017, 0.019)
C_{24}	(0.102, 0.019, 0.021)	(0.082, 0.014, 0.015)	(0.097, 0.017, 0.018)
C_{25}	(0.248, 0.058, 0.063)	(0.18, 0.056, 0.059)	(0.201, 0.031, 0.034)
C_{26}	(0.166, 0.024, 0.025)	(0.125, 0.027, 0.028)	(0.183, 0.017, 0.018)
C_{27}	(0.123, 0.033, 0.037)	(0.15, 0.03, 0.031)	(0.116, 0.025, 0.028)
C_3	(0.312, 0.019, 0.022)	(0.307, 0.036, 0.045)	(0.276, 0.016, 0.018)
C_{31}	(0.139, 0.031, 0.032)	(0.125, 0.031, 0.035)	(0.158, 0.016, 0.017)
C_{32}	(0.133, 0.007, 0.007)	(0.157, 0.015, 0.016)	(0.125, 0.018, 0.02)
C_{33}	(0.209, 0.03, 0.032)	(0.172, 0.046, 0.05)	(0.217, 0.045, 0.049)
C_{34}	(0.174, 0.036, 0.038)	(0.224, 0.041, 0.045)	(0.173, 0.04, 0.043)
C_{35}	(0.126, 0.007, 0.008)	(0.119, 0.006, 0.007)	(0.143, 0.016, 0.02)
C_{36}	(0.115, 0.013, 0.014)	(0.094, 0.014, 0.015)	(0.08, 0.019, 0.021)
C_{37}	(0.104, 0.01, 0.011)	(0.108, 0.01, 0.011)	(0.104, 0.02, 0.021)

Based on the values of subjective weights presented in Table 5 and Equation (14) of Step 6, we can determine the aggregated subjective weights for the main criteria and their sub-criteria. Because of the hierarchical structure of the criteria and sub-criteria, the weights of sub-criteria are local in this step. The global subjective weights of sub-criteria are calculated by multiplying the local weights of them by the weights of the upper level criteria. The local and global subjective weights are represented in Table 6. It should be noted that Equation (15) is used to calculate the global weights.

Table 6. The aggregated local and global subjective weights.

	Aggregated Local Weights	Global Weights of Sub-Criteria
C_1	(0.329, 0.052, 0.06)	—
C_{11}	(0.217, 0.042, 0.046)	(0.071, 0.014, 0.015)
C_{12}	(0.122, 0.022, 0.025)	(0.04, 0.007, 0.008)
C_{13}	(0.119, 0.013, 0.015)	(0.039, 0.004, 0.005)
C_{14}	(0.155, 0.024, 0.026)	(0.051, 0.008, 0.009)
C_{15}	(0.178, 0.021, 0.023)	(0.059, 0.007, 0.008)
C_{16}	(0.209, 0.03, 0.033)	(0.069, 0.01, 0.011)
C_2	(0.372, 0.066, 0.077)	—
C_{21}	(0.116, 0.016,0.018)	(0.043, 0.006, 0.007)
C_{22}	(0.091, 0.009, 0.01)	(0.034, 0.003, 0.004)
C_{23}	(0.203, 0.043, 0.046)	(0.076, 0.016, 0.017)
C_{24}	(0.094, 0.017, 0.018)	(0.035, 0.006, 0.007)
C_{25}	(0.21, 0.048, 0.052)	(0.078, 0.018, 0.019)
C_{26}	(0.158, 0.023, 0.024)	(0.059, 0.009, 0.009)
C_{27}	(0.13, 0.029, 0.032)	(0.048, 0.011, 0.012)
C_3	(0.298, 0.024, 0.028)	—
C_{31}	(0.141, 0.026, 0.028)	(0.042, 0.008, 0.008)
C_{32}	(0.138, 0.013, 0.014)	(0.041, 0.004, 0.004)
C_{33}	(0.199, 0.04, 0.044)	(0.059, 0.012, 0.013)
C_{34}	(0.19, 0.039, 0.042)	(0.057, 0.012, 0.013)
C_{35}	(0.129, 0.01, 0.012)	(0.038, 0.003, 0.004)
C_{36}	(0.096, 0.015, 0.017)	(0.029, 0.004, 0.005)
C_{37}	(0.105, 0.013, 0.014)	(0.031, 0.004, 0.004)

The global weights can be used if we have some alternatives which need to be evaluated with respect to these criteria. Here, the local weights are enough to evaluate the importance of criteria. To show the uncertainty of the subjective weights determined (local subjective weights of sub-criteria) in the form of symmetric interval type-2 fuzzy sets, the graphical representation of them is depicted in Figure 4.

As can be seen in Figure 4 and Table 6, based on the experts' assessments, C_{11} (Individual-level knowledge) and C_{16} (Experience) are the most important and C_{12} (Competence) and C_{13} (Leadership ability) are the least important sub-criteria of C_1 (Human Capital). In the second criterion (Organizational Capital), we can say that C_{25} (Management system) and C_{23} (Working systems) have more importance than the other sub-criteria. Also, we can see the higher weights of C_{33} and C_{34} in the third criterion, which shows that "Market" and "Shareholders" are more important that the other sub-criteria in the "Relational Capital" dimension.

This example shows that the defuzzified subjective weights of some sub-criteria in different main criteria are very close to each other. For example, we can see small difference between the defuzzified subjective weights of C_{12} and C_{13} in the first criterion, C_{22} and C_{24} in the second criterion and C_{31} and C_{32} in the third criterion. However, these sub-criteria can be differentiated by using on the level of uncertainty or their domains based on the results obtained by the proposed approach.

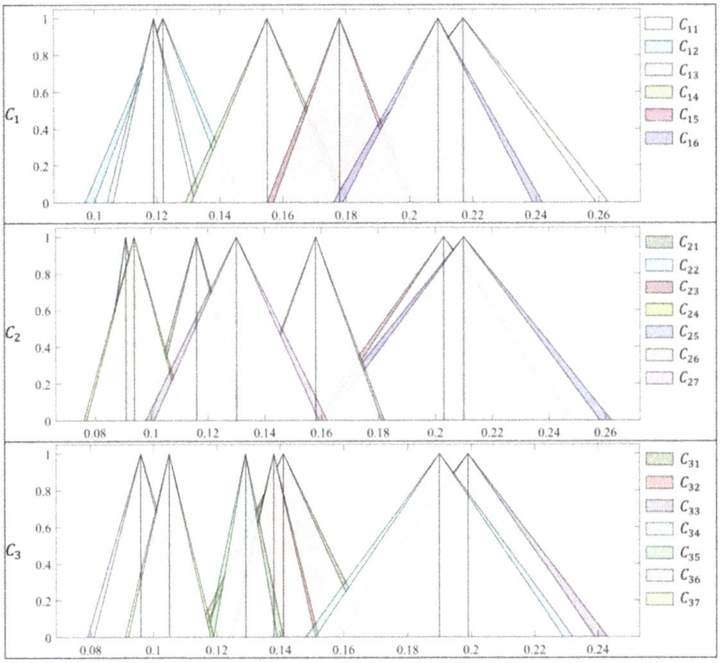

Figure 4. The symmetric IT2FSs related to each sub-criterion.

4. Conclusions

The importance of criteria weights in the process of multi-criteria decision-making has led to developing several methods for determination of them. Some methods have been developed to obtain the objective weights of criteria based on the information of decision-matrices and some other have been devised to determine the subjective weights according to the judgments and preferences of experts or decision-makers. The SWARA method is an efficient method for determination of subjective weights of criteria. Because we are usually confronted with imprecise information given by decision-makers, we need a tool that can be able to handle the uncertainty of information. In this study, we have proposed an extended SWARA method for determination of subjective weights under uncertainty. To model the uncertainty associated with the decision-makers' judgments and preferences, the concept of interval type-2 fuzzy sets has been used. In this regard, a sub-class of IT2FSs—called symmetric interval type-2 fuzzy sets—has been utilized to extend the SWARA method. The symmetrical properties of this sub-class of IT2FSs help us to capture the uncertainty of information flexibly and to make the computations simply. The extended SWARA method has been developed to elicit the preferences of a group of decision-makers under uncertainty and give an aggregated subjective weight for each criterion in the form of symmetric interval type-2 fuzzy sets. The symmetric form of the subjective weights determined by the proposed approach helps us to interpret them in a more efficient way. The application of the extended SWARA method has been presented as a numerical example to determine the weights of dimensions and components of intellectual capital in a company. Future research can apply the proposed approach to different real-world MCDM problems like assessment of green supply chain indicators, evaluation of risk factors and assessment of service quality dimensions. Also, a comparative study of the proposed approach and the AHP method with IT2FSs can be made in future research. Moreover, the proposed approach can be integrated with the other MCDM methods such as TOPSIS, VIKOR, PROMETHEE and EDAS to propose new hybrid approaches to deal with decision-making problems.

Acknowledgments: The authors would like to acknowledge the anonymous reviewers for their valuable comments and suggestions.

Author Contributions: Mehdi Keshavarz-Ghorabaee, Maghsoud Amiri and Edmundas Kazimieras Zavadskas designed the research, analyzed the data and the obtained results and developed the paper. Zenonas Turskis and Jurgita Antucheviciene provided extensive advice throughout the study regarding the research design, methodology, findings and revised the manuscript. All the authors have read and approved the final manuscript.

Conflicts of Interest: The authors declare no conflict of interest.

References

1. Liu, J.; Liu, P.; Liu, S.F.; Zhou, X.Z.; Zhang, T. A study of decision process in MCDM problems with large number of criteria. *Int. Trans. Oper. Res.* **2015**, *22*, 237–264. [CrossRef]
2. Diakoulaki, D.; Mavrotas, G.; Papayannakis, L. Determining objective weights in multiple criteria problems: The CRITIC method. *Comput. Oper. Res.* **1995**, *22*, 763–770. [CrossRef]
3. Goldstein, W.M. Judgments of relative importance in decision making: Global vs local interpretations of subjective weight. *Organ. Behav. Hum. Decis. Process.* **1990**, *47*, 313–336. [CrossRef]
4. Simos, J. *Evaluer L'impact sur L'environnement: Une Approche Originale par L'analyse Multicritère et la Négociation*; Presses Polytechniques et Universitaires Romandes: Lausanne, Switzerland, 1990.
5. Zardari, N.H.; Ahmed, K.; Shirazi, S.M.; Yusop, Z.B. *Weighting Methods and Their Effects on Multi-Criteria Decision Making Model Outcomes in Water Resources Management*; Springer: New York, NY, USA, 2014.
6. Kersuliene, V.; Zavadskas, E.K.; Turskis, Z. Selection of rational dispute resolution method by applying new step-wise weight assessment ratio analysis (SWARA). *J. Bus. Econ. Manag.* **2010**, *11*, 243–258. [CrossRef]
7. Dehnavi, A.; Aghdam, I.N.; Pradhan, B.; Morshed Varzandeh, M.H. A new hybrid model using step-wise weight assessment ratio analysis (SWARA) technique and adaptive neuro-fuzzy inference system (ANFIS) for regional landslide hazard assessment in Iran. *CATENA* **2015**, *135*, 122–148. [CrossRef]
8. Karabasevic, D.; Zavadskas, E.K.; Turskis, Z.; Stanujkic, D. The framework for the selection of personnel based on the SWARA and ARAS methods under uncertainties. *Informatica* **2016**, *27*, 49–65. [CrossRef]
9. Nakhaei, J.; Bitarafan, M.; Lale Arefi, S.; Kaplinski, O. Model for rapid assessment of vulnerability of office buildings to blast using SWARA and SMART methods (a case study of swiss re tower). *J. Civ. Eng. Manag.* **2016**, *22*, 831–843. [CrossRef]
10. Shukla, S.; Mishra, P.; Jain, R.; Yadav, H. An integrated decision making approach for ERP system selection using SWARA and PROMETHEE method. *Int. J. Intell. Enterp.* **2016**, *3*, 120–147. [CrossRef]
11. Işık, A.T.; Adalı, E.A. A new integrated decision making approach based on SWARA and OCRA methods for the hotel selection problem. *Int. J. Adv. Oper. Manag.* **2016**, *8*, 140–151. [CrossRef]

12. Mavi, R.K.; Goh, M.; Zarbakhshnia, N. Sustainable third-party reverse logistic provider selection with fuzzy SWARA and fuzzy MOORA in plastic industry. *Int. J. Adv. Manuf. Technol.* **2017**, *91*, 2401–2418. [CrossRef]
13. Panahi, S.; Khakzad, A.; Afzal, P. Application of stepwise weight assessment ratio analysis (SWARA) for copper prospectivity mapping in the Anarak region, central Iran. *Arab. J. Geosci.* **2017**, *10*, 484. [CrossRef]
14. Stanujkic, D.; Zavadskas, E.K.; Karabasevic, D.; Turskis, Z.; Kersuliene, V. New group decision-making ARCAS approach based on the integration of the SWARA and the ARAS methods adapted for negotiations. *J. Bus. Econ. Manag.* **2017**, *18*, 599–618. [CrossRef]
15. Karabašević, D.; Stanujkić, D.; Urosević, S.; Popović, G.; Maksimović, M. An approach to criteria weights determination by integrating the Delphi and the adapted SWARA methods. *Manag. J. Theory Pract. Manag.* **2017**, *22*, 15–25.
16. Stanujkic, D.; Karabasevic, D.; Zavadskas, E.K. A new approach for selecting alternatives based on the adapted weighted sum and the SWARA methods: A case of personnel selection. *Econ. Comput. Econ. Cybern. Stud. Res.* **2017**, *51*, 39–56.
17. Urosevic, S.; Karabasevic, D.; Stanujkic, D.; Maksimovic, M. An approach to personnel selection in the tourism industry based on the SWARA and the WASPAS methods. *Econ. Comput. Econ. Cybern. Stud. Res.* **2017**, *51*, 75–88.
18. Jamali, G.; Asl, E.K.; Hashemkhani Zolfani, S.; Saparauskas, J. Analysing larg supply chain management competitive strategies in Iranian cement industries. *E+M Ekon. Manag.* **2017**, *20*, 70–83. [CrossRef]
19. Juodagalvienė, B.; Turskis, Z.; Šaparauskas, J.; Endriukaitytė, A. Integrated multi-criteria evaluation of house's plan shape based on the EDAS and SWARA methods. *Eng. Struct. Technol.* **2017**, *9*, 117–125. [CrossRef]
20. Tayyar, N.; Durmu, M. Comparison of Max100, SWARA and pairwise weight elicitation methods. *Int. J. Eng. Res. Appl.* **2017**, *7*, 67–78. [CrossRef]
21. Valipour, A.; Yahaya, N.; Md Noor, N.; Antucheviciene, J.; Tamosaitiene, J. Hybrid SWARA-COPRAS method for risk assessment in deep foundation excavation project: An Iranian case study. *J. Civ. Eng. Manag.* **2017**, *23*, 524–532. [CrossRef]
22. Keshavarz Ghorabaee, M.; Amiri, M.; Zavadskas, E.K.; Antucheviciene, J. A new hybrid fuzzy MCDM approach for evaluation of construction equipment with sustainability considerations. *Arch. Civ. Mech. Eng.* **2018**, *18*, 32–49. [CrossRef]
23. Dahooie, J.H.; Abadi, E.B.J.; Vanaki, A.S.; Firoozfar, H.R. Competency-based IT personnel selection using a hybrid SWARA and ARAS-G methodology. *Hum. Factors Ergon. Manuf. Serv. Ind.* **2018**, *28*, 5–16. [CrossRef]
24. Mardani, A.; Nilashi, M.; Zakuan, N.; Loganathan, N.; Soheilirad, S.; Saman, M.Z.M.; Ibrahim, O. A systematic review and meta-analysis of SWARA and WASPAS methods: Theory and applications with recent fuzzy developments. *Appl. Soft Comput.* **2017**, *57*, 265–292. [CrossRef]
25. Liu, Z.; Qin, K.; Pei, Z. A method for fuzzy soft sets in decision-making based on an ideal solution. *Symmetry* **2017**, *9*, 246. [CrossRef]
26. Alcantud, J.; Rambaud, S.; Torrecillas, M. Valuation fuzzy soft sets: A flexible fuzzy soft set based decision making procedure for the valuation of assets. *Symmetry* **2017**, *9*, 253. [CrossRef]
27. Wang, Z.-X.; Li, J. Correlation coefficients of probabilistic hesitant fuzzy elements and their applications to evaluation of the alternatives. *Symmetry* **2017**, *9*, 259. [CrossRef]
28. Liu, P. Multiple attribute decision-making methods based on normal intuitionistic fuzzy interaction aggregation operators. *Symmetry* **2017**, *9*, 261. [CrossRef]
29. Liu, P.; Mahmood, T.; Khan, Q. Multi-attribute decision-making based on prioritized aggregation operator under hesitant intuitionistic fuzzy linguistic environment. *Symmetry* **2017**, *9*, 270. [CrossRef]
30. Ren, F.L.; Kong, M.M.; Pei, Z. A new hesitant fuzzy linguistic TOPSIS method for group multi-criteria linguistic decision making. *Symmetry* **2017**, *9*, 19. [CrossRef]
31. Garcia, N.; Puente, J.; Fernandez, I.; Priore, P. Suitability of a consensual fuzzy inference system to evaluate suppliers of strategic products. *Symmetry* **2018**, *10*, 20. [CrossRef]
32. Zadeh, L.A. Fuzzy sets. *Inf. Control* **1965**, *8*, 338–353. [CrossRef]
33. Zadeh, L.A. The concept of a linguistic variable and its application to approximate reasoning—I. *Inf. Sci.* **1975**, *8*, 199–249. [CrossRef]
34. Liang, Q.; Mendel, J.M. Interval type-2 fuzzy logic systems: Theory and design. *IEEE Trans. Fuzzy Syst.* **2000**, *8*, 535–550. [CrossRef]

35. Mendel, J.M.; John, R.I.; Liu, F.L. Interval type-2 fuzzy logic systems made simple. *IEEE Trans. Fuzzy Syst.* **2006**, *14*, 808–821. [CrossRef]
36. Mendel, J.M.; Wu, H.W. Type-2 fuzzistics for symmetric interval type-2 fuzzy sets: Part 1, forward problems. *IEEE Trans. Fuzzy Syst.* **2006**, *14*, 781–792. [CrossRef]
37. Keshavarz Ghorabaee, M.; Amiri, M.; Zavadskas, E.K.; Turskis, Z. Multi-criteria group decision-making using an extended EDAS method with interval type-2 fuzzy sets. *E+M Ekon. Manag.* **2017**, *20*, 48–68. [CrossRef]
38. Senturk, S.; Antucheviciene, J. Interval type-2 fuzzy c-control charts: An application in a food company. *Informatica* **2017**, *28*, 269–283. [CrossRef]
39. Soner, O.; Celik, E.; Akyuz, E. Application of AHP and VIKOR methods under interval type 2 fuzzy environment in maritime transportation. *Ocean Eng.* **2017**, *129*, 107–116. [CrossRef]
40. Keshavarz Ghorabaee, M.; Amiri, M.; Zavadskas, E.K.; Turskis, Z.; Antucheviciene, J. A new multi-criteria model based on interval type-2 fuzzy sets and EDAS method for supplier evaluation and order allocation with environmental considerations. *Comput. Ind. Eng.* **2017**, *112*, 156–174. [CrossRef]
41. Qin, J.D.; Liu, X.W.; Pedrycz, W. An extended TODIM multi-criteria group decision making method for green supplier selection in interval type-2 fuzzy environment. *Eur. J. Oper. Res.* **2017**, *258*, 626–638. [CrossRef]
42. Baykasoğlu, A.; Gölcük, İ. Development of an interval type-2 fuzzy sets based hierarchical MADM model by combining DEMATEL and TOPSIS. *Expert Syst. Appl.* **2017**, *70*, 37–51. [CrossRef]
43. Deveci, M.; Demirel, N.C.; Ahmetoglu, E. Airline new route selection based on interval type-2 fuzzy MCDM: A case study of new route between Turkey-North American region destinations. *J. Air Transp. Manag.* **2017**, *59*, 83–99. [CrossRef]
44. Zhong, L.; Yao, L.M. An ELECTRE I-based multi-criteria group decision making method with interval type-2 fuzzy numbers and its application to supplier selection. *Appl. Soft Comput.* **2017**, *57*, 556–576. [CrossRef]
45. Kundu, P.; Kar, S.; Maiti, M. A fuzzy multi-criteria group decision making based on ranking interval type-2 fuzzy variables and an application to transportation mode selection problem. *Soft Comput.* **2017**, *21*, 3051–3062. [CrossRef]
46. Qin, J.D.; Liu, X.W.; Pedrycz, W. A multiple attribute interval type-2 fuzzy group decision making and its application to supplier selection with extended LINMAP method. *Soft Comput.* **2017**, *21*, 3207–3226. [CrossRef]
47. Ju, Y.B.; Ju, D.W.; Wang, A.H.; Ju, M.Y. GRP method for multiple attribute group decision making under trapezoidal interval type-2 fuzzy environment. *J. Intell. Fuzzy Syst.* **2017**, *33*, 3469–3482. [CrossRef]
48. Yao, D.B.; Wang, C.C. Interval type-2 fuzzy information measures and their applications to attribute decision-making approach. *J. Intell. Fuzzy Syst.* **2017**, *33*, 1809–1821. [CrossRef]
49. Liu, K.; Liu, Y.; Qin, J. An integrated ANP-VIKOR methodology for sustainable supplier selection with interval type-2 fuzzy sets. *Granul. Comput.* **2018**. [CrossRef]
50. Wu, T.; Liu, X.W.; Liu, F. An interval type-2 fuzzy TOPSIS model for large scale group decision making problems with social network information. *Inf. Sci.* **2018**, *432*, 392–410. [CrossRef]
51. Celik, E.; Gul, M.; Aydin, N.; Gumus, A.T.; Guneri, A.F. A comprehensive review of multi criteria decision making approaches based on interval type-2 fuzzy sets. *Knowl.-Based Syst.* **2015**, *85*, 329–341. [CrossRef]
52. Hisdal, E. The IF THEN ELSE statement and interval-valued fuzzy sets of higher type. *Int. J. Man Mach. Stud.* **1981**, *15*, 385–455. [CrossRef]
53. Ma, M.; Kandel, A.; Friedman, M. A new approach for defuzzification. *Fuzzy Sets Syst.* **2000**, *111*, 351–356. [CrossRef]
54. Keshavarz Ghorabaee, M. Developing an MCDM method for robot selection with interval type-2 fuzzy sets. *Robot. Comput.-Integr. Manuf.* **2016**, *37*, 221–232. [CrossRef]
55. Bozbura, F.T.; Beskese, A. Prioritization of organizational capital measurement indicators using fuzzy AHP. *Int. J. Approx. Reason.* **2007**, *44*, 124–147. [CrossRef]

symmetry

MDPI

Article

The Recalculation of the Weights of Criteria in MCDM Methods Using the Bayes Approach

Irina Vinogradova [1,*], Valentinas Podvezko [2] and Edmundas Kazimieras Zavadskas [3]

1 Department of Information Technologies, Vilnius Gediminas Technical University, Saulėtekio al. 11, 10223 Vilnius, Lithuania
2 Department of Mathematical Statistics, Vilnius Gediminas Technical University, Saulėtekio al. 11, 10223 Vilnius, Lithuania; valentinas.podvezko@vgtu.lt
3 Laboratory of Operational Research, Vilnius Gediminas Technical University, Saulėtekio al. 11, 10223 Vilnius, Lithuania; edmundas.zavadskas@vgtu.lt
* Correspondence: irina.vinogradova@vgtu.lt; Tel.: +37-052-745-035

Received: 3 May 2018; Accepted: 1 June 2018; Published: 7 June 2018

Abstract: The application of multiple criteria decision-making methods (MCDM) is aimed at choosing the best alternative out of the number of available versions in the absence of the apparently dominant alternative. One of the two major components of multiple criteria decision-making methods is represented by the weights of the criteria describing the considered process. The weights of the criteria quantitatively express their significance and influence on the evaluation result. The criterion weights can be subjective, i.e., based on the estimates assigned by the experts, and the so-called objective, i.e., those which assess the structure of the data array at the time of evaluation. Several groups of experts, representing the opinions of various interested parties may take part in the evaluation of criteria. The evaluation data on the criterion weights also depend on the mathematical methods used for calculations and the estimation scales. In determining the objective weights, several methods, assessing various properties or characteristics of the data array's structure, are usually employed. Therefore, the use of the procedures, improving the accuracy of the evaluation of the weights' values and the integration of the obtained data into a single value, is often required. The present paper offers a new approach to more accurate evaluation of the criteria weights obtained by using various methods based on the idea of the Bayes hypothesis. The performed investigation shows that the suggested method is symmetrical and does not depend on the fact whether a priori or posterior values of the weights are recalculated. This result is the theoretical basis for practical use of the method of combining the weights obtained by various approaches as the geometric mean of various estimates. The ideas suggested by the authors have been repeatedly used in the investigation for combining the objective weights, for recalculating the criteria weights after obtaining the estimates of other groups of experts and for combining the subjective and the objective weights. The recalculated values of the weights of the criteria are used in the work for evaluating the quality of the distant courses taught to the students.

Keywords: MCDM; the criteria of the weights; Bayes' theorem; combining the weights; symmetry of the method; IDOCRIW; FAHP; evaluating the quality of distant courses

---

1. Introduction

The Use of multiple criterion decision-making methods (MCDM) [1–3] allows a decision-maker to choose the best alternative out of a number of the considered alternatives A_1, A_2, ..., A_n or to arrange them according to their importance for the defined purpose. This may be the choice of the best technological process out of the suggested versions, the comparative evaluation of economic, social or ecological situations in particular states or their regions, as well as the performance of banks and enterprises, and the solution of many other similar problems. The MCDM methods are based on using a decision-making matrix $\mathbf{R} = \|r_{ij}\|$ of the values r_{ij} of the criteria $R_1, R_2, \ldots, R_m$, describing the considered process, and the vector $\mathbf{\Omega} = (\omega_j)$ of the significances of these criteria, i.e., their ω_j, where $i = 1, 2, \ldots, n; j = 1, 2, \ldots, m; m$ is the number of criteria and n is the number of the considered alternatives. The values of the criteria r_{ij} can be represented by the statistical data, the estimates assigned by experts and the values of technological or technical characteristics of the considered process. The influence of the criteria on this process and their importance differ to some extent. However, the main idea of the criterion weight evaluation is that, in fact, the most important criterion is assigned the largest weight in any method used for criterion weight evaluation. The obtained weights are usually normalized as follows: $\sum_{j=1}^{m} \omega_j = 1$.

The use of the MCDM methods is based on the integration of the criteria values r_{ij} and their weights ω_j for obtaining the standard of evaluation, which is the criterion of the method. This idea is successfully realized by using the SAW (Simple Additive Weighing) method. The alternatives performance level S_i is calculated by [1]:

$$S_i = \sum_{j=1}^{m} \omega_j \widetilde{r}_{ij}, \tag{1}$$

where ω_j is the weight of the j-th criteria and $\widetilde{r}_{ij}$ is the normalized value of the j-th criteria for the i-th alternative [4,5].

The actual values of the criteria weights have a great influence on determining the importance of the alternatives and choosing the best solution. Therefore, the problems associated with their estimates are widely investigated both in theory and practice of MCDM methods' application.

As mentioned above, the weights of criteria can be subjective and objective. In practice, subjective weights determined by specialists/experts are commonly used. These weights are most commonly used for solving practical problems. A large number of methods to determine the criteria weights based on expert evaluation of their significance (weight) have been developed. These weights are most important for assessing the results because they express the opinions of highly qualified experts with extensive experience. The well-known approaches are the Delphi method [6,7], the expert evaluation method [3], the Analytic Hierarchy Process (AHP) [8–10], the stepwise weight assessment ratio analysis (SWARA) [6,11], the factor relationship (FARE) [12], and KEmeny Median Indicator Ranks Accordance (KEMIRA) [13].

In the process of evaluation, it is also possible to consider the structure of the data array, i.e., the criteria values, and to determine the actual degree of each criterion's dominance, i.e., the so-called objective weights of criteria. In contrast to their subjective counterparts, the objective weights are not so commonly used in practice. They do not play an important role in this process, showing the actual influence of particular criteria at the time of evaluation. The entropy method [14–18], the LINMAP method [1], mathematical programming models for determining the criteria weights [19], the correlation coefficient and standard deviation (CCSD) based on the objective weight determination method [20], as well as the methods of Criterion Impact LOSs (CILOS) and Integrated Determination of Objective CRIteria Weights (IDOCRIW) [4,5,18,21,22], the projection pursuit algorithm [23], a group of correlation methods CRITIC (Criteria Importance Through Intercriteria Correlation) [24], and the least squares' comparison [25] are the well-known practically used methods. Combination weighting is based on the integration of subjective and objective weighting [26–29].

Several groups of experts may take part in determining the weights of the criteria simultaneously. Their estimates represent the opinions of the interested parties. The assigned weights of the considered criteria also depend on the mathematical method used for calculations and the estimation scales.

A number of methods demonstrating the specific features of the data structure (a decision-making matrix) are commonly used simultaneously for determining the objective weights. Therefore, the need arises for improving the accuracy of the obtained weights' values, as well as the integration of the estimates assigned by the experts of various groups and the objective weights obtained by using various methods into an overall estimate. Moreover, to achieve the most accurate evaluation of the criteria weights, the estimates of the objective and subjective weights should be combined.

However, the formal integration of particular weights' estimates into a single value is not correct, because according to the Kendall's theory, the estimates would not be in agreement. This implies that the theoretical grounds for integrating the particular estimates are required.

The authors of the present paper offer a method of weights' recalculation, and the integration of various estimates into a single one, based on the recalculation approach offered by Bayes.

The research procedure is presented in Figure 1, including combination of criteria weights calculated by applying different methods and evaluation of courses by using four MCDM methods.

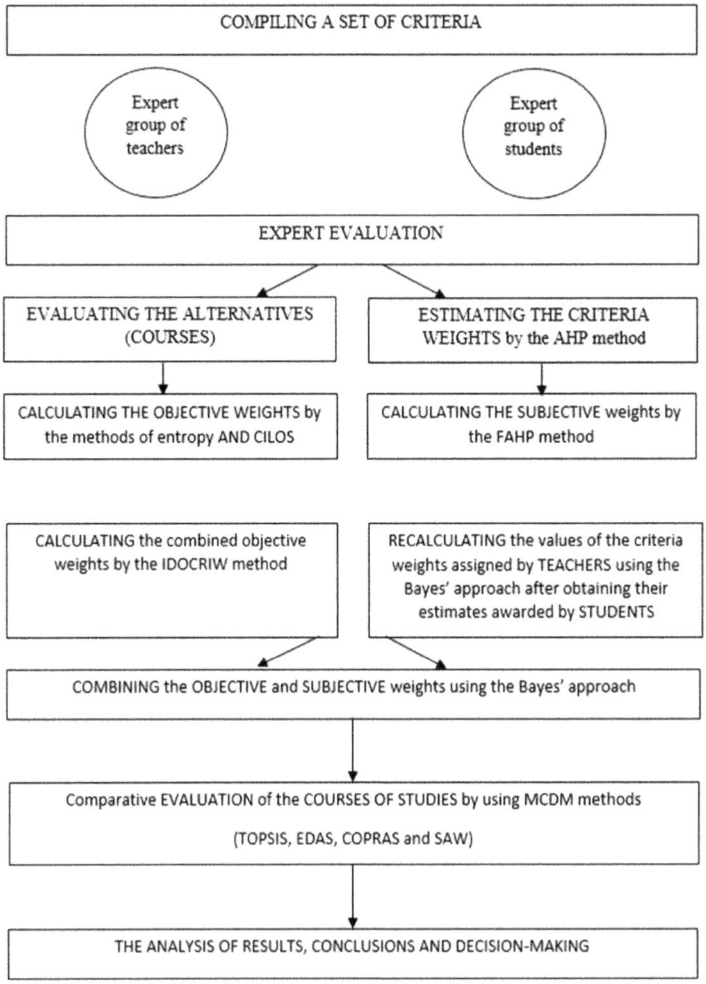

Figure 1. The research procedure.

The ideas suggested in this paper have been repeatedly used in the investigation for combining the subjective and objective weights to evaluate the quality of the distant courses taught to the students. These cases are described to show the potential of the method suggested in the paper. In solving the particular practical problems, one of the procedures suggested in the paper may be used, depending on the type of the problem as follows: the combining of the objective weights calculated by various methods (IDOCRIW), as well as the recalculation of the objective weights assigned by one of the groups of experts after obtaining the estimates of another group or, most often, the combining of the subjective and the objective weights.

2. Integrating the Values of the Weights of the Criteria

The weights of criteria can be considered as random values. The estimates of the criteria weights may vary, depending on the variation in the number of the members in the group of experts and on its decrease or increase. Even one and the same expert can differently assess the criteria weights the day after the first evaluation. The same weights of the criteria can be determined by various groups of experts, who are more or less interested in the obtained results.

The so-called objective weights of the criteria assess the structure of the data array, i.e., the decision matrix.

The elements of the matrix are also either the estimates assigned by the experts to the criteria for the considered alternatives or the statistical data and randomly change in time. The weights of the criteria, as well as the probabilities of the random values, range from 0 to 1. The Bayes' theorem in its application to the criteria weights may be interpreted so that these weights should be recalculated when different criteria weights obtained by another group of experts or by using other evaluation methods become available.

The criteria weights may be considered as a number of random values, making a complete set. In fact, the sum of the criteria weights' values is equal to one: $\sum_{j=1}^{m} \omega_j = 1$.

Besides, the criteria, describing the process evaluated by using multiple criteria methods (MCDM), were chosen in such a way that they could reflect all major aspects and characteristics of this process. Any other criteria were not used for solving the considered problem.

The Bayes' equation [30] used for recalculating the criteria weights in this work was of the form:

$$\omega(R_j/X) = \frac{\omega(R_j)\omega(X/R_j)}{\sum_{j=1}^{m} \omega(R_j)\omega(X/R_j)}, \tag{2}$$

where $\omega(R_j) = \omega_j$ is the initial weight of the j-th criterion R_j; X denotes the event, when new criteria weights are obtained; $\omega(X/R_j) = W_j$ denotes new weights of the criteria calculated by a different method or by another group of experts; $\omega(R_j/X)$ denotes the recalculated criteria weights.

The Equation (2) applied to the weights of the criteria was of the form:

$$\alpha_j = \frac{\omega_j W_j}{\sum_{j=1}^{m} \omega_j W_j}, \tag{3}$$

where α_j denotes the recalculated weights of the criteria.

In using the multiple criteria evaluation (MCDM) methods, the problem of combining the weights of the criteria obtained by using various evaluation methods, or groups of experts, arises.

In these cases, the concept of the geometric mean is commonly used [16,31], though the arithmetic mean or other ideas, helping to combine weights, can also be implemented.

Equation (3) is based on the concept of the geometric mean for integrating weights. It should be noted that Equation (3) is symmetrical, which implies that the result obtained does not depend on the fact, which estimates are original and which ones are the recalculated values.

The same idea was used for calculating the aggregate objective weight by using the IDOCRIW method based on integrating the weights obtained by using the entropy and CILOS methods [18].

3. The Methods Used for Determining the Weights of the Criteria

Various methods are used for determining the subjective weights of the criteria. Experts can assess the importance (significance) of the criteria by using various evaluation techniques. They may include the method of rating the criteria according to their importance and the direct evaluation of criteria, when the sum of the obtained values is equal to one (or 100%), as well as the estimates obtained by using various estimation scales.

The present study is based on using the method of the Analytic Hierarchy Process (AHP) developed by T. Saaty and FAHP, which is an extension of this approach, taking into consideration the uncertainly of the experts' estimates. The methods of entropy, CILOS and IDOCRIW, were used for determining the objective weights. The values of the obtained weights were recalculated using the technique suggested by the authors, which is based on the method developed by Bayes.

3.1. The Method of Fuzzy Analytic Hierarchy Process (FAHP)

The methods AHP and FAHP [8,32,33] were used to determine the weights of the criteria.

The Fuzzy AHP method is suitable for determining the weights of the qualitative criteria, when the experts evaluate the alternatives independently of the judgements of other experts. Each expert performed the evaluation procedure applying a simple AHP method of pairwise comparison. The matrix of the expert's pairwise comparison was verified to check if the expert had not conflicted with his/her own judgment. This facilitated obtaining the weights of the qualitative criteria in a more precise way.

The weights of the criteria were determined by using the FAHP method described below:

Each expert performed pairwise comparison using the scale of the AHP method 1-3-5-7-9. The concordance [8] of the data of the filled in pairwise comparison matrix was checked.

The concordance of the estimates provided by the experts of the whole group was assessed [34].

The matrix $\widetilde{P}$ of the pairwise comparison data obtained from the group of experts, using the FAHP method, was developed based on the particular elements p_{ij}^t of the matrix, constructed using the AHP pairwise comparison data obtained by experts, when $t = 1, 2, \cdots, T$ and T is the number of experts.

The fuzzy triangular numbers $\widetilde{p}_{ij} = (L_{ij}, M_{ij}, U_{ij})$ of the elements of the pairwise comparison matrix $\widetilde{P} = (\widetilde{p}_{ij})$ based on the data provided by the experts' group were calculated by using the offered algorithm as follows [35]:

$$\begin{aligned} M_{ij} &= \frac{\sum_{t=1}^{T} p_{ij}^t}{T}, \\ L_{ij} &= \min_t p_{ij}^t, \\ U_{ij} &= \max_t p_{ij}^t. \end{aligned} \tag{4}$$

Since the matrix is inversely symmetrical, $\widetilde{p}_{ji} = \widetilde{p}_{ij}{}^{-1} = \left(\frac{1}{U_{ij}}, \frac{1}{M_{ij}}, \frac{1}{L_{ij}}\right)$; $\widetilde{p}_{ii} = (1,1,1)$.

$$\widetilde{P} = \begin{pmatrix} (1,1,1) & (L_{12}, M_{12} U_{12}) & \cdots & (L_{1m}, M_{1m} U_{1m}) \\ \left(\frac{1}{U_{12}}, \frac{1}{M_{12}}, \frac{1}{L_{12}}\right) & (1,1,1) & \cdots & (L_{2m}, M_{2m} U_{2m}) \\ \vdots & \vdots & \vdots & \vdots \\ (1/U_{1m}, 1/M_{1m}, 1/L_{1m}) & (1/U_{2m}, 1/M_{2m}, 1/L_{2m}) & \cdots & (1,1,1) \end{pmatrix}. \tag{5}$$

To determine the weights of the criteria based on the matrix of fuzzy numbers, the extent analysis method suggested by Chang [36] was used. The value $\widetilde{S}_j = (l_j, m_j, u_j)$ referred to as the fuzzy synthesis extension was calculated for each criterion:

$$\widetilde{S}_j = \sum_{i=1}^{m} \widetilde{p}_{ij} \otimes \left\{ \sum_{i=1}^{m} \sum_{j=1}^{m} \widetilde{p}_{ij} \right\}^{-1} j = 1, \ldots, m. \tag{6}$$

Each criterion j has the value $\widetilde{S}_j$ expressed by a fuzzy number of the triangle. Then, comparing the criteria (i.e., fuzzy numbers of the triangles), their probability levels (degrees) were determined. The probability level was calculated as follows:

$$V\left(\widetilde{S}_j \geq \widetilde{S}_i\right) = \begin{cases} 1, \text{ if } m_j \geq m_i, \\ \frac{l_i - u_j}{(m_j - u_j) - (m_i - l_i)}, \text{ if } l_i \leq u_j, \quad i, j = 1, \ldots, m; \ i \neq j. \\ 0, \text{ while, in other cases,} \end{cases} \tag{7}$$

The smallest value of the probability level was calculated as follows:

$$V_j = V\left(\widetilde{S}_j \geq \widetilde{S}_1, \widetilde{S}_2, \ldots, \widetilde{S}_{j-1}, \widetilde{S}_{j+1}, \ldots, \widetilde{S}_m\right) = \min_{i \in \{1,\ldots,m; i \neq j\}} V\left(\widetilde{S}_j \geq \widetilde{S}_i\right), \ i = 1, \ldots, m. \tag{8}$$

The vector of the priorities of the fuzzy matrix w_j was calculated by the equation:

$$w_j = \frac{V_j}{\sum_{j=1}^{m} V_j}, \ j = 1, \ldots, m. \tag{9}$$

The objective weights of the criteria were calculated using the Entropy method [17] and the CILOS method [18,37]. The detailed description of these methods and their use are presented in the works of the authors [4,18,21,22].

3.2. The Method Based on Using the Aggregate Objective Weights (IDOCRIW)

Using the idea of combining different weights into an aggregate weight [16,18], it is possible to combine the entropy weights W_j and weights q_j of the criterion impact loss methods as well as connecting them to the objective weights of criteria for assessing the weights ω_j of the structure of the array:

$$\omega_j = \frac{q_j W_j}{\sum_{j=1}^{m} q_j W_j}. \tag{10}$$

These weights emphasize the separation of the particular values of criteria (entropy characteristic), but the impact of these criteria is decreased due to the higher impact loss of other criteria.

The weights calculated in the entropy and the criterion impact loss methods were combined to obtain the aggregate weights and then used in multi-criteria assessment for ranking the alternatives and for selecting the best alternative.

4. The Applied MCDM Methods

For obtaining the relative estimates of the courses and demonstrating the application of MCDM methods, such as TOPSIS (The Technique for Order of Preference by Similarity to Ideal Solution), SAW, COPRAS (Complex Proportional Assessment) and EDAS (Evaluation Based on Distance from Average Solution), which reflect the main ideas of MCDM approaches, were used in the work. They include the calculation of the optimal distance from the best and from the worst alternatives, the combination of the values and weights of the criteria for obtaining the qualitative estimate of the method, determination of the degree of influence of the maximizing and minimizing criteria and taking into consideration the optimal distance from the average estimate. The detailed description of the methods and their use are presented in the works [16,28,38] as well as in the works of the authors of the present paper [39–44].

5. Expert Evaluation of Distance Learning Courses of Studies

As mentioned above, the main constituent parts of the MCDM methods are the values of the criteria for the compared alternatives, i.e., a decision matrix, and the weights of the criteria describing their importance [16,28,33].

Relying on Belton and Stewart's principles of the identification of quality evaluation criteria, these groups of criteria were offered for all stages of the evaluation process. The first group of criteria aims to evaluate the contents of the course of studies. The second group of criteria describes the effective use of tools. The third group or criteria refers to teaching of the course [33]. At each stage, the evaluation was made by a different group of experts. The total number of fifteen criteria was selected.

In the problem of assessing the quality of the courses of studies, the same criteria were considered for each course. The experts, who were specialists in information technologies and lecturers from the respective departments, teaching the courses in particular subjects (which had to be evaluated), as well as students, attending the respective lectures and seminars, had chosen seven criteria for evaluating the quality of the considered courses of studies.

These criteria were as follows: (1) The structure of the course; (2) The relevance of the material of the course; (3) Testing the knowledge of students; (4) Presentation of the material of the course; (5) Communication tools; (6) Readability and accessibility of the material of the course; (7) The practical use of the course of studies.

5.1. Description of the Considered Criteria

Criterion 1. The structure of the course. A general structure of the course is clear. The presentation of the material of the course is consistent. The material is presented in small amounts.

Criterion 2. The relevance of the material of the course. The presented material is relevant and not outdated.

Criterion 3. Testing the knowledge of the students. The presentation of any new topic is followed by the presentation of various tests, helping the students to learn the material. These tests are aimed at checking the knowledge acquired by the students and providing feedback, allowing students to test their knowledge at any time, whenever they wish, without the need for adapting themselves to the timetable of the teachers. A clear and consistent system of testing the knowledge of the students is presented.

Criterion 4. The comprehensiveness of the material of the course. The presented material is easily understood.

Criterion 5. Effective communication tools. Easy and fast access to the learning material is provided by the working group. Communication is secured by the availability of synchronous and asynchronous means of communication. Video conferences present the instrument, allowing all the students to connect to the system simultaneously during the examination period.

Criterion 6. Reading of the material of the course and its accessibility. Effective and fast data communication. The appropriately selected video records' format, quick access to the material, high quality of video record and sound. The material is easily read, using well-known tools, and is accessible without any additional connection sessions.

Criterion 7. The practical use of the course of studies. Having completed the course, the students acquired knowledge, practical skills and competence required for their successful work.

The estimates of the criteria values for each course by teachers and students were used as decision matrices. The subjective weights of the criteria were calculated by using the methods AHP and FAHP, based on the estimates assigned by the teachers and students. The objective weights were determined by using the entropy, CILOS and IDOCRIW methods. Their values were recalculated using the Bayes' method described in the present paper.

5.2. Determining the Subjective Weights

Eight teachers and ten students took part in the evaluation of the criteria weights and quality of the courses of studies. They filled in the AHP matrix of pairwise comparisons of the criteria and the matrix of estimates of the criteria values against the scale of ten points. The values of the AHP matrix filled in by one of the teachers (Table 1).

Table 1. The AHP matrix filled in by one of the teachers.

Criterion	Criterion 1	Criterion 2	Criterion 3	Criterion 4	Criterion 5	Criterion 6	Criterion 7
Criterion 1	1.0000	2.0000	0.5000	0.2500	3.0000	5.0000	4.0000
Criterion 2	0.5000	1.0000	0.3333	0.2500	2.0000	4.0000	3.0000
Criterion 3	2.0000	3.0000	1.0000	0.5000	4.0000	6.0000	5.0000
Criterion 4	4.0000	4.0000	2.0000	1.0000	5.0000	7.0000	6.0000
Criterion 5	0.3333	0.5000	0.2500	0.2000	1.0000	3.0000	2.0000
Criterion 6	0.2000	0.2500	0.1667	0.1429	0.3333	1.0000	0.5000
Criterion 7	0.2500	0.3333	0.2000	0.1667	0.5000	2.0000	1.0000

The concordance degree of the estimates given in each filled in matrix was determined (the concordance coefficient had to be below 0.10 [34]). After evaluation of the criteria describing the course of studies, the ranking of the obtained results and determining of the concordance of the estimates assigned by the experts of each group was made [34]. The judgments of the students were in agreement: the concordance coefficient was $W = 0.57$, while the calculated criterion value $\chi^2 = 34.03$ was larger than the value of the table equal to 12.59 (at the significance level value $\alpha = 0.05$).

The estimates assigned by the teachers were also in agreement: the concordance coefficient was $W = 0.68$, while the calculated criterion value $\chi^2 = 32.41$ was larger than the value of the table equal to 12.59 (at the significance level value $\alpha = 0.05$).

The values of the elements of the matrices calculated by the FAHP method, using Equations (4)–(5) and based on particular AHP matrices for teachers (Table 2) are given below.

Table 2. The matrix filled in by the teachers using the method FAHP.

Criterion	Criterion 1			Criterion 2			Criterion 3			Criterion 4			Criterion 5			Criterion 6			Criterion 7		
Criterion 1	1	1	1	0.3	1.5	3.0	0.3	1.7	4.0	0.2	0.9	3.0	1.0	3.3	7.0	2.0	4.8	6.0	0.3	3.4	5.0
Criterion 2	0.3	0.6	3	1	1	1	0.3	1.4	4.0	0.2	0.7	2.0	0.3	3.0	6.0	3.0	4.6	7.0	0.3	3.4	6.0
Criterion 3	0.3	0.6	4	0.3	0.7	3	1	1	1	0.3	0.6	2.0	0.3	3.2	5.0	3.0	4.8	7.0	0.5	3.3	6.0
Criterion 4	0.3	1.1	5	0.5	1.3	5	0.5	1.7	4	1	1	1	1.0	4.3	6.0	4.0	6.1	7.0	2.0	4.3	6.0
Criterion 5	0.1	0.3	1	0.2	0.3	3	0.2	0.3	3	0.2	0.2	1	1	1	1	0.3	2.8	6.0	0.2	1.9	5.0
Criterion 6	0.2	0.2	0.5	0.1	0.2	0.3	0.1	0.2	0.3	0.1	0.2	0.3	0.2	0.4	3	1	1	1	0.2	0.7	2.0
Criterion 7	0.2	0.3	3	0.2	0.3	4	0.2	0.3	2	0.2	0.2	0.5	0.2	0.5	5	0.5	1.5	6	1	1	1

The values of the parameters of fuzzy sets for teachers (Table 3) were obtained by using Equation (6). The criteria weights assigned by teachers, which were calculated by using the FAHP Methods (7)–(9), are given in Table 4.

Table 3. The matrix of the values $\widetilde{S}_j = (l_j, m_j, u_j)$ for teachers.

Criterion	l_j	m_j	u_j
Criterion 1	0.0303	0.2089	0.9056
Criterion 2	0.0323	0.1883	0.9056
Criterion 3	0.0331	0.1781	0.8744
Criterion 4	0.0553	0.2504	1.0618
Criterion 5	0.0131	0.0865	0.6246
Criterion 6	0.0114	0.0360	0.2316
Criterion 7	0.0142	0.0520	0.6714

Table 4. The values of the subjective weights of the criteria assigned by the teachers.

ω_1	ω_2	ω_3	ω_4	ω_5	ω_6	ω_7
0.1647	0.1610	0.1587	0.1728	0.1341	0.0780	0.1307

In Table 1, the results, obtained using pairwise comparison of seven criteria performed by one of the teachers, are given. The scale 1-3-5-7-9 of the Saaty's approach AHP [8] was used for comparison.

Using the algorithm for developing the matrix (5) of the FAHP method described in Section 3.1 and the estimates assigned by particular teachers, the matrix filled in by a group of teachers based on using the FAHP method was constructed (Table 2).

Based on the data provided in Table 2, the values of $\widetilde{S}_j = (l_j, m_j, u_j)$ were obtained from Equation (6) (Table 3).

The values of the criteria weights assigned by the teachers (Table 3) were obtained using Equations (7)–(9) (Table 4).

According to the teachers, the important criteria, describing the preparation of the course of studies of high quality, include a clear description of the material (ω_4) and the structure of the course material (ω_1), as well as the relevance of the presented material (ω_2). Table 5 provides the results of the pairwise comparison of seven criteria performed by one of the students. The scale 1-3-5-7-9 presented in the AHP method of T. Saaty was used for comparison.

Table 5. The AHP matrix provided by a student.

Criterion	Criterion 1	Criterion 2	Criterion 3	Criterion 4	Criterion 5	Criterion 6	Criterion 7
Criterion 1	1.00	0.50	2.00	0.20	3.00	0.33	0.25
Criterion 2	2.00	1.00	3.00	0.25	4.00	0.50	0.33
Criterion 3	0.50	0.33	1.00	0.17	2.00	0.25	0.20
Criterion 4	5.00	4.00	6.00	1.00	7.00	3.00	2.00
Criterion 5	0.33	0.25	0.50	0.14	1.00	0.20	0.17
Criterion 6	3.00	2.00	4.00	0.33	5.00	1.00	0.50
Criterion 7	4.00	3.00	5.00	0.50	6.00	2.00	1.00

Using the algorithm for developing the matrix (5) of FAHP method, which is described in Section 3.1, and the estimates assigned by individual students, the FAHP matrix of the data provided by a group of students was constructed (Table 6).

Table 6. FAHP matrix of the group of students.

Criterion	Criterion 1			Criterion 2			Criterion 3			Criterion 4			Criterion 5			Criterion 6			Criterion 7		
Criterion 1	1	1	1	0.2	0.4	1.0	0.2	1.1	3.0	0.2	0.2	0.3	0.3	1.8	4.0	0.2	1.4	4.0	0.2	0.3	1.0
Criterion 2	1	2.7	6	1	1	1	1.0	2.9	6.0	0.3	0.6	1.0	0.3	4.0	6.0	0.2	3.3	7.0	0.3	1.0	3.0
Criterion 3	0.3	0.9	5	0.2	0.3	1	1	1	1	0.2	0.4	1.0	0.3	2.3	6.0	0.2	1.8	5.0	0.2	0.4	1.0
Criterion 4	3	4.5	6	1	1.6	4	1	2.7	6	1	1	1	2.0	5.0	7.0	0.5	4.2	7.0	0.3	1.6	4.0
Criterion 5	0.3	0.6	4	0.2	0.3	3	0.2	0.4	3	0.1	0.2	0.5	1	1	1	0.2	1.2	5.0	0.1	0.5	3.0
Criterion 6	0.3	0.7	5	0.1	0.3	5	0.2	0.6	5	0.1	0.2	2	0.2	0.8	5	1	1	1	0.1	0.7	5.0
Criterion 7	1	3.2	6	0.3	1	4	1	2.6	5	0.3	0.6	3	0.3	2	7	0.2	1.3	7	1	1	1

Based on the data presented in Table 6 and using Equation (6), the values of $\widetilde{S}_j = (l_j, m_j, u_j)$ were obtained (Table 7).

Table 7. The matrix of the values $\widetilde{S}_j = (l_j, m_j, u_j)$ for students.

Criterion	l_j	m_j	u_j
Criterion 1	0.0120	0.0884	0.5581
Criterion 2	0.0224	0.2215	1.1682
Criterion 3	0.0133	0.1020	0.7788
Criterion 4	0.0491	0.2966	1.3629
Criterion 5	0.0115	0.0602	0.7593
Criterion 6	0.0116	0.0630	1.0903
Criterion 7	0.0229	0.1683	1.2850

The values of the criteria weights based on the estimates assigned by the students (Table 7) were obtained from Equations (7)–(9) (Table 8). According to the students, the important criteria, describing a high-quality course, include clear presentation of the material (ω_4), the relevance of the material for reading (ω_2) and the practical use of the acquired material (ω_7).

Table 8. The values of the subjective weights of the criteria assigned by the students.

ω_1	ω_2	ω_3	ω_4	ω_5	ω_6	ω_7
0.1201	0.1586	0.1336	0.1692	0.1270	0.1382	0.1533

5.3. The Calculation of the Objective Weights

The objective weights were calculated in the methods of entropy, CILOS and IDOCRIW, based on using a decision matrix, i.e., the values of the criteria obtained for the compared alternatives. In the considered case, the alternatives were the courses of various subjects, using the MOODLE system. These courses were assessed by two different groups of experts: The teachers, who are specialists in the subjects presented in the courses and the students, who learn the materials of these courses.

The objective weights of the criteria assigned by the experts of both groups, as well as their aggregate IDOCRIW weights, were calculated separately and, then, the weights awarded by the teachers were recalculated using the Bayes' equation, when the estimates of the same criteria given by the students, were obtained. Seven criteria were assessed. Similar to the case of the objective weights' evaluation, a group of 8 teachers and a group of 10 students took part in the process. For calculating the objective weights, the average estimates of the courses were used separately for each group. Five courses of studies were evaluated, including discrete mathematics, mathematics 2, integral calculus, operational systems and information technologies. The average estimates of the courses are given in Table 9 (teachers) and in Table 10 (students).

Table 9. The average estimates of five courses by the teachers.

Courses		Criterion 1	Criterion 2	Criterion 3	Criterion 4	Criterion 5	Criterion 6	Criterion 7
Discrete mathematics	Estimate	9	9.875	9.875	9.875	8.375	8.750	9.500
	Place	4.5	1	1	2	4.5	5	4
Mathematics 2	Estimate	9	9.125	8.125	9.625	9.125	10	9.625
	Place	4.5	4.5	5	3.5	1	2	3
Integral calculus	Estimate	10	9.25	9.5	10	8.625	10	9.125
	Place	1	3	2	1	3	2	5
Operational systems	Estimate	9.75	9.75	9	9.5	8.75	10	9.875
	Place	2	2	3.5	5	2	2	1
Information technology	Estimate	9.125	9.125	9	9.625	8.375	9.75	9.75
	Place	3	4.5	3.5	3.5	4.5	4	2

Table 10. The average estimates of five courses by the students.

Courses		Criterion 1	Criterion 2	Criterion 3	Criterion 4	Criterion 5	Criterion 6	Criterion 7
Discrete mathematics	Estimate	9.3	8.6	9.6	9	7.9	8.3	9.2
	Place	3	3	1	5	5	5	2
Mathematics 2	Estimate	8.9	9	9.4	9.1	8	9.7	8.9
	Place	5	2	2	4	4	2.5	3
Integral calculus	Estimate	10	8.1	9	9.2	8.7	9.9	7.2
	Place	1	5	4	3	2	1	5
Operational systems	Estimate	9.4	9.1	9	9.3	8.8	9.6	9.8
	Place	2	1	4	1.5	1	4	1
Information technology	Estimate	9.1	8.5	9	9.3	8.4	9.7	8.8
	Place	4	4	4	1.5	3	2.5	4

Using the data of Tables 5 and 6, the objective weights of the criteria (in the entropy and CILOS methods) provided by the teachers and by the students were calculated. The aggregate objective weights of criteria were obtained by using the method IDOCRIW. It should be noted that using IDOCRIW is equivalent to using the Bayes' Equation (3) for recalculating the entropy weights after the calculation of the weights by the CILOS approach. The values of the weights are presented in Table 11.

Table 11. The objective weights of criteria calculated based on the estimates of the courses awarded by teachers and students.

Experts	Method	Criterion 1	Criterion 2	Criterion 3	Criterion 4	Criterion 5	Criterion 6	Criterion 7
The weights assigned by teachers	Entropy	0.164	0.097	0.351	0.030	0.086	0.213	0.060
	CILOS	0.073	0.173	0.097	0.207	0.153	0.192	0.105
	IDOCRIW	0.093	0.130	0.264	0.047	0.101	0.316	0.049
The weights assigned by students	Entropy	0.079	0.087	0.038	0.008	0.094	0.194	0.501
	CILOS	0.104	0.071	0.158	0.364	0.156	0.114	0.033
	IDOCRIW	0.107	0.081	0.078	0.038	0.191	0.288	0.216

The objective weights of the criteria demonstrate their actual significance at the time of evaluation. The estimates of the courses given by both the teachers and the students show good readability and accessibility of the material of the courses. Based on the judgements of the students, it can be stated that the material of the courses is relevant. According to the estimation of the courses by the teachers, the testing of the students' knowledge of the course material is well-organized.

5.4. The Recalculation of the Values of the Subjective Weights, Assigned by the Teachers, When the Estimates of the Criteria Weights Awarded by the Students Are Obtained

The values of the subjective weights of the criteria assigned by the teachers were recalculated using the Bayes' equation when the weights of the same criteria were obtained by the students. The results of the calculation were the aggregate subjective weights of the teachers and students. In Table 12, the subjective weights assigned by the teachers and the students, as well as the generalized subjective weights, are given. In the case, when the judgements of the groups of experts about the significance of the considered criteria coincide, the overall estimate increases (for example, the estimates of the criteria 4 and 2).

Table 12. The recalculation of the subjective weights' values assigned by the experts based on using the Bayes' method.

Criterion	Teachers	Students	$\omega(R_j)\omega(X/R_j)$	Recalculated Weights
	$\omega(R_j)$	$\omega(X/R_j)$		$\omega(R_j/X)$
Criterion 1	0.16472	0.12010	0.01978	0.13777
Criterion 2	0.16100	0.15858	0.02553	0.17779
Criterion 3	0.15874	0.13360	0.02121	0.14770
Criterion 4	0.17276	0.16922	0.02923	0.20358
Criterion 5	0.13414	0.12697	0.01703	0.11861
Criterion 6	0.07796	0.13822	0.01078	0.07504
Criterion 7	0.13068	0.15331	0.02003	0.13951
-	-	-	0.14360	1.00000

The aggregate objective weights assigned by the teachers and students, using the method IDOCRIW (Table 11) were combined into generalized objective weights using the Bayes' Equation (3). The recalculated values of the objective criteria weights are given in Table 13.

In the case of the recalculated objective weights' values, the significance of the criterion 6, describing the accessibility and readability of the learning material, increased.

The aggregate subjective weights (Table 12) were combined with the objective weights (Table 13) using the Bayes' approach (3). The values of the generalized weights are given in Table 14.

Table 13. The generalized values of the objective criteria weights.

Experts	Criterion 1	Criterion 2	Criterion 3	Criterion 4	Criterion 5	Criterion 6	Criterion 7
The weights assigned by the teachers	0.093	0.130	0.264	0.047	0.101	0.316	0.049
The weights assigned by the students	0.107	0.081	0.078	0.038	0.191	0.288	0.216
The recalculated weights	0.061	0.064	0.126	0.011	0.118	0.556	0.065

Table 14. The recalculation of the values of the subjective criteria weights, taking into consideration the weights of the objective criteria, by using the Bayes' method.

Experts	Criterion 1	Criterion 2	Criterion 3	Criterion 4	Criterion 5	Criterion 6	Criterion 7
Subjective weights	0.061	0.064	0.126	0.011	0.118	0.556	0.065
Objective weights	0.13777	0.17779	0.14770	0.20358	0.11861	0.11861	0.13951
Recalculated weights	0.0797	0.1079	0.1765	0.0212	0.1328	0.3958	0.0860

The subjective and objective weights reflect some different characteristics of the significance of the criteria. Thus, the subjective weights show 'the desired' significance of the criteria, assigned by the experts, while the objective weights reflect the actual weights of criteria at the time of evaluation based on the values of the criteria. The obtained estimates do not usually correlate with each other. It is natural that the most positive effect (or influence) of the criteria weights on the results of the multiple criteria evaluation can be produced in the case of the obvious correlation between these estimates. In this case, a noticeable agreement between the estimates can be observed for the criteria 6, 3 and 5.

6. Evaluating the Quality of the Course of Studies by MCDM Methods Based on the Comparative Analysis

The weights of the criteria are used in MCDM methods for evaluating the compared alternatives. In the present investigation, these methods were used for assessing the courses of studies (the quality of the teaching materials). In the first case, the suggested algorithms for assessing their quality differed because of the averaging of the initial estimates assigned to the courses by the teachers and students and combining the weights of the criteria assigned by them (method of calculation). The averaged data were used as a base for making the calculations by MCDM methods. In the second case, the estimates assigned by the teachers and students were considered separately, while the calculations by using the MCDM methods were made individually for each group, and, then, the obtained estimates were averaged (method of calculation 2).

6.1. Method of Calculation 1

In this case, a decision matrix consists of the averaged estimates of the courses of studies assigned by the students and teachers (Tables 9 and 10). These estimates were combined into their average estimate (Table 15).

Five courses of studies were assessed by using the methods TOPSIS, SAW and EDAS, and the values of the aggregate objective and subjective weights were obtained (Table 14).

The calculation results and the places of courses are given in Table 16.

Table 15. The average estimates of five courses of studies assigned by the students and teachers.

Courses		Criterion 1	Criterion 2	Criterion 3	Criterion 4	Criterion 5	Criterion 6	Criterion 7
Discrete mathematics	Estimate	9.15	9.238	9.738	9.438	8.138	8.525	9.350
	Place	3	3	1	5	5	5	2
Mathematics 2	Estimate	8.950	9.063	8.763	9.363	8.563	9.850	9.263
	Place	5	2	2	4	4	2.5	3
Integral calculus	Estimate	10	8.675	9.250	9.600	8.663	9.95	8.538
	Place	1	5	4	3	2	1	5
Operational systems	Estimate	9.575	9.425	9.0	9.43	8.775	9.8	9.838
	Place	2	1	4	1.5	1	4	1
Information technology	Estimate	9.113	8.8135	9.0	9.463	8.388	9.725	9.275
	Place	4	4	4	1.5	3	2.5	4

Table 16. The estimates of five courses of studies obtained by using method of calculation 1 of MCDM methods.

Courses		TOPSIS	SAW-COPRAS	EDAS	The Average Estimates of the Courses
Discrete mathematics	Estimate	0.2619	0.1928	0.282	5
	Place	5	5	5	
Mathematics 2	Estimate	0.7118	0.2003	0.637	3
	Place	3	4	3	
Integral calculus	Estimate	0.7734	0.2030	0.891	2
	Place	2	2	2	
Operational systems	Estimate	0.7787	0.2045	0.965	1
	Place	1	1	1	
Information technology	Estimate	0.7009	0.1994	0.535	4
	Place	4	3	4	

The results obtained using the methods SAW and COPRAS were the same because all the criteria were maximizing [26].

The estimates of the courses according to method of calculation 1 were the same for all MCDM methods. The highest estimates were assigned to the course of 'Operational systems', while the lowest estimate was awarded to the course of 'Discrete mathematics'. The similarity of the estimates can be attributed to the decrease in the uncertainty of the data due to the averaging of the considered data.

6.2. Method of Calculation 2

This method of calculation of MCDM methods was used to obtain separate estimates of the courses, assigned by the teachers and the students. Their average values were assumed to be the resultant estimate of the courses.

For this purpose, the subjective weights assigned by the teachers were recalculated after the calculation of the objective weights of the same criteria. The values of the recalculated weights assigned by the teachers are given in Table 17.

Based on the estimates of the courses assigned by the teachers (Table 9) and the recalculated weights of the criteria assigned by the teachers (Table 17), five courses of studies were evaluated, using the methods TOPSIS, SAW and EDAS. The calculation results are given in Table 18.

Table 17. The recalculation of the values of the subjective weights of the criteria assigned by the teachers, using the Bayes' approach, taking into consideration the values of the objective weights of criteria.

Experts	Criterion 1	Criterion 2	Criterion 3	Criterion 4	Criterion 5	Criterion 6	Criterion 7
Subjective weights	0.1647	0.1610	0.1587	0.1728	0.1341	0.0780	0.1307
Objective weights	0.093	0.130	0.264	0.047	0.101	0.316	0.049
Recalculated weights	0.117	0.159	0.320	0.062	0.104	0.188	0.049

Table 18. The evaluation of courses of studies by the teachers, using method of calculation 2 of the MCDM methods.

Courses		TOPSIS	SAW-COPRAS	EDAS	The Average Estimates of the Courses
Discrete mathematics	Estimate	0.6860	0.2018	0.700	3
	Place	2	3	3	
Mathematics 2	Estimate	0.2886	0.1934	0.164	5
	Place	5	5	5	
Integral calculus	Estimate	0.7522	0.2048	0.849	1
	Place	1	1	1	
Operational systems	Estimate	0.5720	0.2028	0.706	2
	Place	3	2	2	
Information technology	Estimate	0.4998	0.1972	0.351	4
	Place	4	4	4	

The evaluation of the courses based on the data provided by the students was performed in a similar way. For this purpose, the subjective criteria weights assigned by the students were recalculated after the calculation of the objective weights of the same criteria. The recalculated weights' values assigned to the criteria by the students are given in Table 19.

Table 19. The recalculation of the subjective weights of the criteria assigned by the students, using the Bayes' equation, taking into consideration the objective criteria weights.

Experts	Criterion 1	Criterion 2	Criterion 3	Criterion 4	Criterion 5	Criterion 6	Criterion 7
Subjective weights	0.1201	0.1586	0.1336	0.1692	0.1270	0.1382	0.1533
Objective weights	0.107	0.081	0.078	0.038	0.191	0.288	0.216
Recalculated weights	0.0920	0.0919	0.0746	0.0460	0.1736	0.2849	0.2370

Based on the evaluation of courses by the students (Table 10) and their recalculated criteria weights (Table 19), as well as using the MCDM methods TOPSIS, SAW and EDAS, five courses of studies were evaluated.

The calculation results are given in Table 20.

In the case of using method of calculation 2, the uncertainty of the initial data is much higher because the estimates of the teachers and students differ to some extent. Therefore, the estimates of the courses are different in these groups. However, the average estimates do not differ considerably from those obtained using method of calculation 1. The highest average estimate was awarded to the course of 'Operational systems', while the estimates of the courses of 'Discrete mathematics' and 'Mathematics 2' changed positions.

Table 20. The evaluation of courses of studies by the students, using the MCDM methods, and their overall estimates obtained by using method of calculation 2.

Courses		TOPSIS	SAW-COPRAS	EDAS	The Average Estimates of the Courses	Overall Estimate (Place) of the Course
Discrete mathematics	Estimate	0.4977	0.1937	0.197	5	4
	Place	4	5	5		
Mathematics 2	Estimate	0.6638	0.2008	0.551	2–3	5
	Place	2	2.5	3		
Integral calculus	Estimate	0.4239	0.1955	0.291	4	2
	Place	5	4	4		
Operational systems	Estimate	0.8753	0.2091	0.984	1	1
	Place	1	1	1		
Information technology	Estimate	0.6632	0.2008	0.547	2–3	3
	Place	3	2.5	2		

7. Discussion

The MCDM methods are used for selecting the best alternative or arranging the alternatives in the order of their significance under the condition of the absence of the alternative dominant over others according to all criteria. The MCDM methods take into consideration the influence of the criteria on the evaluated process or object and use scalarization of the criteria and their weights. Therefore, the weight of the criterion plays an important role in making the solution.

Solving the decision-making problems is hardly possible without taking into consideration the judgments of the highly qualified experts.

Therefore, experts in various fields of activities take part in the evaluation. Usually, these experts have different opinions about the considered problems and their interest in the result of the choice of the alternative also differs to great extent. A decision-making person may change his/her estimates, taking into consideration the judgments of other groups of experts. This also applies to the evaluation of the significance of various criteria.

There are two various approaches to determining the weights of criteria. They include the subjective approach, based on the estimates assigned by experts, and the objective approach, assessing the structure of the data array.

A great number of various methods of assessing the subjective and objective weights have been offered. Each method has its specific features because it uses various concepts, mathematical equations and approaches, as well as various estimation scales. However, none of the available methods is universal or the best. Therefore, the need for integrating the estimates obtained by using some particular methods into the overall estimate arises.

The formal integration of particular estimates into a single one is incorrect, because most probably, according to the Kendall's theory of concordance, the results would not be in agreement. Therefore, the theoretical basis for using the method of integrating the particular estimates is required.

The experts' estimates of the criteria weights may be considered as the random values, making a complete set of events, with the sum of values equal to one. Thus, the possibility of using the Bayes' equation for recalculating the values of the criteria weights, when different values, yielded by other evaluation methods or assigned by the experts of another group, are obtained. This provides a theoretical basis for a wide use of the method of combining weights, obtained by employing various methods, as the geometric mean of particular values.

The described method of combining the weights according to the Bayes' approach has been used several times in the present work for recalculating the values of the objective weights and the integration of the values of the subjective and objective weights into a single value in the process of their evaluation. The method offered in the paper may have a wide practical use in solving various decision-making problems.

In solving the particular problems, various methods of combining the weights and the recalculation of their values can be used, depending on the specific nature of the problems.

8. Conclusions

In multiple criteria decision-making methods (MCDM), the weights of the criteria, describing the considered process or object, form one of two components of the evaluation of alternatives. Therefore, in using these methods, the values of the criteria weights have a strong influence on the estimates.

The considered methods use the subjective and objective weights of criteria. The subjective weights of criteria are calculated based on the estimates of the experts, while the objective weights assess the structure of the data array. Each method has its specific features and reflects various characteristics described by the criteria. Their integration for evaluating the significance of the alternatives is of primary importance in the theory and practice of MCDM methods' analysis and application.

In the present study, the authors offer to consider the criteria weights as random values, making a complete set. In this case, the equation offered by Bayes can be used for recalculating the criteria values, when the values of these criteria calculated by a different method are obtained. This allows for combining the weights of the criteria, yielded by various methods and used for assessing their significance, into a single estimate.

The suggested method of combining the criteria weights using the Bayes' approach was repeatedly used in the present study for recalculating the subjective weights of criteria assigned by one of two groups of experts after the weights of these criteria assigned by the experts of the other group were obtained. It was also used for recalculating the objective weights' values and combining the subjective and objective weights for obtaining the overall estimate.

The obtained result provides a theoretical basis for using a widely practically applied method of combining the criteria weights, obtained by using various methods, as a geometric mean of particular values.

In solving the particular decision-making problems, various methods suggested in the paper for combining and recalculating the criteria weights can be used, depending on the specific character of the problem and the available information.

The suggested method of recalculating the criteria weights based on the Bayes' approach was used in the work for assessing the quality of various courses of studies taught to students. The MCDM methods such as SAW, TOPSIS, EDAS and COPRAS were used for evaluation.

The authors believe that the suggested new approach contributes to the solution of various decision-making problems by providing a theoretical basis for combining the weights of criteria obtained by using various MCDM methods and by demonstrating its practical application.

Author Contributions: The idea to use the Bayesian method for weights recalculation offered I.V., V.P. presented the theoretical rationality background for the IDOCRIW method, the idea to apply Bayesian method depends to E.K.Z. and V.P., E.K.Z. is the co-author of the COPRAS, EDAS methods. V.P. and E.K.Z. are authors of the CILOS and IDOCRIW methods. V.P. and I.V. are program's developers, ran the practical calculations. V.P., I.V. and E.K.Z. finally reviewed and edited the manuscript. The discussion was a team task. All authors have read and approved the final manuscript.

Conflicts of Interest: The authors declare no conflict of interest.

References

1. Hwang, C.L.; Lin, M.J. *Group Decision Making under Multiple Criteria: Methods and Applications*; Springer: Berlin, Germany, 1987.
2. Mardani, A.; Jusoh, A.; MD Nor, K.; Khalifah, Z.; Zakwan, N.; Valipour, A. Multiple criteria decision-making techniques and their applications—A review of the literature from 2000 to 2014. *Econ. Res. Ekonomska Istraživanja* **2015**, *28*, 516–571. [CrossRef]
3. Zavadskas, E.K.; Vainiūnas, P.; Turskis, Z.; Tamošaitienė, J. Multiple criteria decision support system for assessment of projects managers in construction. *Int. J. Inf. Technol. Decis. Mak.* **2012**, *11*, 501–520. [CrossRef]
4. Čereška, A.; Zavadskas, E.K.; Bučinskas, V.; Podvezko, V.; Sutinys, E. Analysis of Steel Wire Rope Diagnostic Data Applying Multi-Criteria Methods. *Appl. Sci.* **2018**, *8*, 260. [CrossRef]
5. Zavadskas, E.K.; Cavallaro, F.; Podvezko, V.; Ubarte, I.; Kaklauskas, A. MCDM assessment of a healthy and safe built environment according to sustainable development principles: A practical neighbourhood approach in Vilnius. *Sustainability* **2017**, *9*, 702. [CrossRef]
6. Hashemkhani Zolfani, S.; Saparauskas, J. New Application of SWARA Method in Prioritizing Sustainability Assessment Indicators of Energy System. *Inzinerine Ekonomika Eng. Econ.* **2013**, *24*, 408–414. [CrossRef]
7. Kurilov, J.; Vinogradova, I. Improved fuzzy AHP methodology for evaluating quality of distance learning courses. *Int. J. Eng. Educ.* **2016**, *32*, 1618–1624. [CrossRef]
8. Saaty, T.L. *The Analytic Hierarchy Process*; McGraw-Hill: New York, NY, USA, 1980.
9. Saaty, T.L. *Decision-Making. Analytic Hierarchy Process*; Radio and Communication: Moscow, Russia, 1993.
10. Podvezko, V.; Sivilevicius, H. The use of AHP and rank correlation methods for determining the significance of the interaction between the elements of a transport system having a strong influence on traffic safety. *Transport* **2013**, *28*, 389–403. [CrossRef]

11. Alimardani, M.; Hashemkhani Zolfani, S.; Aghdaie, M.; Tamosaitiene, J. A novel hybrid SWARA and VIKOR methodology for Supplier selection in an agile environment. *Technol. Econ. Dev. Econ.* **2013**, *19*, 533–548. [CrossRef]
12. Ginevicius, R. A new determining method for the criteria weights in multi–criteria evaluation. *Int. J. Inf. Technol. Decis. Mak.* **2011**, *10*, 1067–1095. [CrossRef]
13. Krylovas, A.; Zavadskas, E.K.; Kosareva, N.; Dadelo, S. New KEMIRA method for determining criteria priority and weights in solving MCDM problem. *Int. J. Inf. Technol. Decis. Mak.* **2014**, *13*, 1119–1133. [CrossRef]
14. Han, J.; Li, Y.; Kang, J.; Cai, E.; Tong, Z.; Ouyang, G.; Li, X. Global synchronization of multichannel EEG based on rényi entropy in children with autism spectrum disorder. *Appl. Sci.* **2017**, *7*, 257. [CrossRef]
15. Han, Z.H.; Liu, P.D. A fuzzy multi–attribute decision–making method under risk with unknown attribute weights. *Technol. Econ. Dev. Econ.* **2011**, *17*, 246–258. [CrossRef]
16. Hwang, C.L.; Yoon, K. *Multiple Attribute Decision Making Methods and Applications*; A State of the Art Survey; Springer: Berlin, Germany, 1981.
17. Shannon, C.E. A mathematical theory of communication. *Bell Syst. Tech. J.* **1948**, *27*, 379–423. [CrossRef]
18. Zavadskas, E.K.; Podvezko, V. Integrated determination of objective criteria weights in MCDM. *Int. J. Inf. Technol. Decis. Mak.* **2016**, *15*, 267–283. [CrossRef]
19. Pekelman, D.; Sen, S.K. Mathematical programming models for the determination of attribute weights. *Manag. Sci.* **1974**, *20*, 1217–1229. [CrossRef]
20. Singh, R.K.; Benyoucef, L. A consensus based group decision making methodology for strategic selection problems of supply chain coordination. *Eng. Appl. Artif. Intell.* **2013**, *26*, 122–134. [CrossRef]
21. Čereška, A.; Podvezko, V.; Zavadskas, E.K. Operating characteristics analysis of rotor systems using MCDM methods. *Stud. Inform. Control* **2016**, *25*, 59–68. [CrossRef]
22. Trinkuniene, E.; Podvezko, V.; Zavadskas, E.K.; Joksiene, I.; Vinogradova, I.; Trinkunas, V. Evaluation of quality assurance in contractor contracts by multi-attribute decision-making methods. *Econ. Res. Ekonomska Istrazivanja* **2017**, *30*, 1152–1180. [CrossRef]
23. Su, H.; Qin, P.; Qin, Z. A Method for Evaluating Sea Dike Safety. *Water Resour. Manag.* **2013**, *27*, 5157–5170. [CrossRef]
24. Diakoulaki, D.; Mavrotas, G.; Papayannakis, L. Determining objective weights in multiple criteria problems: The critic method. *Comput. Oper. Res.* **1995**, *22*, 763–770. [CrossRef]
25. Wang, T.C.; Lee, H.D. Developing a fuzzy TOPSIS approach based on subjective weights and objective weights. *Expert Syst. Appl.* **2009**, *36*, 8980–8985. [CrossRef]
26. Lazauskaite, D.; Burinskiene, M.; Podvezko, V. Subjectively and objectively integrated assessment of the quality indices of the suburban residential environment. *Int. J. Strateg. Prop. Manag.* **2015**, *19*, 297–308. [CrossRef]
27. Md Saad, R.; Ahmad, M.Z.; Abu, M.S.; Jusoh, M.S. Hamming Distance Method with Subjective and Objective Weights for Personnel Selection. *Sci. World J. Art.* **2014**, *2014*, 865495. [CrossRef] [PubMed]
28. Parfenova, L.; Pugachev, A.; Podviezko, A. Comparative analysis of tax capacity in regions of Russia. *Technol. Econ. Dev. Econ.* **2016**, *22*, 905–925. [CrossRef]
29. Ustinovichius, L.; Zavadskas, E.K.; Podvezko, V. Application of a quantitative multiple criteria decision making (MCDM-1) approach to the analysis of investments in construction. *Control Cybernet.* **2007**, *36*, 251–268.
30. Jeffreys, H. *Scientific Inference*, 3rd ed.; Cambridge University Press: Cambridge, UK, 1973; p. 31, ISBN 978-0-521-18078-8.
31. Ma, J.; Fan, P.; Huang, L.H. A subjective and objective integrated approach to determine attribute weights. *Eur. J. Oper. Res.* **1999**, *112*, 397–404. [CrossRef]
32. Kou, G.; Lin, C. A cosine maximization method for the priority vector derivation in AHP. *Eur. J. Oper. Res.* **2014**, *235*, 225–232. [CrossRef]
33. Vinogradova, I.; Kliukas, R. Methodology for evaluating the quality of distance learning courses in consecutive stages. *Procedia Soc. Behav. Sci.* **2015**, *191*, 1583–1589. [CrossRef]
34. Kendall, M. *Rank Correlation Methods*; Hafner Publishing House: New York, NY, USA, 1955.

35. Kurilov, J.; Vinogradova, I.; Kubilinskienė, S. New MCEQLS fuzzy AHP methodology for evaluating learning repositories: A tool for technological development of economy. *Technol. Econ. Dev. Econ.* **2016**, *22*, 142–155. [CrossRef]
36. Chang, D.Y. Applications of the extent analysis method on fuzzy AHP. *Eur. J. Oper. Res.* **1996**, *95*, 649–655. [CrossRef]
37. Mirkin, B.G. *Group Choice*; Winston & Sons: Washington, DC, USA, 1979.
38. Stefano, N.M.; Casarotto Filho, N.; Vergara, L.G.L.; Rocha, R.U.G. COPRAS (complex proportional assessment): State of the art research and its applications. *IEEE Latin Am. Trans.* **2015**, *13*, 3899–3906. [CrossRef]
39. Keshavarz Ghorabaee, M.; Zavadskas, E.K.; Olfat, L.; Turskis, Z. Multi-Criteria Inventory Classification Using a New Method of Evaluation Based on Distance from Average Solution (EDAS). *Informatica* **2015**, *26*, 435–451. [CrossRef]
40. Podvezko, V. The Comparative Analysis of MCDM Methods SAW and COPRAS. *Inžinerinė Ekonomika Eng. Econ.* **2011**, *22*, 134–146.
41. Podviezko, A.; Podvezko, V. Absolute and Relative Evaluation of Socio-Economic Objects Based on Multiple Criteria Decision Making Methods. *Inzinerine Ekonomika Eng. Econ.* **2014**, *25*, 522–529. [CrossRef]
42. Turskis, Z.; Keršulienė, V.; Vinogradova, I. A new fuzzy hybrid multi-criteria decision-making approach to solve personnel assessment problems. Case study: Director selection for estates and economy office. *Econ. Comput. Econ. Cybernet. Stud. Res.* **2017**, *51*, 211–229.
43. Zavadskas, E.K.; Kaklauskas, A.; Šarka, V. The new method of multi-criteria complex proportional assessment of projects. *Technol. Econ. Dev. Econ.* **1994**, *1*, 131–139. [CrossRef]
44. Zavadskas, E.K.; Mardani, A.; Turskis, Z.; Jusoh, A.; Nor, K.M.D. Development of TOPSIS method to solve complicated decision-making problems: An overview on developments from 2000 to 2015. *Int. J. Inf. Technol. Decis. Mak.* **2016**, *15*, 645–682. [CrossRef]

symmetry

MDPI

Article

Models for Green Supplier Selection with Some 2-Tuple Linguistic Neutrosophic Number Bonferroni Mean Operators

Jie Wang [1], Guiwu Wei [1,*] and Yu Wei [2,*]

[1] School of Business, Sichuan Normal University, Chengdu 610101, China; JW970326@163.com
[2] School of Finance, Yunnan University of Finance and Economics, Kunming 650221, China
* Correspondence: weiguiwu1973@sicnu.edu.cn (G.W.); ywei@home.swjtu.edu.cn (Y.W.)

Received: 4 April 2018; Accepted: 16 April 2018; Published: 25 April 2018

Abstract: In this paper, we extend the Bonferroni mean (BM) operator, generalized Bonferroni mean (GBM) operator, dual generalized Bonferroni mean (DGBM) operator and dual generalized geometric Bonferroni mean (DGGBM) operator with 2-tuple linguistic neutrosophic numbers (2TLNNs) to propose 2-tuple linguistic neutrosophic numbers weighted Bonferroni mean (2TLNNWBM) operator, 2-tuple linguistic neutrosophic numbers weighted geometric Bonferroni mean (2TLNNWGBM) operator, generalized 2-tuple linguistic neutrosophic numbers weighted Bonferroni mean (G2TLNNWBM) operator, generalized 2-tuple linguistic neutrosophic numbers weighted geometric Bonferroni mean (G2TLNNWGBM) operator, dual generalized 2-tuple linguistic neutrosophic numbers weighted Bonferroni mean (DG2TLNNWBM) operator, and dual generalized 2-tuple linguistic neutrosophic numbers weighted geometric Bonferroni mean (DG2TLNNWGBM) operator. Then, the MADM methods are proposed with these operators. In the end, we utilize an applicable example for green supplier selection in green supply chain management to prove the proposed methods.

Keywords: multiple attribute decision making (MADM); neutrosophic numbers; 2-tuple linguistic neutrosophic numbers set (2TLNNSs); Bonferroni mean (BM) operator; generalized Bonferroni mean (GBM) operator; dual generalized Bonferroni mean (DGBM) operator; dual generalized geometric Bonferroni mean (DGGBM) operator; green supplier selection; green supply chain management

---

1. Introduction

Zadeh [1] introduced a membership function between 0 and 1 instead of traditional crisp value of 0 and 1, and defined the fuzzy set (FS). In order to overcome the insufficiency of FS, Atanassov [2] proposed the concept of an intuitionistic fuzzy set (IFS), which is characterized by its membership function and non-membership function between 0 and 1. Furthermore, Atanassov and Gargov [3] introduced the concept of an interval-valued intuitionistic fuzzy set (IVIFS), which is characterized by its interval membership function and interval non-membership function in the unit interval [0,1]. Because IFSs and IVIFSs cannot depict indeterminate and inconsistent information, Smarandache [4] introduced a neutrosophic set (NS) from a philosophical point of view to express indeterminate and inconsistent information. A NS has more potential power than other modeling mathematical tools, such as fuzzy set [1], IFS [2], and IVIFS [3]. But, it is difficult to apply NSs in solving of real life problems. Therefore, Smarandache [4] and Wang et al. [5,6] defined a single valued neutrosophic set (SVNS) and an interval neutrosophic set (INS), which are characterized by a truth-membership, an indeterminacy membership, and a falsity membership. Ye [7] introduced a simplified neutrosophic set (SNS), including the concepts of SVNS and INS, which are the extension of IFS and IVIFS. Obviously, SNS is a subclass of NS, while SVNS and INS are subclasses of SNS. Ye [8] proposed the correlation and correlation coefficient of single-valued neutrosophic sets (SVNSs) that are based on the extension of the correlation of intuitionistic fuzzy sets and demonstrates that the cosine similarity measure is a special case of the correlation coefficient in SVNS. Broumi and Smarandache [9] extended the correlation coefficient to INSs. Biswas et al. [10] developed a new approach for multi-attribute group decision-making problems by extending the technique for order preference by similarity to ideal solution to single-valued neutrosophic environment. Liu et al. [11] combined Hamacher operations and generalized aggregation operators to NSs, and proposed the generalized neutrosophic number Hamacher weighted averaging (GNNHWA) operator, generalized neutrosophic number Hamacher ordered weighted averaging (GNNHOWA) operator, and generalized neutrosophic number Hamacher hybrid averaging (GNNHHA) operator, and explored some properties of these operators and analyzed some special cases of them. Sahin and Liu [12] developed a maximizing deviation method for solving the multiple attribute decision-making problems with the single-valued neutrosophic information or interval neutrosophic information. Ye [13] defined the Hamming and Euclidean distances between the interval neutrosophic sets (INSs) and proposed the similarity measures between INSs based on the relationship between similarity measures and distances. Zhang et al. [14] defined the operations for INSs and put forward a comparison approach that was based on the related research of interval valued intuitionistic fuzzy sets (IVIFSs) and developed two interval neutrosophic number aggregation operators. Peng et al. [15] developed a new outranking approach for multi-criteria decision-making (MCDM) problems in the context of a simplified neutrosophic environment, where the truth-membership degree, indeterminacy-membership degree, and falsity-membership degree for each element are singleton subsets in [0,1] and defined some outranking relations for simplified neutrosophic number (SNNs) based on ELECTRE (ELimination and Choice Expressing REality), and the properties within the outranking relations are further discussed in detail. Zhang et al. [16] proposed a novel outranking approach for multi-criteria decision-making (MCDM) problems to address situations where there is a set of numbers in the real unit interval and not just a specific number with a neutrosophic set. Liu and Liu [17] proposed the neutrosophic number weighted power averaging (NNWPA) operator, the neutrosophic number weighted geometric power averaging (NNWGPA) operator, the generalized neutrosophic number weighted power averaging (GNNWPA) operator, and studied the properties of above operators are studied, such as idempotency, monotonicity, boundedness, and so on. Peng et al. [18] introduced the multi-valued neutrosophic sets (MVNSs), which allow for the truth-membership, indeterminacy- membership, and falsity-membership degree to have a set of crisp values between zero and one, respectively, and the multi-valued neutrosophic power weighted average (MVNPWA) operator and proposed the multi-valued neutrosophic power weighted geometric (MVNPWG) operator. Zhang et al. [19] presented a new correlation coefficient measure, which satisfies the requirement of this measure equaling one if and only if two interval neutrosophic sets (INSs) are the same and used the proposed weighted correlation coefficient measure of INSs to solve decision-making problems, which take into account the influence of the evaluations' uncertainty and both the objective and subjective weights. Chen and Ye [20] presented the Dombi operations of single-valued neutrosophic numbers (SVNNs) based on the operations of the Dombi T-norm and T-conorm, and then proposed the single-valued neutrosophic Dombi weighted arithmetic average (SVNDWAA) operator and the single-valued neutrosophic Dombi weighted geometric average (SVNDWGA) operator to deal with the aggregation of SVNNs and investigated their properties. Liu and Wang [21] proposed the single-valued neutrosophic normalized weighted Bonferroni mean (SVNNWBM) operator on the basis of Bonferroni mean, the weighted Bonferroni mean (WBM), and the normalized WBM, and developed an approach to solve the multiple attribute decision-making problems with SVNNs that were based on the SVNNWBM operator. Wu et al. [22] defined the prioritized weighted average operator and the prioritized weighted geometric operator for simplified neutrosophic numbers (SNNs) and proposed two novel effective cross-entropy measures for SNSs. Li et al. [23] proposed the improved generalized weighted Heronian mean (IGWHM) operator and the improved generalized weighted geometric Heronian mean (IGWGHM) operator, the single valued neutrosophic number improved generalized weighted Heronian mean (NNIGWHM) operator, and single valued the neutrosophic number improved generalized weighted geometric Heronian mean (NNIGWGHM) operator for multiple attribute group decision making (MAGDM) problems, in which attribute values take the form of SVNNs. Wang et al. [24] combined the generalized weighted BM (GWBM) operator and generalized weighted geometric Bonferroni mean (GWGBM) operator with single valued neutrosophic numbers (SVNNs) to propose the generalized single-valued neutrosophic number weight BM (GSVNNWBM) operator and the generalized single-valued neutrosophic numbers weighted GBM (GSVNNWGBM) operator and developed the MADM methods based on these operators. Wei & Zhang [25] utilized power aggregation operators and the Bonferroni mean to develop some single-valued neutrosophic Bonferroni power aggregation operators and single-valued neutrosophic geometric Bonferroni power aggregation operators. Peng & Dai [26] initiated a new axiomatic definition of single-valued neutrosophic distance measure and similarity measure, which is expressed by a single-valued neutrosophic number that will reduce the information loss and retain more original information.

Although SVNS theory has been successfully applied in some areas, the SVNS is also characterized by the truth-membership degree, indeterminacy-membership degree, and falsity-membership degree information. However, all of the above approaches are unsuitable to describe the truth-membership degree, indeterminacy-membership degree, and falsity-membership degree information of an element to a set by linguistic variables on the basis of the given linguistic term sets, which can reflect the decision maker's confidence level when they are making an evaluation. In order to overcome this limit, we shall propose the concept of 2-tuple linguistic neutrosophic numbers set (2TLNNSs) to solve this problem based on the SVNS [4–6] and the 2-tuple linguistic information processing model [27–42]. Thus, how to aggregate these 2-tuple linguistic neutrosophic numbers is an interesting topic. To solve this issue, in this paper, we shall develop some 2-tuple linguistic neutrosophic information aggregation operators that are based on the traditional Bonferroni mean (BM) operations [43–50]. In order to do so, the remainder of this paper is set out as follows. In the next section, we shall propose the concept of 2TLNNSs. In Section 3, we shall propose some Bonferroni mean (BM) operators with 2TLNNs. In Section 4, we shall propose some generalized Bonferroni mean (GBM) operators with 2TLNNs. In Section 5, we shall propose some dual generalized Bonferroni mean (DGBM) operators with 2TLNNs. In Section 6, we shall present a numerical example for green supplier selection in order to illustrate the method that is proposed in this paper. Section 7 concludes the paper with some remarks.

2. Preliminaries

In this section, we shall propose the concept of 2-tuple linguistic neutrosophic number sets (2TLNNSs) to solve this problem based on the SVNSs [6,7] and 2-tuple linguistic sets (2TLSs) [27,28].

2.1. 2-Tuple Fuzzy Linguistic Representation Model

Definition 1 ([27,28]). *Let $S = \{s_i | i = 0, 1, \ldots, t\}$ be a linguistic term set with odd cardinality. Any label, s_i represents a possible value for a linguistic variable, and S can be defined as:*

$$S = \left\{ \begin{array}{l} s_0 = extremely\ poor, s_1 = very\ poor, s_2 = poor, s_3 = medium, \\ s_4 = good, s_5 = very\ good, s_6 = extremely\ good. \end{array} \right\}$$

Herrera and Martinez [27,28] developed the 2-tuple fuzzy linguistic representation model based on the concept of symbolic translation. It is used for representing the linguistic assessment information by means of a 2-tuple (s_i, ρ_i), where s_i is a linguistic label for predefined linguistic term set S and ρ_i is the value of symbolic translation, and $\rho_i \in [-0.5, 0.5)$.

2.2. SVNSs

Let X be a space of points (objects) with a generic element in fix set X, denoted by x. A single-valued neutrosophic sets (SVNSs) A in X is characterized as following [4–6]:

$$A = \{(x, T_A(x), I_A(x), F_A(x)) | x \in X\} \tag{1}$$

where the truth-membership function $T_A(x)$, indeterminacy-membership $I_A(x)$, and falsity-membership function $F_A(x)$ are single subintervals/subsets in the real standard $[0,1]$, that is, $T_A(x) : X \to [0,1], I_A(x) : X \to [0,1]$ and $F_A(x) : X \to [0,1]$. The sum of $T_A(x)$, $I_A(x)$, and $F_A(x)$ satisfies the condition $0 \le T_A(x) + I_A(x) + F_A(x) \le 3$. Then, a simplification of A is denoted by $A = \{(x, T_A(x), I_A(x), F_A(x)) | x \in X\}$, which is a SVNS.

For a SVNS $\{(x, T_A(x), I_A(x), F_A(x)) | x \in X\}$, the ordered triple components $(T_A(x), I_A(x), F_A(x))$, are described as a single-valued neutrosophic number (SVNN), and each SVNN can be expressed as $A = (T_A, I_A, F_A)$, where $T_A \in [0,1], I_A \in [0,1], F_A \in [0,1]$ $T_A \in [0,1], I_A \in [0,1], F_A \in [0,1]$, and $0 \le T_A + I_A + F_A \le 3$.

2.3. 2TLNNSs

Definition 2. *Assume that* $L = \{l_0, l_1, \ldots, l_t\}$ *is a 2TLSs with odd cardinality* $t + 1$. *If* $l = \langle (s_T, a), (s_I, b), (s_F, c) \rangle$ *is defined for* $(s_T, a), (s_I, b), (s_F, c) \in L$ *and* $a, b, c \in [0, t]$, *where* $(s_T, a), (s_I, b)$ *and* (s_F, c) *express independently the truth degree, indeterminacy degree, and falsity degree by 2TLSs, then 2TLNNSs is defined as follows:*

$$l_j = \left\{ \left(s_{T_j}, a_j\right), \left(s_{I_j}, b_j\right), \left(s_{F_j}, c_j\right) \right\} \tag{2}$$

where $0 \le \Delta^{-1}\left(s_{T_j}, a_j\right) \le t$, $0 \le \Delta^{-1}\left(s_{I_j}, b_j\right) \le t$, $0 \le \Delta^{-1}\left(s_{F_j}, c_j\right) \le t$, *and* $0 \le \Delta^{-1}\left(s_{T_j}, a_j\right) + \Delta^{-1}\left(s_{I_j}, b_j\right) + \Delta^{-1}\left(s_{F_j}, c_j\right) \le 3t$.

Definition 3. *Let* $l_1 = \left\langle (s_{T_1}, a_1), (s_{I_1}, b_1), (s_{F_1}, c_1) \right\rangle$ *be a 2TLNN in* L. *Then the score and accuracy functions of* l_1 *are defined as follows:*

$$S(l_1) = \Delta\left\{ \frac{(2t + \Delta^{-1}(s_{T_1}, a_1) - \Delta^{-1}(s_{I_1}, b_1) - \Delta^{-1}(s_{F_1}, c_1))}{3} \right\}, \Delta^{-1}(S(l_1)) \in [0, t] \tag{3}$$

$$H(l_1) = \Delta\left\{ \frac{t + \Delta^{-1}(s_{T_1}, a_1) - \Delta^{-1}(s_{F_1}, c_1)}{2} \right\}, \Delta^{-1}(H(l_1)) \in [0, t]. \tag{4}$$

Definition 4. *Let* $l_1 = \left\langle (s_{T_1}, a_1), (s_{I_1}, b_1), (s_{F_1}, c_1) \right\rangle$ *and* $l_2 = \left\langle (s_{T_2}, a_2), (s_{I_2}, b_2), (s_{F_2}, c_2) \right\rangle$ *be two 2TLNNs, then*

(1) if $S(l_1) < S(l_2)$, then $l_1 < l_2$;
(2) if $S(l_1) > S(l_2)$, then $l_1 > l_2$;
(3) if $S(l_1) = S(l_2), H(l_1) < H(l_2)$, then $l_1 < l_2$;
(4) if $S(l_1) = S(l_2), H(l_1) > H(l_2)$, then $l_1 > l_2$;
(5) if $S(l_1) = S(l_2), H(l_1) = H(l_2)$, then $l_1 = l_2$.

Definition 5. *Let* $l_1 = \left\langle (s_{T_1}, a_1), (s_{I_1}, b_1), (s_{F_1}, c_1) \right\rangle$ *and* $l_2 = \left\langle (s_{T_2}, a_2), (s_{I_2}, b_2), (s_{F_2}, c_2) \right\rangle$ *be two 2TLNNs, then*

$$(1)\quad l_1 \oplus l_2 = \left\{ \begin{array}{l} \Delta\left(t\left(\frac{\Delta^{-1}(s_{T_1},a_1)}{t} + \frac{\Delta^{-1}(s_{T_2},a_2)}{t} - \frac{\Delta^{-1}(s_{T_1},a_1)}{t} \cdot \frac{\Delta^{-1}(s_{T_2},a_2)}{t}\right)\right), \\ \Delta\left(t\left(\frac{\Delta^{-1}(s_{I_1},b_1)}{t} \cdot \frac{\Delta^{-1}(s_{I_2},b_2)}{t}\right)\right), \Delta\left(t\left(\frac{\Delta^{-1}(s_{F_1},c_1)}{t} \cdot \frac{\Delta^{-1}(s_{F_2},c_2)}{t}\right)\right) \end{array} \right\};$$

$$(2)\quad l_1 \otimes l_2 = \left\{ \begin{array}{l} \Delta\left(t\left(\frac{\Delta^{-1}(s_{T_1},a_1)}{t} \cdot \frac{\Delta^{-1}(s_{T_2},a_2)}{t}\right)\right), \\ \Delta\left(t\left(\frac{\Delta^{-1}(s_{I_1},b_1)}{t} + \frac{\Delta^{-1}(s_{I_2},b_2)}{t} - \frac{\Delta^{-1}(s_{I_1},b_1)}{t} \cdot \frac{\Delta^{-1}(s_{I_2},b_2)}{t}\right)\right), \\ \Delta\left(t\left(\frac{\Delta^{-1}(s_{F_1},c_1)}{t} + \frac{\Delta^{-1}(s_{F_2},c_2)}{t} - \frac{\Delta^{-1}(s_{F_1},c_1)}{t} \cdot \frac{\Delta^{-1}(s_{F_2},c_2)}{t}\right)\right) \end{array} \right\};$$

$$(3)\quad \lambda l_1 = \left\{ \begin{array}{l} \Delta\left(t\left(1-\left(1-\frac{\Delta^{-1}(s_{T_1},a_1)}{t}\right)^{\lambda}\right)\right), \Delta\left(t\left(\frac{\Delta^{-1}(s_{I_1},b_1)}{t}\right)^{\lambda}\right), \\ \Delta\left(t\left(\frac{\Delta^{-1}(s_{F_1},c_1)}{t}\right)^{\lambda}\right) \end{array} \right\}, \lambda > 0;$$

$$(4)\quad (l_1)^{\lambda} = \left\{ \begin{array}{l} \Delta\left(t\left(\frac{\Delta^{-1}(s_{T_1},a_1)}{t}\right)^{\lambda}\right), \Delta\left(t\left(1-\left(1-\frac{\Delta^{-1}(s_{I_1},b_1)}{t}\right)^{\lambda}\right)\right), \\ \Delta\left(t\left(1-\left(1-\frac{\Delta^{-1}(s_{F_1},c_1)}{t}\right)^{\lambda}\right)\right) \end{array} \right\}, \lambda > 0.$$

2.4. BM Operators

Definition 6 ([44]). *Let $p,q > 0$ and $b_i (i = 1, 2, \cdots, n)$ be a collection of nonnegative crisp numbers then the Bonferroni mean (BM) is defined as follows:*

$$\mathrm{BM}^{p,q}(b_1, b_2, \cdots, b_n) = \left(\sum_{i,j=1}^{n} b_i^p b_j^q\right)^{1/(p+q)} \tag{5}$$

Then $\mathrm{BM}^{p,q}$ *is called Bonferroni mean (BM) operator.*

3. 2TLNNWBM and 2TLNNWGBM Operators

3.1. 2TLNNWBM Operator

To consider the attribute weights, the weighted Bonferroni mean (WBM) is defined, as follows.

Definition 7 ([44]). *Let $p,q > 0$ and $b_i (i = 1, 2, \cdots, n)$ be a collection of nonnegative crisp numbers with the weights vector being $\omega = (\omega_1, \omega_2, \cdots \omega_n)^T$, thereby satisfying $\omega_i \in [0,1]$ and $\sum_{i=1}^{n} \omega_i = 1$. The weighted Bonferroni mean (WBM) is defined as follows:*

$$\mathrm{WBM}_{\omega}^{p,q}(b_1, b_2, \cdots, b_n) = \left(\sum_{i,j=1}^{n} \omega_i \omega_j b_i^p b_j^q\right)^{1/(p+q)} \tag{6}$$

Then we extend WBM to fuse the 2TLNNs and propose 2-tuple linguistic neutrosophic number weighted Bonferroni mean (2TLNNWBM) aggregation operator.

Definition 8. *Let $l_i = \langle (s_{T_i}, a_i), (s_{I_i}, b_i), (s_{F_i}, c_i) \rangle$ be a set of 2TLNNs. The 2-tuple linguistic neutrosophic number weighted Bonferroni mean (2TLNNWBM) operator is:*

$$\mathrm{2TLNNWBM}_{\omega}^{p,q}(l_1, l_2, \ldots, l_n) = \left(\bigoplus_{i,j=1}^{n} \left(\omega_i \omega_j \left(l_i^p \otimes l_j^q\right)\right)\right)^{1/(p+q)} \tag{7}$$

Theorem 1. *Let $l_i = \langle (s_{T_i}, a_i), (s_{I_i}, b_i), (s_{F_i}, c_i) \rangle$ be a set of 2TLNNs. The aggregated value by using 2TLNNWBM operators is also a 2TLNN where*

$$\begin{aligned}
&2\text{TLNNWBM}_{\omega}^{p,q}(l_1, l_2, \ldots, l_n) = \left(\underset{i,j=1}{\overset{n}{\oplus}} \left(\omega_i \omega_j \left(l_i^p \otimes l_j^q \right) \right) \right)^{1/(p+q)} \\
&= \left\{ \begin{array}{l}
\Delta\left(t \left(1 - \prod\limits_{i,j=1}^{n} \left(1 - \left(\frac{\Delta^{-1}(s_{T_i}, a_i)}{t} \right)^p \cdot \left(\frac{\Delta^{-1}(s_{T_j}, a_j)}{t} \right)^q \right)^{\omega_i \omega_j} \right)^{1/(p+q)} \right), \\
\Delta\left(t \left(1 - \left(1 - \prod\limits_{i,j=1}^{n} \left(1 - \left(1 - \frac{\Delta^{-1}(s_{I_i}, b_i)}{t} \right)^p \cdot \left(1 - \frac{\Delta^{-1}(s_{I_j}, b_j)}{t} \right)^q \right)^{\omega_i \omega_j} \right)^{1/(p+q)} \right) \right), \\
\Delta\left(t \left(1 - \left(1 - \prod\limits_{i,j=1}^{n} \left(1 - \left(1 - \frac{\Delta^{-1}(s_{F_i}, c_i)}{t} \right)^p \cdot \left(1 - \frac{\Delta^{-1}(s_{F_j}, c_j)}{t} \right)^q \right)^{\omega_i \omega_j} \right)^{1/(p+q)} \right) \right)
\end{array} \right\}
\end{aligned} \tag{8}$$

Proof. According to Definition 5, we can obtain

$$l_i^p = \left\{ \Delta\left(t \left(\frac{\Delta^{-1}(s_{T_i}, a_i)}{t} \right)^p \right), \Delta\left(t \left(1 - \left(1 - \frac{\Delta^{-1}(s_{I_i}, b_i)}{t} \right)^p \right) \right), \Delta\left(t \left(1 - \left(1 - \frac{\Delta^{-1}(s_{F_i}, c_i)}{t} \right)^p \right) \right) \right\} \tag{9}$$

$$l_j^q = \left\{ \Delta\left(t \left(\frac{\Delta^{-1}(s_{T_j}, a_j)}{t} \right)^q \right), \Delta\left(t \left(1 - \left(1 - \frac{\Delta^{-1}(s_{I_j}, b_j)}{t} \right)^q \right) \right), \Delta\left(t \left(1 - \left(1 - \frac{\Delta^{-1}(s_{F_j}, c_j)}{t} \right)^q \right) \right) \right\} \tag{10}$$

Thus,

$$l_i^p \otimes l_j^q = \left\{ \begin{array}{l}
\Delta\left(t \left(\left(\frac{\Delta^{-1}(s_{T_i}, a_i)}{t} \right)^p \cdot \left(\frac{\Delta^{-1}(s_{T_j}, a_j)}{t} \right)^q \right) \right), \\
\Delta\left(t \left(1 - \left(1 - \frac{\Delta^{-1}(s_{I_i}, b_i)}{t} \right)^p \cdot \left(1 - \frac{\Delta^{-1}(s_{I_j}, b_j)}{t} \right)^q \right) \right), \\
\Delta\left(t \left(1 - \left(1 - \frac{\Delta^{-1}(s_{F_i}, c_i)}{t} \right)^p \cdot \left(1 - \frac{\Delta^{-1}(s_{F_j}, c_j)}{t} \right)^q \right) \right)
\end{array} \right\} \tag{11}$$

Thereafter,

$$\begin{aligned}
&\omega_i \omega_j \left(l_i^p \otimes l_j^q \right) \\
&= \left\{ \begin{array}{l}
\Delta\left(t \left(1 - \left(1 - \left(\frac{\Delta^{-1}(s_{T_i}, a_i)}{t} \right)^p \cdot \left(\frac{\Delta^{-1}(s_{T_j}, a_j)}{t} \right)^q \right)^{\omega_i \omega_j} \right) \right), \\
\Delta\left(t \left(1 - \left(1 - \frac{\Delta^{-1}(s_{I_i}, b_i)}{t} \right)^p \cdot \left(1 - \frac{\Delta^{-1}(s_{I_j}, b_j)}{t} \right)^q \right)^{\omega_i \omega_j} \right), \\
\Delta\left(t \left(1 - \left(1 - \frac{\Delta^{-1}(s_{F_i}, c_i)}{t} \right)^p \cdot \left(1 - \frac{\Delta^{-1}(s_{F_j}, c_j)}{t} \right)^q \right)^{\omega_i \omega_j} \right)
\end{array} \right\}
\end{aligned} \tag{12}$$

Furthermore,

$$\begin{aligned}
&\underset{i,j=1}{\overset{n}{\oplus}} \left(\omega_i \omega_j \left(l_i^p \otimes l_j^q \right) \right) \\
&= \left\{ \begin{array}{l}
\Delta\left(t \left(1 - \prod\limits_{i,j=1}^{n} \left(1 - \left(\frac{\Delta^{-1}(s_{T_i}, a_i)}{t} \right)^p \cdot \left(\frac{\Delta^{-1}(s_{T_j}, a_j)}{t} \right)^q \right)^{\omega_i \omega_j} \right) \right), \\
\Delta\left(t \prod\limits_{i,j=1}^{n} \left(1 - \left(1 - \frac{\Delta^{-1}(s_{I_i}, b_i)}{t} \right)^p \cdot \left(1 - \frac{\Delta^{-1}(s_{I_j}, b_j)}{t} \right)^q \right)^{\omega_i \omega_j} \right), \\
\Delta\left(t \prod\limits_{i,j=1}^{n} \left(1 - \left(1 - \frac{\Delta^{-1}(s_{F_i}, c_i)}{t} \right)^p \cdot \left(1 - \frac{\Delta^{-1}(s_{F_j}, c_j)}{t} \right)^q \right)^{\omega_i \omega_j} \right)
\end{array} \right\}
\end{aligned} \tag{13}$$

Therefore,

$$
\begin{aligned}
&2TLNNWBM_{\omega}^{p,q}(l_1, l_2, \ldots, l_n) = \left(\underset{i,j=1}{\overset{n}{\oplus}} \left(\omega_i \omega_j \left(l_i^p \otimes l_j^q \right) \right) \right)^{1/(p+q)} \\
&= \left\{ \begin{array}{l}
\Delta\left(t\left(1 - \prod_{i,j=1}^{n} \left(1 - \left(\frac{\Delta^{-1}(s_{T_i}, a_i)}{t} \right)^p \cdot \left(\frac{\Delta^{-1}(s_{T_j}, a_j)}{t} \right)^q \right)^{\omega_i \omega_j} \right)^{1/(p+q)} \right), \\
\Delta\left(t\left(1 - \left(1 - \prod_{i,j=1}^{n} \left(1 - \left(1 - \frac{\Delta^{-1}(s_{I_i}, b_i)}{t} \right)^p \cdot \left(1 - \frac{\Delta^{-1}(s_{I_j}, b_j)}{t} \right)^q \right)^{\omega_i \omega_j} \right)^{1/(p+q)} \right) \right), \\
\Delta\left(t\left(1 - \left(1 - \prod_{i,j=1}^{n} \left(1 - \left(1 - \frac{\Delta^{-1}(s_{F_i}, c_i)}{t} \right)^p \cdot \left(1 - \frac{\Delta^{-1}(s_{F_j}, c_j)}{t} \right)^q \right)^{\omega_i \omega_j} \right)^{1/(p+q)} \right) \right)
\end{array} \right\}
\end{aligned} \tag{14}
$$

Hence, (8) is kept. □

Then, we need to prove that (8) is a 2TLNN. We need to prove two conditions, as follows:

① $0 \le \Delta^{-1}(s_T, a) \le t, 0 \le \Delta^{-1}(s_I, b) \le t, 0 \le \Delta^{-1}(s_F, c) \le t$
② $0 \le \Delta^{-1}(s_T, a) + \Delta^{-1}(s_I, b) + \Delta^{-1}(s_F, c) \le 3t$

Let

$$
\frac{\Delta^{-1}(s_T, a)}{t} = \left(1 - \prod_{i,j=1}^{n} \left(1 - \left(\frac{\Delta^{-1}(s_{T_i}, a_i)}{t} \right)^p \cdot \left(\frac{\Delta^{-1}(s_{T_j}, a_j)}{t} \right)^q \right)^{\omega_i \omega_j} \right)^{1/(p+q)}
$$

$$
\frac{\Delta^{-1}(s_I, b)}{t} = 1 - \left(1 - \prod_{i,j=1}^{n} \left(1 - \left(1 - \frac{\Delta^{-1}(s_{I_i}, b_i)}{t} \right)^p \cdot \left(1 - \frac{\Delta^{-1}(s_{I_j}, b_j)}{t} \right)^q \right)^{\omega_i \omega_j} \right)^{1/(p+q)}
$$

$$
\frac{\Delta^{-1}(s_F, c)}{t} = 1 - \left(1 - \prod_{i,j=1}^{n} \left(1 - \left(1 - \frac{\Delta^{-1}(s_{F_i}, c_i)}{t} \right)^p \cdot \left(1 - \frac{\Delta^{-1}(s_{F_j}, c_j)}{t} \right)^q \right)^{\omega_i \omega_j} \right)^{1/(p+q)}
$$

Proof. ① Since $0 \le \frac{\Delta^{-1}(s_{T_i}, a_i)}{t} \le 1, 0 \le \frac{\Delta^{-1}(s_{T_j}, a_j)}{t} \le 1$ we get

$$
0 \le 1 - \left(\frac{\Delta^{-1}(s_{T_i}, a_i)}{t} \right)^p \cdot \left(\frac{\Delta^{-1}(s_{T_j}, a_j)}{t} \right)^q \le 1 \tag{15}
$$

Then,

$$
0 \le 1 - \prod_{i,j=1}^{n} \left(1 - \left(\frac{\Delta^{-1}(s_{T_i}, a_i)}{t} \right)^p \cdot \left(\frac{\Delta^{-1}(s_{T_j}, a_j)}{t} \right)^q \right)^{\omega_i \omega_j} \le 1 \tag{16}
$$

$$
0 \le \left(1 - \prod_{i,j=1}^{n} \left(1 - \left(\frac{\Delta^{-1}(s_{T_i}, a_i)}{t} \right)^p \cdot \left(\frac{\Delta^{-1}(s_{T_j}, a_j)}{t} \right)^q \right)^{\omega_i \omega_j} \right)^{1/(p+q)} \le 1 \tag{17}
$$

That means $0 \le \Delta^{-1}(s_T, a) \le t$, so ① is maintained, similarly, we can get $0 \le \Delta^{-1}(s_I, b) \le t$ $0 \le \Delta^{-1}(s_F, c) \le t$. ② Since $0 \le \Delta^{-1}(s_T, a) \le t, 0 \le \Delta^{-1}(s_I, b) \le t, 0 \le \Delta^{-1}(s_F, c) \le t$, we get $0 \le \Delta^{-1}(s_T, a) + \Delta^{-1}(s_I, b) + \Delta^{-1}(s_F, c) \le 3t$. □

Example 1. *Let* $\langle (s_3, 0.4), (s_2, -0.3), (s_4, 0.1) \rangle, \langle (s_2, 0.3), (s_1, 0.2), (s_4, -0.1) \rangle$ *be two 2TLNNs,* $(p, q) = (2, 3)$, $\omega = (0.4, 0.6)$ *according to (8), we have*

$$
\begin{aligned}
&2\mathrm{TLNNWBM}^{(2,3)}_{(0.4,0.6)}\begin{pmatrix} \langle (s_3,0.4),(s_2,-0.3),(s_4,0.1)\rangle, \\ \langle (s_2,0.3),(s_1,0.2),(s_4,-0.1)\rangle \end{pmatrix} \\
&= \left\{ \begin{array}{l}
\Delta\left(t\left(1-\prod\limits_{i,j=1}^{n}\left(1-\left(\frac{\Delta^{-1}(s_{T_i},a_i)}{t}\right)^p\cdot\left(\frac{\Delta^{-1}(s_{T_j},a_j)}{t}\right)^q\right)^{\omega_i\omega_j}\right)^{1/(p+q)}\right), \\
\Delta\left(t\left(1-\left(1-\prod\limits_{i,j=1}^{n}\left(1-\left(1-\frac{\Delta^{-1}(s_{I_i},b_i)}{t}\right)^p\cdot\left(1-\frac{\Delta^{-1}(s_{I_j},b_j)}{t}\right)^q\right)^{\omega_i\omega_j}\right)^{1/(p+q)}\right)\right), \\
\Delta\left(t\left(1-\left(1-\prod\limits_{i,j=1}^{n}\left(1-\left(1-\frac{\Delta^{-1}(s_{F_i},c_i)}{t}\right)^p\cdot\left(1-\frac{\Delta^{-1}(s_{F_j},c_j)}{t}\right)^q\right)^{\omega_i\omega_j}\right)^{1/(p+q)}\right)\right)
\end{array} \right\} \\
&= \left\{ \begin{array}{l}
\Delta\left(6\times\left(1-\left(\begin{array}{l}\left(1-\left(\frac{3.4}{6}\right)^2\times\left(\frac{3.4}{6}\right)^3\right)^{0.4\times0.4}\times\left(1-\left(\frac{3.4}{6}\right)^2\times\left(\frac{2.3}{6}\right)^3\right)^{0.4\times0.6} \\ \times\left(1-\left(\frac{2.3}{6}\right)^2\times\left(\frac{3.4}{6}\right)^3\right)^{0.6\times0.4}\times\left(1-\left(\frac{2.3}{6}\right)^2\times\left(\frac{2.3}{6}\right)^3\right)^{0.6\times0.6}\end{array}\right)\right)^{\frac{1}{2+3}}\right), \\
\Delta\left(6\times\left(1-\left(\begin{array}{l}1-\left(1-\left(1-\frac{1.7}{6}\right)^2\times\left(1-\frac{1.7}{6}\right)^3\right)^{0.4\times0.4}\times\left(1-\left(1-\frac{1.7}{6}\right)^2\times\left(1-\frac{1.2}{6}\right)^3\right)^{0.4\times0.6} \\ \times\left(1-\left(1-\frac{1.2}{6}\right)^2\times\left(1-\frac{1.7}{6}\right)^3\right)^{0.6\times0.4}\times\left(1-\left(1-\frac{1.2}{6}\right)^2\times\left(1-\frac{1.2}{6}\right)^3\right)^{0.6\times0.6}\end{array}\right)^{\frac{1}{2+3}}\right)\right), \\
\Delta\left(6\times\left(1-\left(\begin{array}{l}1-\left(1-\left(1-\frac{4.1}{6}\right)^2\times\left(1-\frac{4.1}{6}\right)^3\right)^{0.4\times0.4}\times\left(1-\left(1-\frac{4.1}{6}\right)^2\times\left(1-\frac{3.9}{6}\right)^3\right)^{0.4\times0.6} \\ \times\left(1-\left(1-\frac{3.9}{6}\right)^2\times\left(1-\frac{4.1}{6}\right)^3\right)^{0.6\times0.4}\times\left(1-\left(1-\frac{3.9}{6}\right)^2\times\left(1-\frac{3.9}{6}\right)^3\right)^{0.6\times0.6}\end{array}\right)^{\frac{1}{2+3}}\right)\right)
\end{array} \right\} \\
&= \langle (s_3,-0.173),(s_1,0.384),(s_4,-0.024)\rangle
\end{aligned}
$$

Then, we will discuss some properties of 2TLNNWBM operator.

Property 1. *(Idempotency) If $l_i = \langle (s_{T_i},a_i),(s_{I_i},b_i),(s_{F_i},c_i)\rangle (i=1,2,\ldots,n)$ are equal, then*

$$2\mathrm{TLNNWBM}^{p,q}_{\omega}(l_1,l_2,\cdots,l_n) = l \tag{18}$$

Proof. Since $l_i = l = \langle (s_T,a),(s_I,b),(s_F,c)\rangle$, then

$$
\begin{aligned}
&2\mathrm{TLNNWBM}^{p,q}_{\omega}(l_1,l_2,\ldots,l_n) = \left(\mathop{\oplus}\limits_{i,j=1}^{n}\left(\omega_i\omega_j(l^p\otimes l^q)\right)\right)^{1/(p+q)} \\
&= \left\{ \begin{array}{l}
\Delta\left(t\left(1-\prod\limits_{i,j=1}^{n}\left(1-\left(\frac{\Delta^{-1}(s_T,a)}{t}\right)^p\cdot\left(\frac{\Delta^{-1}(s_T,a)}{t}\right)^q\right)^{\omega_i\omega_j}\right)^{1/(p+q)}\right), \\
\Delta\left(t\left(1-\left(1-\prod\limits_{i,j=1}^{n}\left(1-\left(1-\frac{\Delta^{-1}(s_I,b)}{t}\right)^p\cdot\left(1-\frac{\Delta^{-1}(s_I,b)}{t}\right)^q\right)^{\omega_i\omega_j}\right)^{1/(p+q)}\right)\right), \\
\Delta\left(t\left(1-\left(1-\prod\limits_{i,j=1}^{n}\left(1-\left(1-\frac{\Delta^{-1}(s_F,c)}{t}\right)^p\cdot\left(1-\frac{\Delta^{-1}(s_F,c)}{t}\right)^q\right)^{\omega_i\omega_j}\right)^{1/(p+q)}\right)\right)
\end{array} \right\} \\
&= \left\{ \begin{array}{l}
\Delta\left(t\left(1-\left(1-\left(\frac{\Delta^{-1}(s_T,a)}{t}\right)^{p+q}\right)^{\sum\limits_{i=1}^{n}\omega_i\sum\limits_{j=1}^{n}\omega_j}\right)^{1/(p+q)}\right), \\
\Delta\left(t\left(1-\left(1-\left(1-\left(1-\frac{\Delta^{-1}(s_I,b)}{t}\right)^{p+q}\right)^{\sum\limits_{i=1}^{n}\omega_i\sum\limits_{j=1}^{n}\omega_j}\right)^{1/(p+q)}\right)\right), \\
\Delta\left(t\left(1-\left(1-\left(1-\left(1-\frac{\Delta^{-1}(s_F,c)}{t}\right)^{p+q}\right)^{\sum\limits_{i=1}^{n}\omega_i\sum\limits_{j=1}^{n}\omega_j}\right)^{1/(p+q)}\right)\right)
\end{array} \right\} \\
&= \langle (s_T,a),(s_I,b),(s_F,c)\rangle = l
\end{aligned}
$$

□

Property 2. *(Monotonicity) Let* $l_{x_i} = \left\langle \left(s_{T_{x_i}}, a_{x_i}\right), \left(s_{I_{x_i}}, b_{x_i}\right), \left(s_{F_{x_i}}, c_{x_i}\right) \right\rangle (i = 1,2,\ldots,n)$ *and* $l_{y_i} = \left\langle \left(s_{T_{y_i}}, a_{y_i}\right), \left(s_{I_{y_i}}, b_{y_i}\right), \left(s_{F_{y_i}}, c_{y_i}\right) \right\rangle (i = 1,2,\ldots,n)$ *be two sets of 2TLNNs. If* $\Delta^{-1}\left(s_{T_{x_i}}, a_{x_i}\right) \leq \Delta^{-1}\left(s_{T_{y_i}}, a_{y_i}\right)$*,*$\Delta^{-1}\left(s_{I_{x_i}}, b_{x_i}\right) \geq \Delta^{-1}\left(s_{I_{y_i}}, b_{y_i}\right)$ *and* $\Delta^{-1}\left(s_{F_{x_i}}, c_{x_i}\right) \geq \Delta^{-1}\left(s_{F_{y_i}}, c_{y_i}\right)$ *hold for all i, then*

$$\text{2TLNNWBM}_{\omega}^{p,q}(l_{x_1}, l_{x_2}, \cdots, l_{x_n}) \leq \text{2TLNNWBM}_{\omega}^{p,q}(l_{y_1}, l_{y_2}, \cdots, l_{y_n}) \tag{19}$$

Proof. Let $\text{2TLNNWBM}_{\omega}^{p,q}(l_{x_1}, l_{x_2}, \cdots, l_{x_n}) = \left\langle \left(s_{T_{x_i}}, a_{x_i}\right), \left(s_{I_{x_i}}, b_{x_i}\right), \left(s_{F_{x_i}}, c_{x_i}\right) \right\rangle (i = 1,2,\ldots,n)$ and $\text{2TLNNWBM}_{\omega}^{p,q}(l_{y_1}, l_{y_2}, \cdots, l_{y_n}) = \left\langle \left(s_{T_{y_i}}, a_{y_i}\right), \left(s_{I_{y_i}}, b_{y_i}\right), \left(s_{F_{y_i}}, c_{y_i}\right) \right\rangle (i = 1,2,\ldots,n)$, given that $\Delta^{-1}\left(s_{T_{x_i}}, a_{x_i}\right) \leq \Delta^{-1}\left(s_{T_{y_i}}, a_{y_i}\right)$, we can obtain

$$\left(\frac{\Delta^{-1}\left(s_{T_{x_i}}, a_{x_i}\right)}{t}\right)^p \cdot \left(\frac{\Delta^{-1}\left(s_{T_{x_j}}, a_{x_j}\right)}{t}\right)^q \leq \left(\frac{\Delta^{-1}\left(s_{T_{y_i}}, a_{y_i}\right)}{t}\right)^p \cdot \left(\frac{\Delta^{-1}\left(s_{T_{y_j}}, a_{y_j}\right)}{t}\right)^q \tag{20}$$

$$\left(1 - \left(\frac{\Delta^{-1}\left(s_{T_{x_i}}, a_{x_i}\right)}{t}\right)^p \cdot \left(\frac{\Delta^{-1}\left(s_{T_{x_j}}, a_{x_j}\right)}{t}\right)^q\right)^{\omega_i\omega_j} \geq \left(1 - \left(\frac{\Delta^{-1}\left(s_{T_{y_i}}, a_{y_i}\right)}{t}\right)^p \cdot \left(\frac{\Delta^{-1}\left(s_{T_{y_j}}, a_{y_j}\right)}{t}\right)^q\right)^{\omega_i\omega_j} \tag{21}$$

Thereafter,

$$1 - \prod_{i,j=1}^{n}\left(1 - \left(\frac{\Delta^{-1}\left(s_{T_{x_i}}, a_{x_i}\right)}{t}\right)^p \cdot \left(\frac{\Delta^{-1}\left(s_{T_{x_j}}, a_{x_j}\right)}{t}\right)^q\right)^{\omega_i\omega_j} \leq 1 - \prod_{i,j=1}^{n}\left(1 - \left(\frac{\Delta^{-1}\left(s_{T_{y_i}}, a_{y_i}\right)}{t}\right)^p \cdot \left(\frac{\Delta^{-1}\left(s_{T_{y_j}}, a_{y_j}\right)}{t}\right)^q\right)^{\omega_i\omega_j} \tag{22}$$

Furthermore,

$$\begin{aligned} &\left(1 - \prod_{i,j=1}^{n}\left(1 - \left(\frac{\Delta^{-1}\left(s_{T_{x_i}}, a_{x_i}\right)}{t}\right)^p \cdot \left(\frac{\Delta^{-1}\left(s_{T_{x_j}}, a_{x_j}\right)}{t}\right)^q\right)^{\omega_i\omega_j}\right)^{1/(p+q)} \\ \leq &\left(1 - \prod_{i,j=1}^{n}\left(1 - \left(\frac{\Delta^{-1}\left(s_{T_{y_i}}, a_{y_i}\right)}{t}\right)^p \cdot \left(\frac{\Delta^{-1}\left(s_{T_{y_j}}, a_{y_j}\right)}{t}\right)^q\right)^{\omega_i\omega_j}\right)^{1/(p+q)} \end{aligned} \tag{23}$$

That means $\Delta^{-1}(s_{T_x}, a_x) \leq \Delta^{-1}\left(s_{T_y}, a_y\right)$. Similarly, we can obtain $\Delta^{-1}(s_{I_x}, b_x) \geq \Delta^{-1}\left(s_{I_y}, b_y\right)$ and $\Delta^{-1}(s_{F_x}, c_x) \geq \Delta^{-1}\left(s_{F_y}, c_y\right)$.

If $\Delta^{-1}(s_{T_x}, a_x) < \Delta^{-1}\left(s_{T_y}, a_y\right)$ and $\Delta^{-1}(s_{I_x}, b_x) \geq \Delta^{-1}\left(s_{I_y}, b_y\right)$ and $\Delta^{-1}(s_{F_x}, c_x) \geq \Delta^{-1}\left(s_{F_y}, c_y\right)$

$$\text{2TLNNWBM}_{\omega}^{p,q}(l_{x_1}, l_{x_2}, \cdots, l_{x_n}) < \text{G2TLNNWBM}_{\omega}^{p,q}(l_{y_1}, l_{y_2}, \cdots, l_{y_n})$$

If $\Delta^{-1}(s_{T_x}, a_x) = \Delta^{-1}\left(s_{T_y}, a_y\right)$ and $\Delta^{-1}(s_{I_x}, b_x) > \Delta^{-1}\left(s_{I_y}, b_y\right)$ and $\Delta^{-1}(s_{F_x}, c_x) > \Delta^{-1}\left(s_{F_y}, c_y\right)$

$$\text{2TLNNWBM}_{\omega}^{p,q}(l_{x_1}, l_{x_2}, \cdots, l_{x_n}) < \text{G2TLNNWBM}_{\omega}^{p,q}(l_{y_1}, l_{y_2}, \cdots, l_{y_n})$$

If $\Delta^{-1}(s_{T_x}, a_x) = \Delta^{-1}\left(s_{T_y}, a_y\right)$ and $\Delta^{-1}(s_{I_x}, b_x) = \Delta^{-1}\left(s_{I_y}, b_y\right)$ and $\Delta^{-1}(s_{F_x}, c_x) = \Delta^{-1}\left(s_{F_y}, c_y\right)$

$$\text{2TLNNWBM}_{\omega}^{p,q}(l_{x_1}, l_{x_2}, \cdots, l_{x_n}) = \text{2TLNNWBM}_{\omega}^{p,q}(l_{y_1}, l_{y_2}, \cdots, l_{y_n})$$

So, Property 2 is right. □

Property 3. *(Boundedness) Let* $l_i = \langle (s_{T_i}, a_i), (s_{I_i}, b_i), (s_{F_i}, c_i) \rangle (i = 1, 2, \ldots, n)$ *be a set of 2TLNNs. If* $l^+ = (\max_i(S_{T_i}, a_i), \min_i(S_{I_i}, b_i), \min_i(S_{F_i}, c_i))$ *and* $l^- = (\min_i(S_{T_i}, a_i), \max_i(S_{I_i}, b_i), \max_i(S_{F_i}, c_i))$ *then*

$$l^- \le \text{2TLNNWBM}_{\omega}^{p,q}(l_1, l_2, \cdots, l_n) \le l^+ \tag{24}$$

From Property 1,

$$\text{2TLNNWBM}_{\omega}^{p,q}(l_1^-, l_2^-, \cdots, l_n^-) = l^-$$
$$\text{2TLNNWBM}_{\omega}^{p,q}(l_1^+, l_2^+, \cdots, l_n^+) = l^+$$

From Property 2,

$$l^- \le \text{2TLNNWBM}_{\omega}^{p,q}(l_1, l_2, \cdots, l_n) \le l^+$$

3.2. 2TLNNWGBM Operator

Similarly to WBM, to consider the attribute weights, the weighted geometric Bonferroni mean (WGBM) is defined, as follows:

Definition 9 ([51]). *Let* $p, q > 0$ *and* $b_i (i = 1, 2, \cdots, n)$ *be a collection of nonnegative crisp numbers with the weights vector being* $\omega = (\omega_1, \omega_2, \cdots \omega_n)^T$*, thereby satisfying* $\omega_i \in [0, 1]$ *and* $\sum_{i=1}^{n} \omega_i = 1$*. If*

$$\text{WGBM}_{\omega}^{p,q}(b_1, b_2, \cdots, b_n) = \frac{1}{p+q} \prod_{i,j=1}^{n} (pb_i + qb_j)^{\omega_i \omega_j} \tag{25}$$

Then we extend WGBM to fuse the 2TLNNs and propose 2-tuple linguistic neutrosophic number weighted geometric Bonferroni mean (2TLNNWGBM) operator.

Definition 10. *Let* $l_i = \langle (s_{T_i}, a_i), (s_{I_i}, b_i), (s_{F_i}, c_i) \rangle (i = 1, 2, \ldots, n)$ *be a set of 2TLNNs with their weight vector be* $w_i = (w_1, w_2, \ldots, w_n)^T$*, thereby satisfying* $w_i \in [0, 1]$ *and* $\sum_{i=1}^{n} w_i = 1$*. If*

$$\text{2TLNNWGBM}_{\omega}^{p,q}(l_1, l_2, \ldots, l_n) = \frac{1}{p+q} \underset{i,j=1}{\overset{n}{\otimes}} (pl_i \oplus ql_j)^{\omega_i \omega_j} \tag{26}$$

Then we called $\text{2TLNNWGBM}_{\omega}^{p,q}$ *the 2-tuple linguistic neutrosophic number weighted geometric BM.*

Theorem 2. *Let* $l_i = \langle (s_{T_i}, a_i), (s_{I_i}, b_i), (s_{F_i}, c_i) \rangle (i = 1, 2, \ldots, n)$ *be a set of 2TLNNs. The aggregated value by using 2TLNNWGBM operators is also a 2TLNN where*

$$\text{2TLNNWGBM}_{\omega}^{p,q}(l_1, l_2, \ldots, l_n) = \frac{1}{p+q} \underset{i,j=1}{\overset{n}{\otimes}} (pl_i \oplus ql_j)^{\omega_i \omega_j}$$
$$= \left\{ \begin{array}{l} \Delta\left(t\left(1-\left(1-\prod_{i,j=1}^{n}\left(1-\left(1-\frac{\Delta^{-1}(s_{T_i},a_i)}{t}\right)^p \cdot \left(1-\frac{\Delta^{-1}(s_{T_j},a_j)}{t}\right)^q\right)^{\omega_i\omega_j}\right)^{\frac{1}{p+q}}\right)\right), \\ \Delta\left(t\left(1-\prod_{i,j=1}^{n}\left(1-\left(\frac{\Delta^{-1}(s_{I_i},b_i)}{t}\right)^p \cdot \left(\frac{\Delta^{-1}(s_{I_j},b_j)}{t}\right)^q\right)^{\omega_i\omega_j}\right)^{\frac{1}{p+q}}\right), \\ \Delta\left(t\left(1-\prod_{i,j=1}^{n}\left(1-\left(\frac{\Delta^{-1}(s_{F_i},c_i)}{t}\right)^p \cdot \left(\frac{\Delta^{-1}(s_{F_j},c_j)}{t}\right)^q\right)^{\omega_i\omega_j}\right)^{\frac{1}{p+q}}\right) \end{array} \right\} \tag{27}$$

Proof. From Definition 5, we can obtain,

$$pl_i = \left\{ \Delta\left(t\left(1-\left(1-\frac{\Delta^{-1}(s_{T_i},a_i)}{t}\right)^p\right)\right), \Delta\left(t\left(\frac{\Delta^{-1}(s_{I_i},b_i)}{t}\right)^p\right), \Delta\left(t\left(\frac{\Delta^{-1}(s_{F_i},c_i)}{t}\right)^p\right) \right\} \tag{28}$$

$$ql_j = \left\{ \Delta\left(t\left(1-\left(1-\frac{\Delta^{-1}(s_{T_j},a_j)}{t}\right)^q\right)\right), \Delta\left(t\left(\frac{\Delta^{-1}(s_{I_j},b_j)}{t}\right)^q\right), \Delta\left(t\left(\frac{\Delta^{-1}(s_{F_j},c_j)}{t}\right)^q\right) \right\} \tag{29}$$

Thus,

$$pl_i \oplus ql_j = \left\{ \begin{array}{l} \Delta\left(t\left(1-(1-)^p\frac{\Delta^{-1}(s_{T_i},a_i)}{t}\cdot\left(1-\frac{\Delta^{-1}(s_{T_j},a_j)}{t}\right)^q\right)\right), \\ \Delta\left(t\left(\frac{\Delta^{-1}(s_{I_i},b_i)}{t}\right)^p\cdot\left(\frac{\Delta^{-1}(s_{I_j},b_j)}{t}\right)^q\right), \Delta\left(t\left(\frac{\Delta^{-1}(s_{F_i},c_i)}{t}\right)^p\cdot\left(\frac{\Delta^{-1}(s_{F_j},c_j)}{t}\right)^q\right) \end{array} \right\} \tag{30}$$

Therefore,

$$\begin{array}{l} (pl_i \oplus ql_j)^{\omega_i\omega_j} \\ = \left\{ \begin{array}{l} \Delta\left(t\left(1-\left(1-\frac{\Delta^{-1}(s_{T_i},a_i)}{t}\right)^p\cdot\left(1-\frac{\Delta^{-1}(s_{T_j},a_j)}{t}\right)^q\right)^{\omega_i\omega_j}\right), \\ \Delta\left(t\left(1-\left(1-\left(\frac{\Delta^{-1}(s_{I_i},b_i)}{t}\right)^p\cdot\left(\frac{\Delta^{-1}(s_{I_j},b_j)}{t}\right)^q\right)^{\omega_i\omega_j}\right)\right), \\ \Delta\left(t\left(1-\left(1-\left(\frac{\Delta^{-1}(s_{F_i},c_i)}{t}\right)^p\cdot\left(\frac{\Delta^{-1}(s_{F_j},c_j)}{t}\right)^q\right)^{\omega_i\omega_j}\right)\right) \end{array} \right\} \end{array} \tag{31}$$

Thereafter,

$$\begin{array}{l} \overset{n}{\underset{i,j=1}{\otimes}} (pl_i \oplus ql_j)^{\omega_i\omega_j} \\ = \left\{ \begin{array}{l} \Delta\left(t\prod_{i,j=1}^{n}\left(1-\left(1-\frac{\Delta^{-1}(s_{T_i},a_i)}{t}\right)^p\cdot\left(1-\frac{\Delta^{-1}(s_{T_j},a_j)}{t}\right)^q\right)^{\omega_i\omega_j}\right), \\ \Delta\left(t\left(1-\prod_{i,j=1}^{n}\left(1-\left(\frac{\Delta^{-1}(s_{I_i},b_i)}{t}\right)^p\cdot\left(\frac{\Delta^{-1}(s_{I_j},b_j)}{t}\right)^q\right)^{\omega_i\omega_j}\right)\right), \\ \Delta\left(t\left(1-\prod_{i,j=1}^{n}\left(1-\left(\frac{\Delta^{-1}(s_{F_i},c_i)}{t}\right)^p\cdot\left(\frac{\Delta^{-1}(s_{F_j},c_j)}{t}\right)^q\right)^{\omega_i\omega_j}\right)\right) \end{array} \right\} \end{array} \tag{32}$$

Furthermore,

$$\begin{array}{l} 2TLNNWGBM_{\omega}^{p,q}(l_1,l_2,\ldots,l_n) = \frac{1}{p+q}\overset{n}{\underset{i,j=1}{\otimes}} (pl_i \oplus ql_j)^{\omega_i\omega_j} \\ = \left\{ \begin{array}{l} \Delta\left(t\left(1-\left(1-\prod_{i,j=1}^{n}\left(1-\left(1-\frac{\Delta^{-1}(s_{T_i},a_i)}{t}\right)^p\cdot\left(1-\frac{\Delta^{-1}(s_{T_j},a_j)}{t}\right)^q\right)^{\omega_i\omega_j}\right)^{\frac{1}{p+q}}\right)\right), \\ \Delta\left(t\left(1-\prod_{i,j=1}^{n}\left(1-\left(\frac{\Delta^{-1}(s_{I_i},b_i)}{t}\right)^p\cdot\left(\frac{\Delta^{-1}(s_{I_j},b_j)}{t}\right)^q\right)^{\omega_i\omega_j}\right)^{\frac{1}{p+q}}\right), \\ \Delta\left(t\left(1-\prod_{i,j=1}^{n}\left(1-\left(\frac{\Delta^{-1}(s_{F_i},c_i)}{t}\right)^p\cdot\left(\frac{\Delta^{-1}(s_{F_j},c_j)}{t}\right)^q\right)^{\omega_i\omega_j}\right)^{\frac{1}{p+q}}\right) \end{array} \right\} \end{array} \tag{33}$$

Hence, (27) is kept. □

Then, we need to prove that (27) is a 2TLNN. We need to prove two conditions, as follows:

① $0 \leq \Delta^{-1}(s_T, a) \leq t, 0 \leq \Delta^{-1}(s_I, b) \leq t, 0 \leq \Delta^{-1}(s_F, c) \leq t$

② $0 \leq \Delta^{-1}(s_T, a) + \Delta^{-1}(s_I, b) + \Delta^{-1}(s_F, c) \leq 3t$

Let

$$\frac{\Delta^{-1}(s_T,a)}{t} = 1 - \left(1 - \prod_{i,j=1}^{n}\left(1 - \left(1 - \frac{\Delta^{-1}(s_{T_i},a_i)}{t}\right)^p \cdot \left(1 - \frac{\Delta^{-1}\left(s_{T_j},a_j\right)}{t}\right)^q\right)^{\omega_i\omega_j}\right)^{\frac{1}{p+q}}$$

$$\frac{\Delta^{-1}(s_I,b)}{t} = \left(1 - \prod_{i,j=1}^{n}\left(1 - \left(\frac{\Delta^{-1}(s_{I_i},b_i)}{t}\right)^p \cdot \left(\frac{\Delta^{-1}\left(s_{I_j},b_j\right)}{t}\right)^q\right)^{\omega_i\omega_j}\right)^{\frac{1}{p+q}}$$

$$\frac{\Delta^{-1}(s_F,c)}{t} = \left(1 - \prod_{i,j=1}^{n}\left(1 - \left(\frac{\Delta^{-1}(s_{F_i},c_i)}{t}\right)^p \cdot \left(\frac{\Delta^{-1}\left(s_{F_j},c_j\right)}{t}\right)^q\right)^{\omega_i\omega_j}\right)^{\frac{1}{p+q}}$$

Proof. ① Since $0 \leq \frac{\Delta^{-1}(s_{T_i},a_i)}{t} \leq 1, 0 \leq \frac{\Delta^{-1}\left(s_{T_j},a_j\right)}{t} \leq 1$ we get

$$0 \leq 1 - \left(1 - \frac{\Delta^{-1}(s_{T_i},a_i)}{t}\right)^p \cdot \left(1 - \frac{\Delta^{-1}\left(s_{T_j},a_j\right)}{t}\right)^q \leq 1 \tag{34}$$

Then,

$$0 \leq \prod_{i,j=1}^{n}\left(1 - \left(1 - \frac{\Delta^{-1}(s_{T_i},a_i)}{t}\right)^p \cdot \left(1 - \frac{\Delta^{-1}\left(s_{T_j},a_j\right)}{t}\right)^q\right)^{\omega_i\omega_j} \leq 1 \tag{35}$$

$$0 \leq 1 - \left(1 - \prod_{i,j=1}^{n}\left(1 - \left(1 - \frac{\Delta^{-1}(s_{T_i},a_i)}{t}\right)^p \cdot \left(1 - \frac{\Delta^{-1}\left(s_{T_j},a_j\right)}{t}\right)^q\right)^{\omega_i\omega_j}\right)^{\frac{1}{p+q}} \leq 1 \tag{36}$$

That means $0 \leq \Delta^{-1}(s_T, a) \leq t$, so ① is maintained, similarly, we can get $0 \leq \Delta^{-1}(s_I, b) \leq t, 0 \leq \Delta^{-1}(s_F, c) \leq t$. ② Since $0 \leq \Delta^{-1}(s_T, a) \leq t, 0 \leq \Delta^{-1}(s_I, b) \leq t, 0 \leq \Delta^{-1}(s_F, c) \leq t$, we get $0 \leq \Delta^{-1}(s_T, a) + \Delta^{-1}(s_I, b) + \Delta^{-1}(s_F, c) \leq 3t$. □

Example 2. *Let* $\langle (s_3, 0.4), (s_2, -0.3), (s_4, 0.1)\rangle, \langle (s_2, 0.3), (s_1, 0.2), (s_4, -0.1)\rangle$ *be two 2TLNNs,* $(p,q) = (2,3), \omega = (0.4, 0.6)$ *according to (27), we have*

$$
\begin{aligned}
&\text{2TLNNWGBM}^{(2,3)}_{(0.4,0.6)}\begin{pmatrix}\langle (s_3, 0.4), (s_2, -0.3), (s_4, 0.1)\rangle, \\ \langle (s_2, 0.3), (s_1, 0.2), (s_4, -0.1)\rangle\end{pmatrix} \\
&= \left\{\begin{array}{l}
\Delta\left(t\left(1-\left(1-\prod\limits_{i,j=1}^{n}\left(1-\left(1-\frac{\Delta^{-1}(s_{T_i},a_i)}{t}\right)^{p}\cdot\left(1-\frac{\Delta^{-1}(s_{T_j},a_j)}{t}\right)^{q}\right)^{\omega_i\omega_j}\right)^{\frac{1}{p+q}}\right)\right), \\
\Delta\left(t\left(1-\prod\limits_{i,j=1}^{n}\left(1-\left(\frac{\Delta^{-1}(s_{I_i},b_i)}{t}\right)^{p}\cdot\left(\frac{\Delta^{-1}(s_{I_j},b_j)}{t}\right)^{q}\right)^{\omega_i\omega_j}\right)^{\frac{1}{p+q}}\right), \\
\Delta\left(t\left(1-\prod\limits_{i,j=1}^{n}\left(1-\left(\frac{\Delta^{-1}(s_{F_i},c_i)}{t}\right)^{p}\cdot\left(\frac{\Delta^{-1}(s_{F_j},c_j)}{t}\right)^{q}\right)^{\omega_i\omega_j}\right)^{\frac{1}{p+q}}\right)
\end{array}\right\} \\
&= \left\{\begin{array}{l}
\Delta\left(6\times\left(1-\left(\begin{array}{l}1-\left(1-\left(1-\frac{3.4}{6}\right)^2\times\left(1-\frac{3.4}{6}\right)^3\right)^{0.4\times0.4}\times\left(1-\left(1-\frac{3.4}{6}\right)^2\times\left(1-\frac{2.3}{6}\right)^3\right)^{0.4\times0.6} \\ \times\left(1-\left(1-\frac{2.3}{6}\right)^2\times\left(1-\frac{3.4}{6}\right)^3\right)^{0.6\times0.4}\times\left(1-\left(1-\frac{2.3}{6}\right)^2\times\left(1-\frac{2.3}{6}\right)^3\right)^{0.6\times0.6}\end{array}\right)^{\frac{1}{2+3}}\right)\right), \\
\Delta\left(6\times\left(1-\left(\begin{array}{l}\left(1-\left(\frac{1.7}{6}\right)^2\times\left(\frac{1.7}{6}\right)^3\right)^{0.4\times0.4}\times\left(1-\left(\frac{1.7}{6}\right)^2\times\left(\frac{1.2}{6}\right)^3\right)^{0.4\times0.6} \\ \times\left(1-\left(\frac{1.2}{6}\right)^2\times\left(\frac{1.7}{6}\right)^3\right)^{0.6\times0.4}\times\left(1-\left(\frac{1.2}{6}\right)^2\times\left(\frac{1.2}{6}\right)^3\right)^{0.6\times0.6}\end{array}\right)\right)^{\frac{1}{2+3}}\right), \\
\Delta\left(6\times\left(1-\left(\begin{array}{l}\left(1-\left(\frac{4.1}{6}\right)^2\times\left(\frac{4.1}{6}\right)^3\right)^{0.4\times0.4}\times\left(1-\left(\frac{4.1}{6}\right)^2\times\left(\frac{3.9}{6}\right)^3\right)^{0.4\times0.6} \\ \times\left(1-\left(\frac{3.9}{6}\right)^2\times\left(\frac{4.1}{6}\right)^3\right)^{0.6\times0.4}\times\left(1-\left(\frac{3.9}{6}\right)^2\times\left(\frac{3.9}{6}\right)^3\right)^{0.6\times0.6}\end{array}\right)\right)^{\frac{1}{2+3}}\right)
\end{array}\right\} \\
&= \langle (s_3, -0.334), (s_1, 0.434), (s_4, -0.018)\rangle
\end{aligned}
$$

Similar to 2TLNNWBM, the 2TLNNWGBM has the same properties, as follows. The proof are omitted here to save space.

Property 4. *(Idempotency) If* $l_i = \langle (s_{T_i}, a_i), (s_{I_i}, b_i), (s_{F_i}, c_i)\rangle (i = 1, 2, \ldots, n)$ *are equal, then*

$$\text{2TLNNWGBM}^{p,q}_{\omega}(l_1, l_2, \cdots, l_n) = l \tag{37}$$

Property 5. *(Monotonicity) Let* $l_{x_i} = \left\langle \left(s_{T_{x_i}}, a_{x_i}\right), \left(s_{I_{x_i}}, b_{x_i}\right), \left(s_{F_{x_i}}, c_{x_i}\right)\right\rangle (i = 1, 2, \ldots, n)$ *and* $l_{y_i} = \left\langle \left(s_{T_{y_i}}, a_{y_i}\right), \left(s_{I_{y_i}}, b_{y_i}\right), \left(s_{F_{y_i}}, c_{y_i}\right)\right\rangle (i = 1, 2, \ldots, n)$ *be two sets of 2TLNNs. If* $\Delta^{-1}\left(s_{T_{x_i}}, a_{x_i}\right) \leq \Delta^{-1}\left(s_{T_{y_i}}, a_{y_i}\right), \Delta^{-1}\left(s_{I_{x_i}}, b_{x_i}\right) \geq \Delta^{-1}\left(s_{I_{y_i}}, b_{y_i}\right)$ *and* $\Delta^{-1}\left(s_{F_{x_i}}, c_{x_i}\right) \geq \Delta^{-1}\left(s_{F_{y_i}}, c_{y_i}\right)$ *hold for all i, then*

$$\text{2TLNNWGBM}^{p,q}_{\omega}(l_{x_1}, l_{x_2}, \cdots, l_{x_n}) \leq \text{2TLNNWGBM}^{p,q}_{\omega}(l_{y_1}, l_{y_2}, \cdots, l_{y_n}) \tag{38}$$

Property 6. *(Boundedness) Let* $l_i = \langle (s_{T_i}, a_i), (s_{I_i}, b_i), (s_{F_i}, c_i)\rangle (i = 1, 2, \ldots, n)$ *be a set of 2TLNNs. If* $l^+ = (\max_i(S_{T_i}, a_i), \min_i(S_{I_i}, b_i), \min_i(S_{F_i}, c_i))$ *and* $l^- = (\min_i(S_{T_i}, a_i), \max_i(S_{I_i}, b_i), \max_i(S_{F_i}, c_i))$ *then*

$$l^- \leq \text{2TLNNWGBM}^{p,q}_{\omega}(l_1, l_2, \cdots, l_n) \leq l^+ \tag{39}$$

4. G2TLNNWBM and G2TLNNWGBM Operators

4.1. G2TLNNWBM Operator

The primary advantage of BM is that it can determine the interrelationship between arguments. However, the traditional BM can only consider the correlations of any two aggregated arguments. Thereafter, Beliakov et al. [43] extended the BM and introduced the generalized BM(GBM) operator. Zhu et al. [51] introduced the generalized weighted BM(GWBM) operator, as follows.

Definition 11 ([51]). *Let $p, q, r > 0$ and $b_i (i = 1, 2, \cdots, n)$ be a collection of nonnegative crisp numbers with the weights vector being $\omega = (\omega_1, \omega_2, \cdots \omega_n)^T$, thereby satisfying $\omega_i \in [0, 1]$ and $\sum_{i=1}^{n} \omega_i = 1$. The generalized weighted Bonferroni mean (GWBM) is defined as follows:*

$$\mathrm{GWBM}_{\omega}^{p,q,r}(b_1, b_2, \cdots, b_n) = \left(\sum_{i,j,k=1}^{n} \omega_i \omega_j \omega_k b_i^p b_j^q b_k^r \right)^{1/(p+q+r)} \tag{40}$$

Then we extend GWBM to fuse the 2TLNNs and propose generalized 2-tuple linguistic neutrosophic number weighted Bonferroni mean (G2TLNNWBM) aggregation operator.

Definition 12. *Let $l_i = \langle (s_{T_i}, a_i), (s_{I_i}, b_i), (s_{F_i}, c_i) \rangle$ be a set of 2TLNNs. The generalized 2-tuple linguistic neutrosophic number weighted Bonferroni mean (G2TLNNWBM) operator is:*

$$\mathrm{G2TLNNWBM}_{\omega}^{p,q,r}(l_1, l_2, \ldots, l_n) = \left(\underset{i,j,k=1}{\overset{n}{\oplus}} \left(\omega_i \omega_j \omega_k \left(l_i^p \otimes l_j^q \otimes l_k^r \right) \right) \right)^{1/(p+q+r)} \tag{41}$$

Theorem 3. *Let $l_i = \langle (s_{T_i}, a_i), (s_{I_i}, b_i), (s_{F_i}, c_i) \rangle$ be a set of 2TLNNs. The aggregated value by using G2TLNNWBM operators is also a 2TLNN where*

$$\begin{aligned}
&\mathrm{G2TLNNWBM}_{\omega}^{p,q,r}(l_1, l_2, \ldots, l_n) = \left(\underset{i,j,k=1}{\overset{n}{\oplus}} \left(\omega_i \omega_j \omega_k \left(l_i^p \otimes l_j^q \otimes l_k^r \right) \right) \right)^{1/(p+q+r)} \\
&= \left\{ \begin{array}{l}
\Delta\left(t \left(1 - \prod_{i,j,k=1}^{n} \left(1 - \left(\frac{\Delta^{-1}(s_{T_i}, a_i)}{t} \right)^p \cdot \left(\frac{\Delta^{-1}(s_{T_j}, a_j)}{t} \right)^q \cdot \left(\frac{\Delta^{-1}(s_{T_k}, a_k)}{t} \right)^r \right)^{\omega_i \omega_j \omega_k} \right)^{1/(p+q+r)} \right), \\
\Delta\left(t \left(1 - \left(1 - \prod_{i,j,k=1}^{n} \left(1 - \left(1 - \frac{\Delta^{-1}(s_{I_i}, b_i)}{t} \right)^p \cdot \left(1 - \frac{\Delta^{-1}(s_{I_j}, b_j)}{t} \right) \cdot \left(1 - \frac{\Delta^{-1}(s_{I_k}, b_k)}{t} \right)^r \right)^{\omega_i \omega_j \omega_k} \right)^{1/(p+q+r)} \right) \right), \\
\Delta\left(t \left(1 - \left(1 - \prod_{i,j,k=1}^{n} \left(1 - \left(1 - \frac{\Delta^{-1}(s_{F_i}, c_i)}{t} \right)^p \cdot \left(1 - \frac{\Delta^{-1}(s_{F_j}, c_j)}{t} \right)^q \cdot \left(1 - \frac{\Delta^{-1}(s_{F_k}, c_k)}{t} \right)^r \right)^{\omega_i \omega_j \omega_k} \right)^{1/(p+q+r)} \right) \right)
\end{array} \right\}
\end{aligned} \tag{42}$$

Proof. According to Definition 5, we can obtain

$$l_i^p = \left\{ \Delta\left(t \left(\frac{\Delta^{-1}(s_{T_i}, a_i)}{t} \right)^p \right), \Delta\left(t \left(1 - \left(1 - \frac{\Delta^{-1}(s_{I_i}, b_i)}{t} \right)^p \right) \right), \Delta\left(t \left(1 - \left(1 - \frac{\Delta^{-1}(s_{F_i}, c_i)}{t} \right)^p \right) \right) \right\} \tag{43}$$

$$l_j^q = \left\{ \Delta\left(t \left(\frac{\Delta^{-1}(s_{T_j}, a_j)}{t} \right)^q \right), \Delta\left(t \left(1 - \left(1 - \frac{\Delta^{-1}(s_{I_j}, b_j)}{t} \right)^q \right) \right), \Delta\left(t \left(1 - \left(1 - \frac{\Delta^{-1}(s_{F_j}, c_j)}{t} \right)^q \right) \right) \right\} \tag{44}$$

$$l_k^r = \left\{ \Delta\left(t \left(\frac{\Delta^{-1}(s_{T_k}, a_k)}{t} \right)^r \right), \Delta\left(t \left(1 - \left(1 - \frac{\Delta^{-1}(s_{I_k}, b_k)}{t} \right)^r \right) \right), \Delta\left(t \left(1 - \left(1 - \frac{\Delta^{-1}(s_{F_k}, c_k)}{t} \right)^r \right) \right) \right\} \tag{45}$$

Thus,

$$l_i^p \otimes l_j^q \otimes l_k^r = \left\{ \begin{array}{l}
\Delta\left(t \left(\left(\frac{\Delta^{-1}(s_{T_i}, a_i)}{t} \right)^p \cdot \left(\frac{\Delta^{-1}(s_{T_j}, a_j)}{t} \right)^q \cdot \left(\frac{\Delta^{-1}(s_{T_k}, a_k)}{t} \right)^r \right) \right), \\
\Delta\left(t \left(1 - \left(1 - \frac{\Delta^{-1}(s_{I_i}, b_i)}{t} \right)^p \cdot \left(1 - \frac{\Delta^{-1}(s_{I_j}, b_j)}{t} \right)^q \cdot \left(1 - \frac{\Delta^{-1}(s_{I_k}, b_k)}{t} \right)^r \right) \right), \\
\Delta\left(t \left(1 - \left(1 - \frac{\Delta^{-1}(s_{F_i}, c_i)}{t} \right)^p \cdot \left(1 - \frac{\Delta^{-1}(s_{F_j}, c_j)}{t} \right) \cdot \left(1 - \frac{\Delta^{-1}(s_{F_k}, c_k)}{t} \right)^r \right) \right)
\end{array} \right\} \tag{46}$$

Thereafter,

$$
\omega_i\omega_j\omega_k\left(l_i^p \otimes l_j^q \otimes l_k^r\right)
= \left\{ \begin{array}{l}
\Delta\left(t\left(1-\left(1-\left(\frac{\Delta^{-1}\left(s_{T_i},a_i\right)}{t}\right)^p \cdot \left(\frac{\Delta^{-1}\left(s_{T_j},a_j\right)}{t}\right)^q \cdot \left(\frac{\Delta^{-1}\left(s_{T_k},a_k\right)}{t}\right)^r\right)^{\omega_i\omega_j\omega_k}\right)\right), \\
\Delta\left(t\left(1-\left(1-\frac{\Delta^{-1}\left(s_{I_i},b_i\right)}{t}\right)^p \cdot \left(1-\frac{\Delta^{-1}\left(s_{I_j},b_j\right)}{t}\right)^q \cdot \left(1-\frac{\Delta^{-1}\left(s_{I_k},b_k\right)}{t}\right)^r\right)^{\omega_i\omega_j\omega_k}\right), \\
\Delta\left(t\left(1-\left(1-\frac{\Delta^{-1}\left(s_{F_i},c_i\right)}{t}\right)^p \cdot \left(1-\frac{\Delta^{-1}\left(s_{F_j},c_j\right)}{t}\right)^q \cdot \left(1-\frac{\Delta^{-1}\left(s_{F_k},c_k\right)}{t}\right)^r\right)^{\omega_i\omega_j\omega_k}\right)
\end{array} \right\} \tag{47}
$$

Furthermore,

$$
\underset{i,j,k=1}{\overset{n}{\oplus}}\left(\omega_i\omega_j\omega_k\left(l_i^p \otimes l_j^q \otimes l_k^r\right)\right)
= \left\{ \begin{array}{l}
\Delta\left(t\left(1-\prod_{i,j,k=1}^{n}\left(1-\left(\frac{\Delta^{-1}\left(s_{T_i},a_i\right)}{t}\right)^p \cdot \left(\frac{\Delta^{-1}\left(s_{T_j},a_j\right)}{t}\right)^q \cdot \left(\frac{\Delta^{-1}\left(s_{T_k},a_k\right)}{t}\right)^r\right)^{\omega_i\omega_j\omega_k}\right)\right), \\
\Delta\left(t\prod_{i,j,k=1}^{n}\left(1-\left(1-\frac{\Delta^{-1}\left(s_{I_i},b_i\right)}{t}\right)^p \cdot \left(1-\frac{\Delta^{-1}\left(s_{I_j},b_j\right)}{t}\right)^q \cdot \left(1-\frac{\Delta^{-1}\left(s_{I_k},b_k\right)}{t}\right)^r\right)^{\omega_i\omega_j\omega_k}\right), \\
\Delta\left(t\prod_{i,j,k=1}^{n}\left(1-\left(1-\frac{\Delta^{-1}\left(s_{F_i},c_i\right)}{t}\right)^p \cdot \left(1-\frac{\Delta^{-1}\left(s_{F_j},c_j\right)}{t}\right)^q \cdot \left(1-\frac{\Delta^{-1}\left(s_{F_k},c_k\right)}{t}\right)^r\right)^{\omega_i\omega_j\omega_k}\right)
\end{array} \right\} \tag{48}
$$

Therefore,

$$
\mathrm{G2TLNNWBM}_{\omega}^{p,q,r}(l_1,l_2,\ldots,l_n)=\left(\underset{i,j,k=1}{\overset{n}{\oplus}}\left(\omega_i\omega_j\omega_k\left(l_i^p \otimes l_j^q \otimes l_k^r\right)\right)\right)^{1/(p+q+r)}
$$

$$
= \left\{ \begin{array}{l}
\Delta\left(t\left(1-\prod_{i,j,k=1}^{n}\left(1-\left(\frac{\Delta^{-1}\left(s_{T_i},a_i\right)}{t}\right)^p \cdot \left(\frac{\Delta^{-1}\left(s_{T_j},a_j\right)}{t}\right)^q \cdot \left(\frac{\Delta^{-1}\left(s_{T_k},a_k\right)}{t}\right)^r\right)^{\omega_i\omega_j\omega_k}\right)^{1/(p+q+r)}\right), \\
\Delta\left(t\left(1-\left(1-\prod_{i,j,k=1}^{n}\left(1-\left(1-\frac{\Delta^{-1}\left(s_{I_i},b_i\right)}{t}\right)^p \cdot \left(1-\frac{\Delta^{-1}\left(s_{I_j},b_j\right)}{t}\right)^q \cdot \left(1-\frac{\Delta^{-1}\left(s_{I_k},b_k\right)}{t}\right)^r\right)^{\omega_i\omega_j\omega_k}\right)^{1/(p+q+r)}\right)\right), \\
\Delta\left(t\left(1-\left(1-\prod_{i,j,k=1}^{n}\left(1-\left(1-\frac{\Delta^{-1}\left(s_{F_i},c_i\right)}{t}\right)^p \cdot \left(1-\frac{\Delta^{-1}\left(s_{F_j},c_j\right)}{t}\right)^q \cdot \left(1-\frac{\Delta^{-1}\left(s_{F_k},c_k\right)}{t}\right)^r\right)^{\omega_i\omega_j\omega_k}\right)^{1/(p+q+r)}\right)\right)
\end{array} \right\} \tag{49}
$$

Hence, (42) is kept. □

Then, we need to prove that (42) is a 2TLNN. We need to prove two conditions as follows:

① $0 \le \Delta^{-1}(s_T,a) \le t, 0 \le \Delta^{-1}(s_I,b) \le t, 0 \le \Delta^{-1}(s_F,c) \le t$
② $0 \le \Delta^{-1}(s_T,a) + \Delta^{-1}(s_I,b) + \Delta^{-1}(s_F,c) \le 3t$

Let

$$
\frac{\Delta^{-1}(s_T,a)}{t}=\left(1-\prod_{i,j,k=1}^{n}\left(1-\left(\frac{\Delta^{-1}\left(s_{T_i},a_i\right)}{t}\right)^p \cdot \left(\frac{\Delta^{-1}\left(s_{T_j},a_j\right)}{t}\right)^q \cdot \left(\frac{\Delta^{-1}\left(s_{T_k},a_k\right)}{t}\right)^r\right)^{\omega_i\omega_j\omega_k}\right)^{1/(p+q+r)}
$$

$$
\frac{\Delta^{-1}(s_I,b)}{t}=1-\left(1-\prod_{i,j,k=1}^{n}\left(1-\left(1-\frac{\Delta^{-1}\left(s_{I_i},b_i\right)}{t}\right)^p \cdot \left(1-\frac{\Delta^{-1}\left(s_{I_j},b_j\right)}{t}\right)^q \cdot \left(1-\frac{\Delta^{-1}\left(s_{I_k},b_k\right)}{t}\right)^r\right)^{\omega_i\omega_j\omega_k}\right)^{1/(p+q+r)}
$$

$$
\frac{\Delta^{-1}(s_F,c)}{t}=1-\left(1-\prod_{i,j,k=1}^{n}\left(1-\left(1-\frac{\Delta^{-1}\left(s_{F_i},c_i\right)}{t}\right)^p \cdot \left(1-\frac{\Delta^{-1}\left(s_{F_j},c_j\right)}{t}\right)^q \cdot \left(1-\frac{\Delta^{-1}\left(s_{F_k},c_k\right)}{t}\right)^r\right)^{\omega_i\omega_j\omega_k}\right)^{1/(p+q+r)}
$$

Proof. ① Since $0 \le \frac{\Delta^{-1}\left(s_{T_i}, a_i\right)}{t} \le 1, 0 \le \frac{\Delta^{-1}\left(s_{T_j}, a_j\right)}{t} \le 1, 0 \le \frac{\Delta^{-1}\left(s_{T_k}, a_k\right)}{t} \le 1$ we get

$$0 \le 1 - \left(\frac{\Delta^{-1}\left(s_{T_i}, a_i\right)}{t}\right)^p \cdot \left(\frac{\Delta^{-1}\left(s_{T_j}, a_j\right)}{t}\right)^q \cdot \left(\frac{\Delta^{-1}\left(s_{T_k}, a_k\right)}{t}\right)^r \le 1 \tag{50}$$

Then,

$$0 \le 1 - \prod_{i,j,k=1}^{n}\left(1 - \left(\frac{\Delta^{-1}\left(s_{T_i}, a_i\right)}{t}\right)^p \cdot \left(\frac{\Delta^{-1}\left(s_{T_j}, a_j\right)}{t}\right)^q \cdot \left(\frac{\Delta^{-1}\left(s_{T_k}, a_k\right)}{t}\right)^r\right)^{\omega_i\omega_j\omega_k} \le 1 \tag{51}$$

$$0 \le \left(1 - \prod_{i,j,k=1}^{n}\left(1 - \left(\frac{\Delta^{-1}\left(s_{T_i}, a_i\right)}{t}\right)^p \cdot \left(\frac{\Delta^{-1}\left(s_{T_j}, a_j\right)}{t}\right)^q \cdot \left(\frac{\Delta^{-1}\left(s_{T_k}, a_k\right)}{t}\right)^r\right)^{\omega_i\omega_j\omega_k}\right)^{1/(p+q+r)} \le 1 \tag{52}$$

That means $0 \le \Delta^{-1}(s_T, a) \le t$, so ① is maintained, similarly, we can get $0 \le \Delta^{-1}(s_I, b) \le t, 0 \le \Delta^{-1}(s_F, c) \le t$. ② Since $0 \le \Delta^{-1}(s_T, a) \le t, 0 \le \Delta^{-1}(s_I, b) \le t, 0 \le \Delta^{-1}(s_F, c) \le t$, we get $0 \le \Delta^{-1}(s_T, a) + \Delta^{-1}(s_I, b) + \Delta^{-1}(s_F, c) \le 3t$. □

Then we will discuss some properties of G2TLNNWBM operator.

Property 7. *(Idempotency) If $l_i = \left\langle \left(s_{T_i}, a_i\right), \left(s_{I_i}, b_i\right), \left(s_{F_i}, c_i\right) \right\rangle (i = 1, 2, \ldots, n)$ are equal, then*

$$\text{G2TLNNWBM}_{\omega}^{p,q,r}(l_1, l_2, \cdots, l_n) = l \tag{53}$$

Proof. Since $l_i = l = \langle (s_T, a), (s_I, b), (s_F, c) \rangle$, then

$$\text{G2TLNNWBM}_{\omega}^{p,q,r}(l_1, l_2, \ldots, l_n) = \left(\underset{i,j,k=1}{\overset{n}{\oplus}}\left(\omega_i\omega_j\omega_k\left(l_i^p \otimes l_j^q \otimes l_k^r\right)\right)\right)^{1/(p+q+r)}$$

$$= \left\{\begin{array}{l} \Delta\left(t\left(1 - \prod_{i,j,k=1}^{n}\left(1 - \left(\frac{\Delta^{-1}\left(s_{T_i}, a_i\right)}{t}\right)^p \cdot \left(\frac{\Delta^{-1}\left(s_{T_j}, a_j\right)}{t}\right)^q \cdot \left(\frac{\Delta^{-1}\left(s_{T_k}, a_k\right)}{t}\right)^r\right)^{\omega_i\omega_j\omega_k}\right)^{1/(p+q+r)}\right), \\ \Delta\left(t\left(1 - \left(1 - \prod_{i,j,k=1}^{n}\left(1 - \left(1 - \frac{\Delta^{-1}\left(s_{I_i}, b_i\right)}{t}\right)^p \cdot \left(1 - \frac{\Delta^{-1}\left(s_{I_j}, b_j\right)}{t}\right)^q \cdot \left(1 - \frac{\Delta^{-1}\left(s_{I_k}, b_k\right)}{t}\right)^r\right)^{\omega_i\omega_j\omega_k}\right)^{1/(p+q+r)}\right)\right), \\ \Delta\left(t\left(1 - \left(1 - \prod_{i,j,k=1}^{n}\left(1 - \left(1 - \frac{\Delta^{-1}\left(s_{F_i}, c_i\right)}{t}\right)^p \cdot \left(1 - \frac{\Delta^{-1}\left(s_{F_j}, c_j\right)}{t}\right)^q \cdot \left(1 - \frac{\Delta^{-1}\left(s_{F_k}, c_k\right)}{t}\right)^r\right)^{\omega_i\omega_j\omega_k}\right)^{1/(p+q+r)}\right)\right) \end{array}\right\}$$

$$= \left\{\begin{array}{l} \Delta\left(t\left(1 - \left(1 - \left(\frac{\Delta^{-1}(s_T, a)}{t}\right)^{p+q+r}\right)^{\sum_{i=1}^{n}\omega_i\sum_{j=1}^{n}\omega_j\sum_{k=1}^{n}\omega_k}\right)^{1/(p+q+r)}\right), \\ \Delta\left(t\left(1 - \left(1 - \left(1 - \left(1 - \frac{\Delta^{-1}(s_I, b)}{t}\right)^{p+q+r}\right)^{\sum_{i=1}^{n}\omega_i\sum_{j=1}^{n}\omega_j\sum_{k=1}^{n}\omega_k}\right)^{1/(p+q+r)}\right)\right), \\ \Delta\left(t\left(1 - \left(1 - \left(1 - \left(1 - \frac{\Delta^{-1}(s_F, c)}{t}\right)^{p+q+r}\right)^{\sum_{i=1}^{n}\omega_i\sum_{j=1}^{n}\omega_j\sum_{k=1}^{n}\omega_k}\right)^{1/(p+q+r)}\right)\right) \end{array}\right\}$$

$$= \langle (s_T, a), (s_I, b), (s_F, c) \rangle = l$$

□

Property 8. *(Monotonicity) Let* $l_{x_i} = \left\langle \left(s_{T_{x_i}}, a_{x_i}\right), \left(s_{I_{x_i}}, b_{x_i}\right), \left(s_{F_{x_i}}, c_{x_i}\right) \right\rangle (i = 1, 2, \ldots, n)$ *and* $l_{y_i} = \left\langle \left(s_{T_{y_i}}, a_{y_i}\right), \left(s_{I_{y_i}}, b_{y_i}\right), \left(s_{F_{y_i}}, c_{y_i}\right) \right\rangle (i = 1, 2, \ldots, n)$ *be two sets of 2TLNNs. If* $\Delta^{-1}\left(s_{T_{x_i}}, a_{x_i}\right) \le \Delta^{-1}\left(s_{T_{y_i}}, a_{y_i}\right)$,$\Delta^{-1}\left(s_{I_{x_i}}, b_{x_i}\right) \ge \Delta^{-1}\left(s_{I_{y_i}}, b_{y_i}\right)$ *and* $\Delta^{-1}\left(s_{F_{x_i}}, c_{x_i}\right) \ge \Delta^{-1}\left(s_{F_{y_i}}, c_{y_i}\right)$ *hold for all i, then*

$$\text{G2TLNNWBM}_{\omega}^{p,q,r}(l_{x_1}, l_{x_2}, \cdots, l_{x_n}) \le \text{G2TLNNWBM}_{\omega}^{p,q,r}(l_{y_1}, l_{y_2}, \cdots, l_{y_n}) \tag{54}$$

Proof. Let $\text{G2TLNNWBM}_{\omega}^{p,q,r}(l_{x_1}, l_{x_2}, \cdots, l_{x_n}) = \left\langle \left(s_{T_{x_i}}, a_{x_i}\right), \left(s_{I_{x_i}}, b_{x_i}\right), \left(s_{F_{x_i}}, c_{x_i}\right) \right\rangle (i = 1, 2, \ldots, n)$ and $\text{G2TLNNWBM}_{\omega}^{p,q,r}(l_{y_1}, l_{y_2}, \cdots, l_{y_n}) = \left\langle \left(s_{T_{y_i}}, a_{y_i}\right), \left(s_{I_{y_i}}, b_{y_i}\right), \left(s_{F_{y_i}}, c_{y_i}\right) \right\rangle (i = 1, 2, \ldots, n)$, given that $\Delta^{-1}\left(s_{T_{x_i}}, a_{x_i}\right) \le \Delta^{-1}\left(s_{T_{y_i}}, a_{y_i}\right)$, we can obtain

$$\begin{aligned} &\left(\frac{\Delta^{-1}\left(s_{T_{x_i}}, a_{x_i}\right)}{t}\right)^p \cdot \left(\frac{\Delta^{-1}\left(s_{T_{x_j}}, a_{x_j}\right)}{t}\right)^q \cdot \left(\frac{\Delta^{-1}\left(s_{T_{x_k}}, a_{x_k}\right)}{t}\right)^r \\ &\le \left(\frac{\Delta^{-1}\left(s_{T_{y_i}}, a_{y_i}\right)}{t}\right)^p \cdot \left(\frac{\Delta^{-1}\left(s_{T_{y_j}}, a_{y_j}\right)}{t}\right)^q \cdot \left(\frac{\Delta^{-1}\left(s_{T_{y_k}}, a_{y_k}\right)}{t}\right)^r \end{aligned} \tag{55}$$

$$\begin{aligned} &\left(1 - \left(\frac{\Delta^{-1}\left(s_{T_{x_i}}, a_{x_i}\right)}{t}\right)^p \cdot \left(\frac{\Delta^{-1}\left(s_{T_{x_j}}, a_{x_j}\right)}{t}\right)^q \cdot \left(\frac{\Delta^{-1}\left(s_{T_{x_k}}, a_{x_k}\right)}{t}\right)^r\right)^{\omega_i \omega_j \omega_k} \\ &\ge \left(1 - \left(\frac{\Delta^{-1}\left(s_{T_{y_i}}, a_{y_i}\right)}{t}\right)^p \cdot \left(\frac{\Delta^{-1}\left(s_{T_{y_j}}, a_{y_j}\right)}{t}\right)^q \cdot \left(\frac{\Delta^{-1}\left(s_{T_{y_k}}, a_{y_k}\right)}{t}\right)^r\right)^{\omega_i \omega_j \omega_k} \end{aligned} \tag{56}$$

Thereafter,

$$\begin{aligned} &1 - \prod_{i,j,k=1}^{n} \left(1 - \left(\frac{\Delta^{-1}\left(s_{T_{x_i}}, a_{x_i}\right)}{t}\right)^p \cdot \left(\frac{\Delta^{-1}\left(s_{T_{x_j}}, a_{x_j}\right)}{t}\right)^q \cdot \left(\frac{\Delta^{-1}\left(s_{T_{x_k}}, a_{x_k}\right)}{t}\right)^r\right)^{\omega_i \omega_j \omega_k} \\ &\le 1 - \prod_{i,j,k=1}^{n} \left(1 - \left(\frac{\Delta^{-1}\left(s_{T_{y_i}}, a_{y_i}\right)}{t}\right)^p \cdot \left(\frac{\Delta^{-1}\left(s_{T_{y_j}}, a_{y_j}\right)}{t}\right)^q \cdot \left(\frac{\Delta^{-1}\left(s_{T_{y_k}}, a_{y_k}\right)}{t}\right)^r\right)^{\omega_i \omega_j \omega_k} \end{aligned} \tag{57}$$

Furthermore,

$$\begin{aligned} &t\left(1 - \prod_{i,j,k=1}^{n} \left(1 - \left(\frac{\Delta^{-1}\left(s_{T_{x_i}}, a_{x_i}\right)}{t}\right)^p \cdot \left(\frac{\Delta^{-1}\left(s_{T_{x_j}}, a_{x_j}\right)}{t}\right)^q \cdot \left(\frac{\Delta^{-1}\left(s_{T_{x_k}}, a_{x_k}\right)}{t}\right)^r\right)^{\omega_i \omega_j \omega_k}\right)^{1/(p+q+r)} \\ &\le t\left(1 - \prod_{i,j,k=1}^{n} \left(1 - \left(\frac{\Delta^{-1}\left(s_{T_{y_i}}, a_{y_i}\right)}{t}\right)^p \cdot \left(\frac{\Delta^{-1}\left(s_{T_{y_j}}, a_{y_j}\right)}{t}\right)^q \cdot \left(\frac{\Delta^{-1}\left(s_{T_{y_k}}, a_{y_k}\right)}{t}\right)^r\right)^{\omega_i \omega_j \omega_k}\right)^{1/(p+q+r)} \end{aligned} \tag{58}$$

That means $\Delta^{-1}(s_{T_x}, a_x) \le \Delta^{-1}\left(s_{T_y}, a_y\right)$. Similarly, we can obtain $\Delta^{-1}(s_{I_x}, b_x) \ge \Delta^{-1}\left(s_{I_y}, b_y\right)$ and $\Delta^{-1}(s_{F_x}, c_x) \ge \Delta^{-1}\left(s_{F_y}, c_y\right)$.

If $\Delta^{-1}(s_{T_x}, a_x) < \Delta^{-1}\left(s_{T_y}, a_y\right)$ and $\Delta^{-1}(s_{I_x}, b_x) \ge \Delta^{-1}\left(s_{I_y}, b_y\right)$ and $\Delta^{-1}(s_{F_x}, c_x) \ge \Delta^{-1}\left(s_{F_y}, c_y\right)$

$$\text{G2TLNNWBM}_{\omega}^{p,q,r}(l_{x_1}, l_{x_2}, \cdots, l_{x_n}) < \text{G2LNNWBM}_{\omega}^{p,q,r}(l_{y_1}, l_{y_2}, \cdots, l_{y_n})$$

If $\Delta^{-1}(s_{T_x}, a_x) = \Delta^{-1}\left(s_{T_y}, a_y\right)$ and $\Delta^{-1}(s_{I_x}, b_x) > \Delta^{-1}\left(s_{I_y}, b_y\right)$ and $\Delta^{-1}(s_{F_x}, c_x) > \Delta^{-1}\left(s_{F_y}, c_y\right)$

$$\text{G2TLNNWBM}_{\omega}^{p,q,r}(l_{x_1}, l_{x_2}, \cdots, l_{x_n}) < \text{DG2LNNWBM}_{\omega}^{p,q,r}(l_{y_1}, l_{y_2}, \cdots, l_{y_n})$$

If $\Delta^{-1}(s_{T_x}, a_x) = \Delta^{-1}\left(s_{T_y}, a_y\right)$ and $\Delta^{-1}(s_{I_x}, b_x) = \Delta^{-1}\left(s_{I_y}, b_y\right)$ and $\Delta^{-1}(s_{F_x}, c_x) = \Delta^{-1}\left(s_{F_y}, c_y\right)$

$$\text{G2TLNNWBM}_{\omega}^{p,q,r}(l_{x_1}, l_{x_2}, \cdots, l_{x_n}) = \text{G2TLNNWBM}_{\omega}^{p,q,r}(l_{y_1}, l_{y_2}, \cdots, l_{y_n})$$

So Property 8 is right. □

Property 9. *(Boundedness) Let* $l_i = \langle (s_{T_i}, a_i), (s_{I_i}, b_i), (s_{F_i}, c_i) \rangle (i = 1, 2, \ldots, n)$ *be a set of 2TLNNs. If* $l^+ = (\max_i (S_{T_i}, a_i), \min_i (S_{I_i}, b_i), \min_i (S_{F_i}, c_i))$ *and* $l^- = (\min_i (S_{T_i}, a_i), \max_i (S_{I_i}, b_i), \max_i (S_{F_i}, c_i))$ *then*

$$l^- \leq \text{G2TLNNWBM}_{\omega}^{p,q,r}(l_1, l_2, \cdots, l_n) \leq l^+ \tag{59}$$

From Property 7,

$$\begin{aligned} \text{G2TLNNWBM}_{\omega}^{p,q,r}(l_1^-, l_2^-, \cdots, l_n^-) = l^- \\ \text{G2TLNNWBM}_{\omega}^{p,q,r}(l_1^+, l_2^+, \cdots, l_n^+) = l^+ \end{aligned}$$

From Property 8,

$$l^- \leq \text{G2TLNNWBM}_{\omega}^{p,q,r}(l_1, l_2, \cdots, l_n) \leq l^+$$

4.2. G2TLNNWGBM Operator

Similarly to GWBM, to consider the attribute weights, the generalized weighted geometric Bonferroni mean (GWGBM) is defined, as follows.

Definition 13 ([51]). *Let* $p, q, r > 0$ *and* $b_i (i = 1, 2, \cdots, n)$ *be a collection of nonnegative crisp numbers with the weights vector being* $\omega = (\omega_1, \omega_2, \cdots \omega_n)^T$, *thereby satisfying* $\omega_i \in [0, 1]$ *and* $\sum_{i=1}^{n} \omega_i = 1$. *If*

$$\text{GWGBM}_{\omega}^{p,q,r}(b_1, b_2, \cdots, b_n) = \frac{1}{p+q+r} \prod_{i,j,k=1}^{n} (pb_i + qb_j + rb_k)^{\omega_i \omega_j \omega_k} \tag{60}$$

Then we extend GWGBM to fuse the 2TLNNs and propose generalized 2-tuple linguistic neutrosophic number weighted geometric Bonferroni mean (G2TLNNWGBM) aggregation operator.

Definition 14. *Let* $l_i = \langle (s_{T_i}, a_i), (s_{I_i}, b_i), (s_{F_i}, c_i) \rangle (i = 1, 2, \ldots, n)$ *be a set of 2TLNNs with their weight vector be* $w_i = (w_1, w_2, \ldots, w_n)^T$, *thereby satisfying* $w_i \in [0, 1]$ *and* $\sum_{i=1}^{n} w_i = 1$. *If*

$$\text{G2TLNNWGBM}_{\omega}^{p,q,r}(l_1, l_2, \ldots, l_n) = \frac{1}{p+q+r} \bigotimes_{i,j,k=1}^{n} (pl_i \oplus ql_j \oplus rl_k)^{\omega_i \omega_j \omega_k} \tag{61}$$

Then we called $\text{G2TLNNWGBM}_{\omega}^{p,q,r}$ *the generalized 2-tuple linguistic neutrosophic number weighted geometric BM.*

Theorem 4. *Let $l_i = \langle (s_{T_i}, a_i), (s_{I_i}, b_i), (s_{F_i}, c_i) \rangle (i = 1, 2, \ldots, n)$ be a set of 2TLNNs. The aggregated value by using G2TLNNWGBM operators is also a 2TLNN where*

$$\mathrm{G2TLNNWGBM}_{\omega}^{p,q,r}(l_1, l_2, \ldots, l_n) = \frac{1}{p+q+r} \underset{i,j,k=1}{\overset{n}{\otimes}} (pl_i \oplus ql_j \oplus rl_k)^{\omega_i \omega_j \omega_k}$$
$$= \left\{ \begin{array}{l} \Delta\left(t\left(1-\left(1-\prod_{i,j,k=1}^{n}\left(1-\left(1-\frac{\Delta^{-1}(s_{T_i},a_i)}{t}\right)^p \cdot \left(1-\frac{\Delta^{-1}(s_{T_j},a_j)}{t}\right)^q \cdot \left(1-\frac{\Delta^{-1}(s_{T_k},a_k)}{t}\right)^r\right)^{\omega_i\omega_j\omega_k}\right)^{\frac{1}{p+q+r}}\right)\right), \\ \Delta\left(t\left(1-\prod_{i,j,k=1}^{n}\left(1-\left(\frac{\Delta^{-1}(s_{I_i},b_i)}{t}\right)^p \cdot \left(\frac{\Delta^{-1}(s_{I_j},b_j)}{t}\right)^q \cdot \left(\frac{\Delta^{-1}(s_{I_k},b_k)}{t}\right)^r\right)^{\omega_i\omega_j\omega_k}\right)^{\frac{1}{p+q+r}}\right), \\ \Delta\left(t\left(1-\prod_{i,j,k=1}^{n}\left(1-\left(\frac{\Delta^{-1}(s_{F_i},c_i)}{t}\right)^p \cdot \left(\frac{\Delta^{-1}(s_{F_j},c_j)}{t}\right)^q \cdot \left(\frac{\Delta^{-1}(s_{F_k},c_k)}{t}\right)^r\right)^{\omega_i\omega_j\omega_k}\right)^{\frac{1}{p+q+r}}\right) \end{array} \right\} \tag{62}$$

Proof. From Definition 5, we can obtain,

$$pl_i = \left\{\Delta\left(t\left(1-\left(1-\frac{\Delta^{-1}(s_{T_i},a_i)}{t}\right)^p\right)\right), \Delta\left(t\left(\frac{\Delta^{-1}(s_{I_i},b_i)}{t}\right)^p\right), \Delta\left(t\left(\frac{\Delta^{-1}(s_{F_i},c_i)}{t}\right)^p\right)\right\} \tag{63}$$

$$ql_j = \left\{\Delta\left(t\left(1-\left(1-\frac{\Delta^{-1}(s_{T_j},a_j)}{t}\right)^q\right)\right), \Delta\left(t\left(\frac{\Delta^{-1}(s_{I_j},b_j)}{t}\right)^q\right), \Delta\left(t\left(\frac{\Delta^{-1}(s_{F_j},c_j)}{t}\right)^q\right)\right\} \tag{64}$$

$$rl_k = \left\{\Delta\left(t\left(1-\left(1-\frac{\Delta^{-1}(s_{T_k},a_k)}{t}\right)^r\right)\right), \Delta\left(t\left(\frac{\Delta^{-1}(s_{I_k},b_k)}{t}\right)^r\right), \Delta\left(t\left(\frac{\Delta^{-1}(s_{F_k},c_k)}{t}\right)^r\right)\right\} \tag{65}$$

Thus,

$$pl_i \oplus ql_j \oplus rl_k$$
$$= \left\{ \begin{array}{l} \Delta\left(t\left(1-\left(1-\frac{\Delta^{-1}(s_{T_i},a_i)}{t}\right)^p \cdot \left(1-\frac{\Delta^{-1}(s_{T_j},a_j)}{t}\right)^q \cdot \left(1-\frac{\Delta^{-1}(s_{T_k},a_k)}{t}\right)^r\right)\right), \\ \Delta\left(t\left(\frac{\Delta^{-1}(s_{I_i},b_i)}{t}\right)^p \cdot \left(\frac{\Delta^{-1}(s_{I_j},b_j)}{t}\right)^q \cdot \left(\frac{\Delta^{-1}(s_{I_k},b_k)}{t}\right)^r\right), \\ \Delta\left(t\left(\frac{\Delta^{-1}(s_{F_i},c_i)}{t}\right)^p \cdot \left(\frac{\Delta^{-1}(s_{F_j},c_j)}{t}\right)^q \cdot \left(\frac{\Delta^{-1}(s_{F_k},c_k)}{t}\right)^r\right) \end{array} \right\} \tag{66}$$

Therefore,

$$(pl_i \oplus ql_j \oplus rl_k)^{\omega_i\omega_j\omega_k}$$
$$= \left\{ \begin{array}{l} \Delta\left(t\left(1-\left(1-\frac{\Delta^{-1}(s_{T_i},a_i)}{t}\right)^p \cdot \left(1-\frac{\Delta^{-1}(s_{T_j},a_j)}{t}\right)^q \cdot \left(1-\frac{\Delta^{-1}(s_{T_k},a_k)}{t}\right)^r\right)^{\omega_i\omega_j\omega_k}\right), \\ \Delta\left(t\left(1-\left(1-\left(\frac{\Delta^{-1}(s_{I_i},b_i)}{t}\right)^p \cdot \left(\frac{\Delta^{-1}(s_{I_j},b_j)}{t}\right)^q \cdot \left(\frac{\Delta^{-1}(s_{I_k},b_k)}{t}\right)^r\right)^{\omega_i\omega_j\omega_k}\right)\right), \\ \Delta\left(t\left(1-\left(1-\left(\frac{\Delta^{-1}(s_{F_i},c_i)}{t}\right)^p \cdot \left(\frac{\Delta^{-1}(s_{F_j},c_j)}{t}\right)^q \cdot \left(\frac{\Delta^{-1}(s_{F_k},c_k)}{t}\right)^r\right)^{\omega_i\omega_j\omega_k}\right)\right) \end{array} \right\} \tag{67}$$

Thereafter,

$$\underset{i,j,k=1}{\overset{n}{\otimes}}(pl_i \oplus ql_j \oplus rl_k)^{\omega_i\omega_j\omega_k}$$
$$= \left\{ \begin{array}{l} \Delta\left(t\prod_{i,j,k=1}^{n}\left(1-\left(1-\frac{\Delta^{-1}(s_{T_i},a_i)}{t}\right)^p\cdot\left(1-\frac{\Delta^{-1}(s_{T_j},a_j)}{t}\right)^q\cdot\left(1-\frac{\Delta^{-1}(s_{T_k},a_k)}{t}\right)^r\right)^{\omega_i\omega_j\omega_k}\right), \\ \Delta\left(t\left(1-\prod_{i,j,k=1}^{n}\left(1-\left(\frac{\Delta^{-1}(s_{I_i},b_i)}{t}\right)^p\cdot\left(\frac{\Delta^{-1}(s_{I_j},b_j)}{t}\right)^q\cdot\left(\frac{\Delta^{-1}(s_{I_k},b_k)}{t}\right)^r\right)^{\omega_i\omega_j\omega_k}\right)\right), \\ \Delta\left(t\left(1-\prod_{i,j,k=1}^{n}\left(1-\left(\frac{\Delta^{-1}(s_{F_i},c_i)}{t}\right)^p\cdot\left(\frac{\Delta^{-1}(s_{F_j},c_j)}{t}\right)^q\cdot\left(\frac{\Delta^{-1}(s_{F_k},c_k)}{t}\right)^r\right)^{\omega_i\omega_j\omega_k}\right)\right) \end{array} \right\} \tag{68}$$

Furthermore,

$$G2TLNNWGBM_{\omega}^{p,q,r}(l_1,l_2,\ldots,l_n) = \frac{1}{p+q+r}\underset{i,j,k=1}{\overset{n}{\otimes}}(pl_i \oplus ql_j \oplus rl_k)^{\omega_i\omega_j\omega_k}$$
$$= \left\{ \begin{array}{l} \Delta\left(t\left(1-\left(1-\prod_{i,j,k=1}^{n}\left(1-\left(1-\frac{\Delta^{-1}(s_{T_i},a_i)}{t}\right)^p\cdot\left(1-\frac{\Delta^{-1}(s_{T_j},a_j)}{t}\right)^q\cdot\left(1-\frac{\Delta^{-1}(s_{T_k},a_k)}{t}\right)^r\right)^{\omega_i\omega_j\omega_k}\right)^{\frac{1}{p+q+r}}\right)\right), \\ \Delta\left(t\left(1-\prod_{i,j,k=1}^{n}\left(1-\left(\frac{\Delta^{-1}(s_{I_i},b_i)}{t}\right)^p\cdot\left(\frac{\Delta^{-1}(s_{I_j},b_j)}{t}\right)^q\cdot\left(\frac{\Delta^{-1}(s_{I_k},b_k)}{t}\right)^r\right)^{\omega_i\omega_j\omega_k}\right)^{\frac{1}{p+q+r}}\right), \\ \Delta\left(t\left(1-\prod_{i,j,k=1}^{n}\left(1-\left(\frac{\Delta^{-1}(s_{F_i},c_i)}{t}\right)^p\cdot\left(\frac{\Delta^{-1}(s_{F_j},c_j)}{t}\right)^q\cdot\left(\frac{\Delta^{-1}(s_{F_k},c_k)}{t}\right)^r\right)^{\omega_i\omega_j\omega_k}\right)^{\frac{1}{p+q+r}}\right) \end{array} \right\} \tag{69}$$

Hence, (62) is kept. □

Then, we need to prove that (62) is a 2TLNN. We need to prove two conditions, as follows:

① $0 \le \Delta^{-1}(s_T,a) \le t, 0 \le \Delta^{-1}(s_I,b) \le t, 0 \le \Delta^{-1}(s_F,c) \le t$
② $0 \le \Delta^{-1}(s_T,a) + \Delta^{-1}(s_I,b) + \Delta^{-1}(s_F,c) \le 3t$

Let

$$\frac{\Delta^{-1}(s_T,a)}{t} = 1-\left(1-\prod_{i,j,k=1}^{n}\left(1-\left(1-\frac{\Delta^{-1}(s_{T_i},a_i)}{t}\right)^p\cdot\left(1-\frac{\Delta^{-1}(s_{T_j},a_j)}{t}\right)^q\cdot\left(1-\frac{\Delta^{-1}(s_{T_k},a_k)}{t}\right)^r\right)^{\omega_i\omega_j\omega_k}\right)^{\frac{1}{p+q+r}}$$
$$\frac{\Delta^{-1}(s_I,b)}{t} = \left(1-\prod_{i,j,k=1}^{n}\left(1-\left(\frac{\Delta^{-1}(s_{I_i},b_i)}{t}\right)^p\cdot\left(\frac{\Delta^{-1}(s_{I_j},b_j)}{t}\right)^q\cdot\left(\frac{\Delta^{-1}(s_{I_k},b_k)}{t}\right)^r\right)^{\omega_i\omega_j\omega_k}\right)^{\frac{1}{p+q+r}}$$
$$\frac{\Delta^{-1}(s_F,c)}{t} = \left(1-\prod_{i,j,k=1}^{n}\left(1-\left(\frac{\Delta^{-1}(s_{F_i},c_i)}{t}\right)^p\cdot\left(\frac{\Delta^{-1}(s_{F_j},c_j)}{t}\right)^q\cdot\left(\frac{\Delta^{-1}(s_{F_k},c_k)}{t}\right)^r\right)^{\omega_i\omega_j\omega_k}\right)^{\frac{1}{p+q+r}}$$

Proof. ① Since $0 \le \frac{\Delta^{-1}(s_{T_i},a_i)}{t} \le 1, 0 \le \frac{\Delta^{-1}(s_{T_j},a_j)}{t} \le 1, 0 \le \frac{\Delta^{-1}(s_{T_k},a_k)}{t} \le 1$ we get

$$0 \le 1-\left(1-\frac{\Delta^{-1}(s_{T_i},a_i)}{t}\right)^p\cdot\left(1-\frac{\Delta^{-1}(s_{T_j},a_j)}{t}\right)^q\cdot\left(1-\frac{\Delta^{-1}(s_{T_k},a_k)}{t}\right)^r \le 1 \tag{70}$$

Then,

$$0 \le \prod_{i,j,k=1}^{n}\left(1-\left(1-\frac{\Delta^{-1}(s_{T_i},a_i)}{t}\right)^p\cdot\left(1-\frac{\Delta^{-1}(s_{T_j},a_j)}{t}\right)^q\cdot\left(1-\frac{\Delta^{-1}(s_{T_k},a_k)}{t}\right)^r\right)^{\omega_i\omega_j\omega_k} \le 1 \tag{71}$$

$$0 \le 1-\left(1-\prod_{i,j,k=1}^{n}\left(1-\left(1-\frac{\Delta^{-1}(s_{T_i},a_i)}{t}\right)^{p}\cdot\left(1-\frac{\Delta^{-1}(s_{T_j},a_j)}{t}\right)^{q}\cdot\left(1-\frac{\Delta^{-1}(s_{T_k},a_k)}{t}\right)^{r}\right)^{\omega_i\omega_j\omega_k}\right)^{\frac{1}{p+q+r}} \le 1 \quad (72)$$

That means $0 \le \Delta^{-1}(s_T,a) \le t$, so ① is maintained, similarly, we can get $0 \le \Delta^{-1}(s_I,b) \le t, 0 \le \Delta^{-1}(s_F,c) \le t$. ② Since $0 \le \Delta^{-1}(s_T,a) \le t, 0 \le \Delta^{-1}(s_I,b) \le t, 0 \le \Delta^{-1}(s_F,c) \le t$, we get $0 \le \Delta^{-1}(s_T,a) + \Delta^{-1}(s_I,b) + \Delta^{-1}(s_F,c) \le 3t$. □

Similar to G2TLNNWBM, the G2TLNNWGBM has the same properties as follows. The proof are omitted here to save space.

Property 10. *(Idempotency) If* $l_i = \langle (s_{T_i},a_i),(s_{I_i},b_i),(s_{F_i},c_i)\rangle (i = 1,2,\ldots,n)$ *are equal, then*

$$\text{G2TLNNWGBM}_{\omega}^{p,q,r}(l_1,l_2,\cdots,l_n) = l \quad (73)$$

Property 11. *(Monotonicity) Let* $l_{x_i} = \left\langle \left(s_{T_{x_i}},a_{x_i}\right),\left(s_{I_{x_i}},b_{x_i}\right),\left(s_{F_{x_i}},c_{x_i}\right)\right\rangle (i = 1,2,\ldots,n)$ *and* $l_{y_i} = \left\langle \left(s_{T_{y_i}},a_{y_i}\right),\left(s_{I_{y_i}},b_{y_i}\right),\left(s_{F_{y_i}},c_{y_i}\right)\right\rangle (i = 1,2,\ldots,n)$ *be two sets of 2TLNNs. If* $\Delta^{-1}\left(s_{T_{x_i}},a_{x_i}\right) \le \Delta^{-1}\left(s_{T_{y_i}},a_{y_i}\right), \Delta^{-1}\left(s_{I_{x_i}},b_{x_i}\right) \ge \Delta^{-1}\left(s_{I_{y_i}},b_{y_i}\right)$ *and* $\Delta^{-1}\left(s_{F_{x_i}},c_{x_i}\right) \ge \Delta^{-1}\left(s_{F_{y_i}},c_{y_i}\right)$ *hold for all i, then*

$$\text{G2TLNNWGBM}_{\omega}^{p,q,r}(l_{x_1},l_{x_2},\cdots,l_{x_n}) \le \text{G2TLNNWGBM}_{\omega}^{p,q,r}(l_{y_1},l_{y_2},\cdots,l_{y_n}) \quad (74)$$

Property 12. *(Boundedness) Let* $l_i = \langle (s_{T_i},a_i),(s_{I_i},b_i),(s_{F_i},c_i)\rangle (i = 1,2,\ldots,n)$ *be a set of 2TLNNs. If* $l^+ = (\max_i(S_{T_i},a_i),\min_i(S_{I_i},b_i),\min_i(S_{F_i},c_i))$ *and* $l^- = (\min_i(S_{T_i},a_i),\max_i(S_{I_i},b_i),\max_i(S_{F_i},c_i))$ *then*

$$l^- \le \text{G2TLNNWGBM}_{\omega}^{p,q,r}(l_1,l_2,\cdots,l_n) \le l^+ \quad (75)$$

5. DG2TLNNWBM and DG2TLNNWGBM Operators

5.1. DG2TLNNWBM Operator

However, the GBM still has some drawbacks, GBWM and GWGBM can only consider the interrelationship between any three aggregated arguments. So, Zhang et al. [52] introduced a new generalization of the traditional BM because the correlations are ubiquitous among all of the arguments. The new generalization of the traditional BM is called the dual GWBM (DGBM). The DGWBM is defined, as follows.

Definition 15 ([52]). *Let* $b_i(i = 1,2,\cdots,n)$ *be a collection of nonnegative crisp numbers with the weights vector being* $\omega = (\omega_1,\omega_2,\cdots\omega_n)^T$*, thereby satisfying* $\omega_i \in [0,1]$ *and* $\sum_{i=1}^{n}\omega_i = 1$*. The dual generalized weighted Bonferroni mean (DGWBM) is defined as follows:*

$$\text{DGWBM}_{w}^{R}(b_1,b_2,\ldots,b_n) = \left(\sum_{i_1,i_2,\ldots,i_n=1}^{n}\left(\prod_{j=1}^{n} w_{i_j} b_{i_j}^{r_j}\right)\right)^{1/\sum_{j=1}^{n} r_j} \quad (76)$$

where $R = (r_1,r_2,\ldots,r_n)^T$ *is the parameter vector with* $r_i \ge 0\ (i = 1,2,\ldots,n)$*.*

Then we extend DGWBM to fuse the 2TLNNs and propose dual generalized 2-tuple linguistic neutrosophic number weighted Bonferroni mean (DG2TLNNWBM) operator.

Definition 16. *Let* $l_i = \langle (s_{T_i}, a_i), (s_{I_i}, b_i), (s_{F_i}, c_i) \rangle$ *be a set of 2TLNNs. The dual generalized 2-tuple linguistic neutrosophic number weighted Bonferroni mean (DG2TLNNWBM) operator is:*

$$\mathrm{DG2TLNNWBM}_w^R(l_1, l_2, \ldots, l_n) = \left(\mathop{\oplus}_{i_1, i_2, \ldots, i_n = 1}^{n} \left(\mathop{\otimes}_{j=1}^{n} w_{i_j} l_{ij}^{r_j} \right) \right)^{1/\sum_{i=1}^{n} r_j} \tag{77}$$

Theorem 5. *Let* $l_i = \langle (s_{T_i}, a_i), (s_{I_i}, b_i), (s_{F_i}, c_i) \rangle$ *be a set of 2TLNNs. The aggregated value by using DG2TLNNWBM operators is also a 2TLNN where*

$$\begin{aligned} &\mathrm{DG2TLNNWBM}_w^R(l_1, l_2, \ldots, l_n) = \left(\mathop{\oplus}_{i_1, i_2, \ldots, i_n = 1}^{n} \left(\mathop{\otimes}_{j=1}^{n} w_{i_j} l_{i_j}^{r_j} \right) \right)^{1/\sum_{i=1}^{n} r_j} \\ &= \left\{ \begin{array}{l} \Delta\left(t\left(1 - \prod_{i_1, i_2, \ldots, i_n = 1}^{n} \left(1 - \prod_{j=1}^{n} \left(1 - \left(1 - \left(\frac{\Delta^{-1}\left(s_{T_{i_j}}, a_{i_j}\right)}{t} \right)^{r_j} \right)^{w_{i_j}} \right) \right) \right)^{1/\sum_{i=1}^{n} r_j} \right), \\ \Delta\left(t\left(1 - \left(1 - \prod_{i_1, i_2, \ldots, i_n = 1}^{n} \left(1 - \prod_{j=1}^{n} \left(1 - \left(1 - \left(1 - \frac{\Delta^{-1}\left(s_{I_{i_j}}, b_{i_j}\right)}{t} \right)^{r_j} \right)^{w_{i_j}} \right) \right) \right)^{1/\sum_{i=1}^{n} r_j} \right) \right), \\ \Delta\left(t\left(1 - \left(1 - \prod_{i_1, i_2, \ldots, i_n = 1}^{n} \left(1 - \prod_{j=1}^{n} \left(1 - \left(1 - \left(1 - \frac{\Delta^{-1}\left(s_{F_{i_j}}, c_{i_j}\right)}{t} \right)^{r_j} \right)^{w_{i_j}} \right) \right) \right)^{1/\sum_{i=1}^{n} r_j} \right) \right) \end{array} \right\} \end{aligned} \tag{78}$$

Proof. According to Definition 5, we can obtain

$$l_{i_j}^{r_j} = \left\{ \Delta\left(t\left(\frac{\Delta^{-1}\left(s_{T_{i_j}}, a_{i_j}\right)}{t} \right)^{r_j} \right), \Delta\left(t\left(1 - \left(1 - \frac{\Delta^{-1}\left(s_{I_{i_j}}, b_{i_j}\right)}{t} \right)^{r_j} \right) \right), \Delta\left(t\left(1 - \left(1 - \frac{\Delta^{-1}\left(s_{F_{i_j}}, c_{i_j}\right)}{t} \right)^{r_j} \right) \right) \right\} \tag{79}$$

Thus,

$$w_{i_j} l_{i_j}^{r_j} = \left\{ \begin{array}{l} \Delta\left(t\left(1 - \left(1 - \left(\frac{\Delta^{-1}\left(s_{T_{i_j}}, a_{i_j}\right)}{t} \right)^{r_j} \right)^{w_{i_j}} \right) \right), \Delta\left(t\left(1 - \left(1 - \frac{\Delta^{-1}\left(s_{I_{i_j}}, b_{i_j}\right)}{t} \right)^{r_j} \right)^{w_{i_j}} \right), \\ \Delta\left(t\left(1 - \left(1 - \frac{\Delta^{-1}\left(s_{F_{i_j}}, c_{i_j}\right)}{t} \right)^{r_j} \right)^{w_{i_j}} \right) \end{array} \right\} \tag{80}$$

Thereafter,

$$\mathop{\otimes}_{j=1}^{n} w_{i_j} l_{i_j}^{r_j} = \left\{ \begin{array}{l} \Delta\left(t \prod_{j=1}^{n} \left(1 - \left(1 - \left(\frac{\Delta^{-1}\left(s_{T_{i_j}}, a_{i_j}\right)}{t} \right)^{r_j} \right)^{w_{i_j}} \right) \right), \\ \Delta\left(t\left(1 - \prod_{j=1}^{n} \left(1 - \left(1 - \left(1 - \frac{\Delta^{-1}\left(s_{I_{i_j}}, b_{i_j}\right)}{t} \right)^{r_j} \right)^{w_{i_j}} \right) \right) \right), \\ \Delta\left(t\left(1 - \prod_{j=1}^{n} \left(1 - \left(1 - \left(1 - \frac{\Delta^{-1}\left(s_{F_{i_j}}, c_{i_j}\right)}{t} \right)^{r_j} \right)^{w_{i_j}} \right) \right) \right) \end{array} \right\} \tag{81}$$

Furthermore,

$$\begin{aligned}
&\overset{n}{\underset{i_1,i_2,\ldots,i_n=1}{\oplus}}\left(\overset{n}{\underset{j=1}{\otimes}} w_{i_j} l_{i_j}^{r_j}\right) \\
&= \left\{\begin{array}{l}
\Delta\left(t\left(1-\prod\limits_{i_1,i_2,\ldots,i_n=1}^{n}\left(1-\prod\limits_{j=1}^{n}\left(1-\left(1-\left(\frac{\Delta^{-1}\left(s_{T_{i_j}},a_{i_j}\right)}{t}\right)^{r_j}\right)^{w_{i_j}}\right)\right)\right)\right), \\
\Delta\left(t\prod\limits_{i_1,i_2,\ldots,i_n=1}^{n}\left(1-\prod\limits_{j=1}^{n}\left(1-\left(1-\left(1-\frac{\Delta^{-1}\left(s_{I_{i_j}},b_{i_j}\right)}{t}\right)^{r_j}\right)^{w_{i_j}}\right)\right)\right), \\
\Delta\left(t\prod\limits_{i_1,i_2,\ldots,i_n=1}^{n}\left(1-\prod\limits_{j=1}^{n}\left(1-\left(1-\left(1-\frac{\Delta^{-1}\left(s_{F_{i_j}},c_{i_j}\right)}{t}\right)^{r_j}\right)^{w_{i_j}}\right)\right)\right)
\end{array}\right\}
\end{aligned} \tag{82}$$

Therefore,

$$\begin{aligned}
&\text{DG2TLNNWBM}_w^R(l_1,l_2,\ldots,l_n) = \left(\overset{n}{\underset{i_1,i_2,\ldots,i_n=1}{\oplus}}\left(\overset{n}{\underset{j=1}{\otimes}} w_{i_j} l_{i_j}^{r_j}\right)\right)^{1/\sum_{i=1}^n r_j} \\
&= \left\{\begin{array}{l}
\Delta\left(t\left(1-\prod\limits_{i_1,i_2,\ldots,i_n=1}^{n}\left(1-\prod\limits_{j=1}^{n}\left(1-\left(1-\left(\frac{\Delta^{-1}\left(s_{T_{i_j}},a_{i_j}\right)}{t}\right)^{r_j}\right)^{w_{i_j}}\right)\right)\right)^{1/\sum_{i=1}^n r_j}\right), \\
\Delta\left(t\left(1-\left(1-\prod\limits_{i_1,i_2,\ldots,i_n=1}^{n}\left(1-\prod\limits_{j=1}^{n}\left(1-\left(1-\left(1-\frac{\Delta^{-1}\left(s_{I_{i_j}},b_{i_j}\right)}{t}\right)^{r_j}\right)^{w_{i_j}}\right)\right)\right)^{1/\sum_{i=1}^n r_j}\right)\right), \\
\Delta\left(t\left(1-\left(1-\prod\limits_{i_1,i_2,\ldots,i_n=1}^{n}\left(1-\prod\limits_{j=1}^{n}\left(1-\left(1-\left(1-\frac{\Delta^{-1}\left(s_{F_{i_j}},c_{i_j}\right)}{t}\right)^{r_j}\right)^{w_{i_j}}\right)\right)\right)^{1/\sum_{i=1}^n r_j}\right)\right)
\end{array}\right\}
\end{aligned} \tag{83}$$

Hence, (78) is kept. □

Then, we need to prove that (78) is a 2TLNN. We need to prove two conditions as follows:

① $0 \le \Delta^{-1}(s_T,a) \le t, 0 \le \Delta^{-1}(s_I,b) \le t, 0 \le \Delta^{-1}(s_F,c) \le t$

② $0 \le \Delta^{-1}(s_T,a) + \Delta^{-1}(s_I,b) + \Delta^{-1}(s_F,c) \le 3t$

Let

$$\frac{\Delta^{-1}(s_T,a)}{t} = \left(1-\prod_{i_1,i_2,\ldots,i_n=1}^{n}\left(1-\prod_{j=1}^{n}\left(1-\left(1-\left(\frac{\Delta^{-1}\left(s_{T_{i_j}},a_{i_j}\right)}{t}\right)^{r_j}\right)^{w_{i_j}}\right)\right)\right)^{1/\sum_{i=1}^n r_j}$$

$$\frac{\Delta^{-1}(s_I,b)}{t} = 1-\left(1-\prod_{i_1,i_2,\ldots,i_n=1}^{n}\left(1-\prod_{j=1}^{n}\left(1-\left(1-\left(1-\frac{\Delta^{-1}\left(s_{I_{i_j}},b_{i_j}\right)}{t}\right)^{r_j}\right)^{w_{i_j}}\right)\right)\right)^{1/\sum_{i=1}^n r_j}$$

$$\frac{\Delta^{-1}(s_F,c)}{t} = 1-\left(1-\prod_{i_1,i_2,\ldots,i_n=1}^{n}\left(1-\prod_{j=1}^{n}\left(1-\left(1-\left(1-\frac{\Delta^{-1}\left(s_{I_{i_j}},b_{i_j}\right)}{t}\right)^{r_j}\right)^{w_{i_j}}\right)\right)\right)^{1/\sum_{i=1}^n r_j}$$

Proof. ① Since $0 \le \frac{\Delta^{-1}\left(s_{T_{i_j}}, a_{i_j}\right)}{t} \le 1$, we get

$$0 \le 1 - \left(\frac{\Delta^{-1}\left(s_{T_{i_j}}, a_{i_j}\right)}{t}\right)^{r_j} \le 1 \tag{84}$$

Then,

$$0 \le 1 - \prod_{j=1}^{n}\left(1 - \left(1 - \left(\frac{\Delta^{-1}\left(s_{T_{i_j}}, a_{i_j}\right)}{t}\right)^{r_j}\right)^{w_{i_j}}\right) \le 1 \tag{85}$$

Furthermore,

$$0 \le 1 - \prod_{i_1, i_2, \ldots, i_n = 1}^{n}\left(1 - \prod_{j=1}^{n}\left(1 - \left(1 - \left(\frac{\Delta^{-1}\left(s_{T_{i_j}}, a_{i_j}\right)}{t}\right)^{r_j}\right)^{w_{i_j}}\right)\right) \le 1 \tag{86}$$

Therefore,

$$0 \le \left(1 - \prod_{i_1, i_2, \ldots, i_n = 1}^{n}\left(1 - \prod_{j=1}^{n}\left(1 - \left(1 - \left(\frac{\Delta^{-1}\left(s_{T_{i_j}}, a_{i_j}\right)}{t}\right)^{r_j}\right)^{w_{i_j}}\right)\right)\right)^{1/\sum_{i=1}^{n} r_j} \le 1 \tag{87}$$

That means $0 \le \Delta^{-1}(s_T, a) \le t$, so ① is maintained, similarly, we can get $0 \le \Delta^{-1}(s_I, b) \le t, 0 \le \Delta^{-1}(s_F, c) \le t$. ② Since $0 \le \Delta^{-1}(s_T, a) \le t, 0 \le \Delta^{-1}(s_I, b) \le t, 0 \le \Delta^{-1}(s_F, c) \le t$, we get $0 \le \Delta^{-1}(s_T, a) + \Delta^{-1}(s_I, b) + \Delta^{-1}(s_F, c) \le 3t$. □

Then, we will discuss some properties of the DG2TLNNWBM operator.

Property 13. *(Monotonicity) Let* $l_{x_i} = \left\langle \left(s_{T_{x_i}}, a_{x_i}\right), \left(s_{I_{x_i}}, b_{x_i}\right), \left(s_{F_{x_i}}, c_{x_i}\right) \right\rangle (i = 1, 2, \ldots, n)$ *and* $l_{y_i} = \left\langle \left(s_{T_{y_i}}, a_{y_i}\right), \left(s_{I_{y_i}}, b_{y_i}\right), \left(s_{F_{y_i}}, c_{y_i}\right) \right\rangle (i = 1, 2, \ldots, n)$ *be two sets of 2TLNNs. If* $\Delta^{-1}\left(s_{T_{x_i}}, a_{x_i}\right) \le \Delta^{-1}\left(s_{T_{y_i}}, a_{y_i}\right), \Delta^{-1}\left(s_{I_{x_i}}, b_{x_i}\right) \ge \Delta^{-1}\left(s_{I_{y_i}}, b_{y_i}\right)$ *and* $\Delta^{-1}\left(s_{F_{x_i}}, c_{x_i}\right) \ge \Delta^{-1}\left(s_{F_{y_i}}, c_{y_i}\right)$ *hold for all i, then*

$$\text{DG2TLNNWBM}_w^R(l_{x_1}, l_{x_2}, \cdots, l_{x_n}) \le \text{DG2TLNNWBM}_w^R(l_{y_1}, l_{y_2}, \cdots, l_{y_n}) \tag{88}$$

Proof. Let $\text{DG2TLNNWBM}_w^R(l_{x_1}, l_{x_2}, \cdots, l_{x_n}) = \left\langle \left(s_{T_{x_i}}, a_{x_i}\right), \left(s_{I_{x_i}}, b_{x_i}\right), \left(s_{F_{x_i}}, c_{x_i}\right) \right\rangle (i = 1, 2, \ldots, n)$ and $\text{DG2TLNNWBM}_w^R(l_{y_1}, l_{y_2}, \cdots, l_{y_n}) = \left\langle \left(s_{T_{y_i}}, a_{y_i}\right), \left(s_{I_{y_i}}, b_{y_i}\right), \left(s_{F_{y_i}}, c_{y_i}\right) \right\rangle (i = 1, 2, \ldots, n)$, given that $\Delta^{-1}\left(s_{T_{x_i}}, a_{x_i}\right) \le \Delta^{-1}\left(s_{T_{y_i}}, a_{y_i}\right)$, we can obtain

$$\left(\frac{\Delta^{-1}\left(s_{T_{x_{i_j}}}, a_{x_{i_j}}\right)}{t}\right)^{r_j} \le \left(\frac{\Delta^{-1}\left(s_{T_{y_{i_j}}}, a_{y_{i_j}}\right)}{t}\right)^{r_j} \tag{89}$$

$$\left(1 - \left(\frac{\Delta^{-1}\left(s_{T_{x_{i_j}}}, a_{x_{i_j}}\right)}{t}\right)^{r_j}\right)^{w_{i_j}} \ge \left(1 - \left(\frac{\Delta^{-1}\left(s_{T_{y_{i_j}}}, a_{y_{i_j}}\right)}{t}\right)^{r_j}\right)^{w_{i_j}} \tag{90}$$

Thereafter,

$$\prod_{j=1}^{n}\left(1-\left(1-\left(\frac{\Delta^{-1}\left(s_{Tx_{i_j}},a_{x_{i_j}}\right)}{t}\right)\right)^{r_j}\right)^{w_{i_j}} \leq \prod_{j=1}^{n}\left(1-\left(1-\left(\frac{\Delta^{-1}\left(s_{Ty_{i_j}},a_{y_{i_j}}\right)}{t}\right)\right)^{r_j}\right)^{w_{i_j}} \tag{91}$$

Furthermore,

$$\prod_{i_1,i_2,\ldots,i_n=1}^{n}\left(1-\prod_{j=1}^{n}\left(1-\left(1-\left(\frac{\Delta^{-1}\left(s_{Tx_{i_j}},a_{x_{i_j}}\right)}{t}\right)\right)^{r_j}\right)^{w_{i_j}}\right) \geq \prod_{i_1,i_2,\ldots,i_n=1}^{n}\left(1-\prod_{j=1}^{n}\left(1-\left(1-\left(\frac{\Delta^{-1}\left(s_{Ty_{i_j}},a_{y_{i_j}}\right)}{t}\right)\right)^{r_j}\right)^{w_{i_j}}\right) \tag{92}$$

$$\left(1-\prod_{i_1,i_2,\ldots,i_n=1}^{n}\left(1-\prod_{j=1}^{n}\left(1-\left(1-\left(\frac{\Delta^{-1}\left(s_{Tx_{i_j}},a_{x_{i_j}}\right)}{t}\right)\right)^{r_j}\right)^{w_{i_j}}\right)\right)^{1/\sum_{i=1}^{n}r_j}$$
$$\leq\left(1-\prod_{i_1,i_2,\ldots,i_n=1}^{n}\left(1-\prod_{j=1}^{n}\left(1-\left(1-\left(\frac{\Delta^{-1}\left(s_{Ty_{i_j}},a_{y_{i_j}}\right)}{t}\right)\right)^{r_j}\right)^{w_{i_j}}\right)\right)^{1/\sum_{i=1}^{n}r_j}$$

That means $\Delta^{-1}(s_{T_x},a_x) \leq \Delta^{-1}\left(s_{T_y},a_y\right)$. Similarly, we can obtain $\Delta^{-1}(s_{I_x},b_x) \geq \Delta^{-1}\left(s_{I_y},b_y\right)$ and $\Delta^{-1}(s_{F_x},c_x) \geq \Delta^{-1}\left(s_{F_y},c_y\right)$.

If $\Delta^{-1}(s_{T_x},a_x) < \Delta^{-1}\left(s_{T_y},a_y\right)$ and $\Delta^{-1}(s_{I_x},b_x) \geq \Delta^{-1}\left(s_{I_y},b_y\right)$ and $\Delta^{-1}(s_{F_x},c_x) \geq \Delta^{-1}\left(s_{F_y},c_y\right)$

$$\text{DG2TLNNWBM}_w^R(l_{x_1},l_{x_2},\cdots,l_{x_n}) < \text{DG2TLNNWBM}_w^R(l_{y_1},l_{y_2},\cdots,l_{y_n})$$

If $\Delta^{-1}(s_{T_x},a_x) = \Delta^{-1}\left(s_{T_y},a_y\right)$ and $\Delta^{-1}(s_{I_x},b_x) > \Delta^{-1}\left(s_{I_y},b_y\right)$ and $\Delta^{-1}(s_{F_x},c_x) > \Delta^{-1}\left(s_{F_y},c_y\right)$

$$\text{DG2TLNNWBM}_w^R(l_{x_1},l_{x_2},\cdots,l_{x_n}) < \text{DG2TLNNWBM}_w^R(l_{y_1},l_{y_2},\cdots,l_{y_n})$$

If $\Delta^{-1}(s_{T_x},a_x) = \Delta^{-1}\left(s_{T_y},a_y\right)$ and $\Delta^{-1}(s_{I_x},b_x) = \Delta^{-1}\left(s_{I_y},b_y\right)$ and $\Delta^{-1}(s_{F_x},c_x) = \Delta^{-1}\left(s_{F_y},c_y\right)$

$$\text{DG2TLNNWBM}_w^R(l_{x_1},l_{x_2},\cdots,l_{x_n}) = \text{DG2TLNNWBM}_w^R(l_{y_1},l_{y_2},\cdots,l_{y_n})$$

So Property 13 is right. □

Property 14. *(Boundedness) Let*$l_i = \left\langle (s_{T_i},a_i),(s_{I_i},b_i),(s_{F_i},c_i)\right\rangle (i = 1,2,\ldots,n)$ *be a set of 2TLNNs. If* $l^+ = \left(\max_i(S_{T_i},a_i),\min_i(S_{I_i},b_i),\min_i(S_{F_i},c_i)\right)$ *and* $l^- = \left(\min_i(S_{T_i},a_i),\max_i(S_{I_i},b_i),\max_i(S_{F_i},c_i)\right)$ *then*

$$l^- \leq \text{DG2TLNNWBM}_w^R(l_1,l_2,\cdots,l_n) \leq l^+ \tag{93}$$

From Theorem 5, we can obtain

$$\mathrm{DG2TLNNWBM}_w^R(l_1^-, l_2^-, \cdots, l_n^-)$$
$$= \left\{ \begin{array}{l} \Delta\left(t\left(1 - \prod\limits_{i_1,i_2,\ldots,i_n=1}^{n}\left(1 - \prod\limits_{j=1}^{n}\left(1 - \left(1 - \left(\frac{\min\Delta^{-1}\left(s_{T_{i_j}}, a_{i_j}\right)}{t}\right)\right)^{r_j}\right)^{w_{i_j}}\right)\right)^{1/\sum_{i=1}^{n} r_j}\right), \\ \Delta\left(t\left(1 - \left(1 - \prod\limits_{i_1,i_2,\ldots,i_n=1}^{n}\left(1 - \prod\limits_{j=1}^{n}\left(1 - \left(1 - \left(1 - \frac{\max\Delta^{-1}\left(s_{I_{i_j}}, b_{i_j}\right)}{t}\right)\right)^{r_j}\right)^{w_{i_j}}\right)\right)^{1/\sum_{i=1}^{n} r_j}\right)\right), \\ \Delta\left(t\left(1 - \left(1 - \prod\limits_{i_1,i_2,\ldots,i_n=1}^{n}\left(1 - \prod\limits_{j=1}^{n}\left(1 - \left(1 - \left(1 - \frac{\max\Delta^{-1}\left(s_{F_{i_j}}, c_{i_j}\right)}{t}\right)\right)^{r_j}\right)^{w_{i_j}}\right)\right)^{1/\sum_{i=1}^{n} r_j}\right)\right) \end{array} \right\}$$

$$\mathrm{DG2TLNNWBM}_w^R(l_1^+, l_2^+, \cdots, l_n^+)$$
$$= \left\{ \begin{array}{l} \Delta\left(t\left(1 - \prod\limits_{i_1,i_2,\ldots,i_n=1}^{n}\left(1 - \prod\limits_{j=1}^{n}\left(1 - \left(1 - \left(\frac{\max\Delta^{-1}\left(s_{T_{i_j}}, a_{i_j}\right)}{t}\right)\right)^{r_j}\right)^{w_{i_j}}\right)\right)^{1/\sum_{i=1}^{n} r_j}\right), \\ \Delta\left(t\left(1 - \left(1 - \prod\limits_{i_1,i_2,\ldots,i_n=1}^{n}\left(1 - \prod\limits_{j=1}^{n}\left(1 - \left(1 - \left(1 - \frac{\min\Delta^{-1}\left(s_{I_{i_j}}, b_{i_j}\right)}{t}\right)\right)^{r_j}\right)^{w_{i_j}}\right)\right)^{1/\sum_{i=1}^{n} r_j}\right)\right), \\ \Delta\left(t\left(1 - \left(1 - \prod\limits_{i_1,i_2,\ldots,i_n=1}^{n}\left(1 - \prod\limits_{j=1}^{n}\left(1 - \left(1 - \left(1 - \frac{\min\Delta^{-1}\left(s_{F_{i_j}}, c_{i_j}\right)}{t}\right)\right)^{r_j}\right)^{w_{i_j}}\right)\right)^{1/\sum_{i=1}^{n} r_j}\right)\right) \end{array} \right\}$$

From Property 13,

$$l^- \leq \mathrm{DG2TLNNWBM}_w^R(l_1, l_2, \cdots, l_n) \leq l^+$$

5.2. DG2TLNNWGBM Operator

Similarly to DGWBM, to consider the attribute weights, the dual generalized weighted geometric Bonferroni mean (DGWGBM) is defined, as follows.

Definition 17 ([52]). *Let and $b_i (i = 1, 2, \cdots, n)$ be a collection of nonnegative crisp numbers with the weights vector being $\omega = (\omega_1, \omega_2, \cdots \omega_n)^T$, thereby satisfying $\omega_i \in [0, 1]$ and $\sum\limits_{i=1}^{n} \omega_i = 1$. If*

$$\mathrm{DGWGBM}_w^R(b_1, b_2, \cdots, b_n) = \frac{1}{\sum_{j=1}^{n} r_j}\left(\prod_{i_1,i_2,\ldots,i_n=1}^{n}\left(\sum_{j=1}^{n}\left(r_j l_{i_j}\right)\right)\right)^{\prod_{j=1}^{n} w_{i_j}} \tag{94}$$

where $R = (r_1, r_2, \ldots, r_n)^T$ is the parameter vector with $r_i \geq 0\ (i = 1, 2, \ldots, n)$.

Then we extend DGWGBM to fuse the 2TLNNs and propose dual generalized 2-tuple linguistic neutrosophic number weighted geometric Bonferroni mean (DG2TLNNWGBM) aggregation operator.

Definition 18. *Let $l_i = \langle (s_{T_i}, a_i), (s_{I_i}, b_i), (s_{F_i}, c_i) \rangle (i = 1, 2, \ldots, n)$be a set of 2TLNNs with their weight vector be $w_i = (w_1, w_2, \ldots, w_n)^T$, thereby satisfying $w_i \in [0, 1]$ and $\sum_{i=1}^{n} w_i = 1$. If*

$$\mathrm{DG2TLNNWGBM}_w^R(l_1, l_2, \ldots, l_n) = \frac{1}{\sum_{j=1}^{n} r_j}\left(\bigotimes_{i_1,i_2,\ldots,i_n=1}^{n}\left(\bigoplus_{j=1}^{n}\left(r_j l_{i_j}\right)\right)^{\prod_{j=1}^{n} w_{i_j}}\right) \tag{95}$$

Then we called $\mathrm{DG2TLNNWGBM}_w^R$ the dual generalized 2-tuple linguistic neutrosophic number weighted geometric BM.

Theorem 6. *Let* $l_i = \langle (s_{T_i}, a_i), (s_{I_i}, b_i), (s_{F_i}, c_i) \rangle (i = 1, 2, \ldots, n)$ *be a set of 2TLNNs. The aggregated value by using DG2TLNNWGBM operators is also a 2TLNN where*

$$\begin{aligned}
&\mathrm{DG2TLNNWGBM}_w^R(l_1, l_2, \ldots, l_n) \\
&= \frac{1}{\sum_{j=1}^n r_j}\left(\underset{i_1,i_2,\ldots,i_n=1}{\overset{n}{\otimes}}\left(\underset{j=1}{\overset{n}{\oplus}}\left(r_j l_{i_j}\right)\right)^{\prod_{j=1}^n w_{i_j}}\right) \\
&= \left\{\begin{array}{l}
\Delta\left(t\left(1-\left(1-\prod_{i_1,i_2,\ldots,i_n=1}^{n}\left(1-\prod_{j=1}^{n}\left(1-\frac{\Delta^{-1}\left(s_{T_{i_j}}, a_{i_j}\right)}{t}\right)^{r_j}\right)^{\prod_{j=1}^n w_{i_j}}\right)^{\frac{1}{\sum_{j=1}^n r_j}}\right)\right), \\
\Delta\left(t\left(1-\prod_{i_1,i_2,\ldots,i_n=1}^{n}\left(1-\prod_{j=1}^{n}\left(\frac{\Delta^{-1}\left(s_{I_{i_j}}, b_{i_j}\right)}{t}\right)^{r_j}\right)^{\prod_{j=1}^n w_{i_j}}\right)^{\frac{1}{\sum_{j=1}^n r_j}}\right), \\
\Delta\left(t\left(1-\prod_{i_1,i_2,\ldots,i_n=1}^{n}\left(1-\prod_{j=1}^{n}\left(\frac{\Delta^{-1}\left(s_{F_{i_j}}, c_{i_j}\right)}{t}\right)^{r_j}\right)^{\prod_{j=1}^n w_{i_j}}\right)^{\frac{1}{\sum_{j=1}^n r_j}}\right)
\end{array}\right\}
\end{aligned} \tag{96}$$

Proof. From Definition 5, we can obtain,

$$r_j l_{i_j} = \left\{\begin{array}{l}
\Delta\left(t\left(1-\left(1-\frac{\Delta^{-1}\left(s_{T_{i_j}}, a_{i_j}\right)}{t}\right)^{r_j}\right)\right), \Delta\left(t\left(1-\left(1-\frac{\Delta^{-1}\left(s_{T_{i_j}}, a_{i_j}\right)}{t}\right)^{r_j}\right)\right), \\
\Delta\left(t\left(\frac{\Delta^{-1}\left(s_{F_{i_j}}, c_{i_j}\right)}{t}\right)^{r_j}\right)
\end{array}\right\} \tag{97}$$

Thus,

$$\underset{j=1}{\overset{n}{\oplus}}\left(r_j l_{i_j}\right) = \left\{\begin{array}{l}
\Delta\left(t\left(1-\prod_{j=1}^{n}\left(1-\frac{\Delta^{-1}\left(s_{T_{i_j}}, a_{i_j}\right)}{t}\right)^{r_j}\right)\right), \Delta\left(t\prod_{j=1}^{n}\left(\frac{\Delta^{-1}\left(s_{I_{i_j}}, b_{i_j}\right)}{t}\right)^{r_j}\right), \\
\Delta\left(t\prod_{j=1}^{n}\left(\frac{\Delta^{-1}\left(s_{F_{i_j}}, c_{i_j}\right)}{t}\right)^{r_j}\right)
\end{array}\right\} \tag{98}$$

Therefore,

$$\begin{aligned}
&\left(\underset{j=1}{\overset{n}{\oplus}}\left(r_j l_{i_j}\right)\right)^{\prod_{j=1}^n w_{i_j}} \\
&= \left\{\begin{array}{l}
\Delta\left(t\left(1-\prod_{j=1}^{n}\left(1-\frac{\Delta^{-1}\left(s_{T_{i_j}}, a_{i_j}\right)}{t}\right)^{r_j}\right)^{\prod_{j=1}^n w_{i_j}}\right), \Delta\left(t\left(1-\left(1-\prod_{j=1}^{n}\left(\frac{\Delta^{-1}\left(s_{I_{i_j}}, b_{i_j}\right)}{t}\right)^{r_j}\right)^{\prod_{j=1}^n w_{i_j}}\right)\right), \\
\Delta\left(t\left(1-\left(1-\prod_{j=1}^{n}\left(\frac{\Delta^{-1}\left(s_{F_{i_j}}, c_{i_j}\right)}{t}\right)^{r_j}\right)^{\prod_{j=1}^n w_{i_j}}\right)\right)
\end{array}\right\}
\end{aligned} \tag{99}$$

Thereafter,

$$\underset{i_1,i_2,\ldots,i_n=1}{\overset{n}{\otimes}}\left(\underset{j=1}{\overset{n}{\oplus}}\left(r_j l_{i_j}\right)\right)^{\prod_{j=1}^{n} w_{i_j}} = \left\{\begin{array}{l} \Delta\left(t\prod_{i_1,i_2,\ldots,i_n=1}^{n}\left(1-\prod_{j=1}^{n}\left(1-\frac{\Delta^{-1}\left(s_{T_{i_j}},a_{i_j}\right)}{t}\right)^{r_j}\right)^{\prod_{j=1}^{n} w_{i_j}}\right), \\ \Delta\left(t\left(1-\prod_{i_1,i_2,\ldots,i_n=1}^{n}\left(1-\prod_{j=1}^{n}\left(\frac{\Delta^{-1}\left(s_{I_{i_j}},b_{i_j}\right)}{t}\right)^{r_j}\right)^{\prod_{j=1}^{n} w_{i_j}}\right)\right), \\ \Delta\left(t\left(1-\prod_{i_1,i_2,\ldots,i_n=1}^{n}\left(1-\prod_{j=1}^{n}\left(\frac{\Delta^{-1}\left(s_{F_{i_j}},c_{i_j}\right)}{t}\right)^{r_j}\right)^{\prod_{j=1}^{n} w_{i_j}}\right)\right) \end{array}\right\} \quad (100)$$

Furthermore,

$$\begin{aligned} & \text{DG2TLNNWGBM}_w^R(l_1,l_2,\ldots,l_n) \\ & = \frac{1}{\sum_{j=1}^{n} r_j}\left(\underset{i_1,i_2,\ldots,i_n=1}{\overset{n}{\otimes}}\left(\underset{j=1}{\overset{n}{\oplus}}\left(r_j l_{i_j}\right)\right)^{\prod_{j=1}^{n} w_{i_j}}\right) \\ & = \left\{\begin{array}{l} \Delta\left(t\left(1-\left(1-\prod_{i_1,i_2,\ldots,i_n=1}^{n}\left(1-\prod_{j=1}^{n}\left(1-\frac{\Delta^{-1}\left(s_{T_{i_j}},a_{i_j}\right)}{t}\right)^{r_j}\right)^{\prod_{j=1}^{n} w_{i_j}}\right)^{\frac{1}{\sum_{j=1}^{n} r_j}}\right)\right), \\ \Delta\left(t\left(1-\prod_{i_1,i_2,\ldots,i_n=1}^{n}\left(1-\prod_{j=1}^{n}\left(\frac{\Delta^{-1}\left(s_{I_{i_j}},b_{i_j}\right)}{t}\right)^{r_j}\right)^{\prod_{j=1}^{n} w_{i_j}}\right)^{\frac{1}{\sum_{j=1}^{n} r_j}}\right), \\ \Delta\left(t\left(1-\prod_{i_1,i_2,\ldots,i_n=1}^{n}\left(1-\prod_{j=1}^{n}\left(\frac{\Delta^{-1}\left(s_{F_{i_j}},c_{i_j}\right)}{t}\right)^{r_j}\right)^{\prod_{j=1}^{n} w_{i_j}}\right)^{\frac{1}{\sum_{j=1}^{n} r_j}}\right) \end{array}\right\} \end{aligned} \quad (101)$$

Hence, (96) is kept. □

Then, we need to prove that (96) is a 2TLNN. We need to prove two conditions, as follows:

① $0 \le \Delta^{-1}(s_T,a) \le t, 0 \le \Delta^{-1}(s_I,b) \le t, 0 \le \Delta^{-1}(s_F,c) \le t$
② $0 \le \Delta^{-1}(s_T,a) + \Delta^{-1}(s_I,b) + \Delta^{-1}(s_F,c) \le 3t$

Let

$$\frac{\Delta^{-1}(s_T,a)}{t} = 1-\left(1-\prod_{i_1,i_2,\ldots,i_n=1}^{n}\left(1-\prod_{j=1}^{n}\left(1-\frac{\Delta^{-1}\left(s_{T_{i_j}},a_{i_j}\right)}{t}\right)^{r_j}\right)^{\prod_{j=1}^{n} w_{i_j}}\right)^{\frac{1}{\sum_{j=1}^{n} r_j}}$$

$$\frac{\Delta^{-1}(s_I,b)}{t} = \left(1-\prod_{i_1,i_2,\ldots,i_n=1}^{n}\left(1-\prod_{j=1}^{n}\left(\frac{\Delta^{-1}\left(s_{I_{i_j}},b_{i_j}\right)}{t}\right)^{r_j}\right)^{\prod_{j=1}^{n} w_{i_j}}\right)^{\frac{1}{\sum_{j=1}^{n} r_j}}$$

$$\frac{\Delta^{-1}(s_F,c)}{t} = \left(1-\prod_{i_1,i_2,\ldots,i_n=1}^{n}\left(1-\prod_{j=1}^{n}\left(\frac{\Delta^{-1}\left(s_{F_{i_j}},c_{i_j}\right)}{t}\right)^{r_j}\right)^{\prod_{j=1}^{n} w_{i_j}}\right)^{\frac{1}{\sum_{j=1}^{n} r_j}}$$

Proof. ① Since $0 \le \frac{\Delta^{-1}\left(s_{T_{i_j}}, a_{i_j}\right)}{t} \le 1$, we get

$$0 \le 1 - \prod_{j=1}^{n}\left(1 - \frac{\Delta^{-1}\left(s_{T_{i_j}}, a_{i_j}\right)}{t}\right)^{r_j} \le 1 \quad (102)$$

Then,

$$0 \le \prod_{i_1,i_2,\ldots,i_n=1}^{n}\left(1 - \prod_{j=1}^{n}\left(1 - \frac{\Delta^{-1}\left(s_{T_{i_j}}, a_{i_j}\right)}{t}\right)^{r_j}\right)^{\prod_{j=1}^{n} w_{i_j}} \le 1 \quad (103)$$

$$0 \le 1 - \left(1 - \prod_{i_1,i_2,\ldots,i_n=1}^{n}\left(1 - \prod_{j=1}^{n}\left(1 - \frac{\Delta^{-1}\left(s_{T_{i_j}}, a_{i_j}\right)}{t}\right)^{r_j}\right)^{\prod_{j=1}^{n} w_{i_j}}\right)^{\frac{1}{\sum_{j=1}^{n} r_j}} \le 1 \quad (104)$$

That means $0 \le \Delta^{-1}(s_T, a) \le t$, so ① is maintained, similarly, we can get $0 \le \Delta^{-1}(s_I, b) \le t, 0 \le \Delta^{-1}(s_F, c) \le t$. ② Since $0 \le \Delta^{-1}(s_T, a) \le t, 0 \le \Delta^{-1}(s_I, b) \le t, 0 \le \Delta^{-1}(s_F, c) \le t$, we get $0 \le \Delta^{-1}(s_T, a) + \Delta^{-1}(s_I, b) + \Delta^{-1}(s_F, c) \le 3t$. □

Similar to DG2TLNNWBM, the DG2TLNNWGBM has the same properties, as follows. The proof are omitted here to save space.

Property 15. *(Monotonicity) Let* $l_{x_i} = \left\langle \left(s_{T_{x_i}}, a_{x_i}\right), \left(s_{I_{x_i}}, b_{x_i}\right), \left(s_{F_{x_i}}, c_{x_i}\right) \right\rangle (i = 1, 2, \ldots, n)$ *and* $l_{y_i} = \left\langle \left(s_{T_{y_i}}, a_{y_i}\right), \left(s_{I_{y_i}}, b_{y_i}\right), \left(s_{F_{y_i}}, c_{y_i}\right) \right\rangle (i = 1, 2, \ldots, n)$ *be two sets of 2TLNNs. If* $\Delta^{-1}\left(s_{T_{x_i}}, a_{x_i}\right) \le \Delta^{-1}\left(s_{T_{y_i}}, a_{y_i}\right), \Delta^{-1}\left(s_{I_{x_i}}, b_{x_i}\right) \ge \Delta^{-1}\left(s_{I_{y_i}}, b_{y_i}\right)$ *and* $\Delta^{-1}\left(s_{F_{x_i}}, c_{x_i}\right) \ge \Delta^{-1}\left(s_{F_{y_i}}, c_{y_i}\right)$ *hold for all i, then*

$$\text{DG2TLNNWGBM}_w^R(l_{x_1}, l_{x_2}, \cdots, l_{x_n}) \le \text{DG2TLNNWGBM}_w^R(l_{y_1}, l_{y_2}, \cdots, l_{y_n}) \quad (105)$$

Property 16. *(Boundedness) Let* $l_i = \langle (s_{T_i}, a_i), (s_{I_i}, b_i), (s_{F_i}, c_i) \rangle (i = 1, 2, \ldots, n)$ *be a set of 2TLNNs. If* $l^+ = (\max_i(S_{T_i}, a_i), \min_i(S_{I_i}, b_i), \min_i(S_{F_i}, c_i))$ *and* $l^- = (\min_i(S_{T_i}, a_i), \max_i(S_{I_i}, b_i), \max_i(S_{F_i}, c_i))$ *then*

$$l^- \le \text{DG2TLNNWGBM}_w^R(l_1, l_2, \cdots, l_n) \le l^+ \quad (106)$$

6. Numerical Example and Comparative Analysis

6.1. Numerical Example

Given the rise in environmental and resource conservation importance, green supply chain management has seen growth within industry. In addition, balancing economic development and environmental development is one of the critical issues faced by managers to help organizations maintain a strategically competitive position. Green supplier management as one of important part of green supply chain is also critical issue for effective green supply chain management. Thus, in this section, we shall present a numerical example to select green suppliers in green supply chain management with 2TLNNs in order to illustrate the method that is proposed in this paper. There is a panel with five possible green suppliers in green supply chain management A_i $(i = 1, 2, 3, 4, 5)$ to select. The experts selects four attribute to evaluate the five possible green suppliers: ① G_1 is the product quality factor; ② G_2 is the environmental factors; ③ G_3 is the delivery factor; and, ④ G_4 is the price factor. The five possible green suppliers A_i $(i = 1, 2, 3, 4, 5)$ are to be evaluated using the 2TLNNs by the three decision maker under the above four attributes (whose weighting vector $\omega = (0.35, 0.10, 0.25, 0.30)$, expert weighting vector, which are listed in Tables 1–3.

Table 1. 2-tuple linguistic neutrosophic number (2TLNN) decision matrix (R_1).

	G_1	G_2	G_3	G_4
A_1	$<(s_3, 0), (s_2, 0)\ (s_1, 0)>$	$<(s_3, 0), (s_4, 0)\ (s_2, 0)>$	$<(s_4, 0), (s_4, 0)\ (s_2, 0)>$	$<(s_4, 0), (s_4, 0)\ (s_2, 0)>$
A_2	$<(s_4, 0), (s_3, 0)\ (s_2, 0)>$	$<(s_5, 0), (s_4, 0)\ (s_4, 0)>$	$<(s_4, 0), (s_4, 0)\ (s_2, 0)>$	$<(s_4, 0), (s_3, 0)\ (s_3, 0)>$
A_3	$<(s_4, 0), (s_3, 0)\ (s_4, 0)>$	$<(s_3, 0), (s_3, 0)\ (s_2, 0)>$	$<(s_5, 0), (s_3, 0)\ (s_2, 0)>$	$<(s_4, 0), (s_3, 0)\ (s_4, 0)>$
A_4	$<(s_5, 0), (s_5, 0)\ (s_4, 0)>$	$<(s_4, 0), (s_3, 0)\ (s_3, 0)>$	$<(s_5, 0), (s_4, 0)\ (s_5, 0)>$	$<(s_3, 0), (s_4, 0)\ (s_1, 0)>$
A_5	$<(s_3, 0), (s_4, 0)\ (s_2, 0)>$	$<(s_4, 0), (s_5, 0)\ (s_2, 0)>$	$<(s_3, 0), (s_4, 0)\ (s_1, 0)>$	$<(s_4, 0), (s_3, 0)\ (s_2, 0)>$

Table 2. 2TLNN decision matrix (R_2).

	G_1	G_2	G_3	G_4
A_1	$<(s_5, 0), (s_4, 0)\ (s_3, 0)>$	$<(s_3, 0), (s_5, 0)\ (s_2, 0)>$	$<(s_3, 0), (s_1, 0)\ (s_2, 0)>$	$<(s_4, 0), (s_1, 0)\ (s_3, 0)>$
A_2	$<(s_2, 0), (s_3, 0)\ (s_3, 0)>$	$<(s_3, 0), (s_3, 0)\ (s_3, 0)>$	$<(s_3, 0), (s_4, 0)\ (s_2, 0)>$	$<(s_5, 0), (s_4, 0)\ (s_3, 0)>$
A_3	$<(s_5, 0), (s_3, 0)\ (s_3, 0)>$	$<(s_3, 0), (s_2, 0)\ (s_2, 0)>$	$<(s_2, 0), (s_3, 0)\ (s_4, 0)>$	$<(s_3, 0), (s_2, 0)\ (s_4, 0)>$
A_4	$<(s_3, 0), (s_5, 0)\ (s_2, 0)>$	$<(s_4, 0), (s_2, 0)\ (s_3, 0)>$	$<(s_3, 0), (s_4, 0)\ (s_5, 0)>$	$<(s_5, 0), (s_1, 0)\ (s_4, 0)>$
A_5	$<(s_3, 0), (s_3, 0)\ (s_1, 0)>$	$<(s_3, 0), (s_4, 0)\ (s_5, 0)>$	$<(s_4, 0), (s_5, 0)\ (s_1, 0)>$	$<(s_5, 0), (s_3, 0)\ (s_2, 0)>$

Table 3. 2TLNN decision matrix (R_3).

	G_1	G_2	G_3	G_4
A_1	$<(s_5, 0), (s_3, 0)\ (s_1, 0)>$	$<(s_4, 0), (s_2, 0)\ (s_1, 0)>$	$<(s_4, 0), (s_4, 0)\ (s_3, 0)>$	$<(s_4, 0), (s_1, 0)\ (s_3, 0)>$
A_2	$<(s_4, 0), (s_2, 0)\ (s_2, 0)>$	$<(s_4, 0), (s_5, 0)\ (s_4, 0)>$	$<(s_3, 0), (s_2, 0)\ (s_3, 0)>$	$<(s_2, 0), (s_1, 0)\ (s_3, 0)>$
A_3	$<(s_2, 0), (s_1, 0)\ (s_4, 0)>$	$<(s_3, 0), (s_2, 0)\ (s_2, 0)>$	$<(s_4, 0), (s_5, 0)\ (s_2, 0)>$	$<(s_2, 0), (s_4, 0)\ (s_4, 0)>$
A_4	$<(s_5, 0), (s_4, 0)\ (s_4, 0)>$	$<(s_5, 0), (s_4, 0)\ (s_2, 0)>$	$<(s_3, 0), (s_4, 0)\ (s_5, 0)>$	$<(s_5, 0), (s_3, 0)\ (s_1, 0)>$
A_5	$<(s_3, 0), (s_3, 0)\ (s_2, 0)>$	$<(s_4, 0), (s_2, 0)\ (s_2, 0)>$	$<(s_4, 0), (s_2, 0)\ (s_3, 0)>$	$<(s_5, 0), (s_3, 0)\ (s_4, 0)>$

In the following, we utilize the approach that was developed to select green suppliers in green supply chain management.

Definition 19. *Let* $l_j = \left\langle \left(s_{T_j}, a_j\right), \left(s_{I_j}, b_j\right), \left(s_{F_j}, c_j\right) \right\rangle (j = 1, 2, \ldots, n)$ *be a set of 2TLNNs with their weight vector be* $w_i = (w_1, w_2, \ldots, w_n)^T$*, thereby satisfying* $w_i \in [0, 1]$ *and* $\sum_{i=1}^{n} w_i = 1$*, then we can obtain*

$$\mathrm{2TLNNWAA}(l_1, l_2, \ldots, l_n) = \sum_{j=1}^{n} w_j l_j$$
$$= \left\{ \begin{array}{l} \Delta\left(t\left(1 - \prod_{j=1}^{n}\left(1 - \frac{\Delta^{-1}\left(s_{T_j}, a_j\right)}{t}\right)^{w_j}\right)\right), \Delta\left(t\prod_{j=1}^{n}\left(\frac{\Delta^{-1}\left(s_{I_j}, b_j\right)}{t}\right)^{w_j}\right), \\ \Delta\left(t\prod_{j=1}^{n}\left(\frac{\Delta^{-1}\left(s_{F_j}, c_j\right)}{t}\right)^{w_j}\right) \end{array} \right\} \tag{107}$$

$$\mathrm{2TLNNWGA}(l_1, l_2, \ldots, l_n) = \sum_{j=1}^{n} (l_j)^{w_j}$$
$$= \left\{ \begin{array}{l} \Delta\left(t\prod_{j=1}^{n}\left(\frac{\Delta^{-1}\left(s_{T_j}, a_j\right)}{t}\right)^{w_j}\right), \Delta\left(t\left(1 - \prod_{j=1}^{n}\left(1 - \frac{\Delta^{-1}\left(s_{I_j}, b_j\right)}{t}\right)^{w_j}\right)\right), \\ \Delta\left(t\left(1 - \prod_{j=1}^{n}\left(1 - \frac{\Delta^{-1}\left(s_{F_j}, c_j\right)}{t}\right)^{w_j}\right)\right) \end{array} \right\} \tag{108}$$

Step 1. According to 2TLNNs $r_{ij} (i = 1, 2, 3, 4, 5, j = 1, 2, 3, 4)$, we can aggregate all of the 2TLNNs r_{ij} by using the 2TLNNWAA (2TLNNWGA) operator to get the overall 2TLNNs A_i $(i = 1, 2, 3, 4, 5)$ of the green suppliers A_i. Then the aggregating results are shown in Table 4.

Table 4. The aggregating results by the 2TLNNWAA operator.

	G_1	G_2
A_1	<(s_5, 0.000), (s_3, 0.464), (s_2, −0.268)>	<(s_4, −0.449), (s_3, 0.162), (s_1, 0.414)>
A_2	<(s_3, 0.172), (s_2, 0.449), (s_2, 0.449)>	<(s_4, −0.449), (s_4, −0.127), (s_3, 0.464)>
A_3	<(s_4, 0.000), (s_2, −0.268), (s_3, 0.464)>	< (s_3, 0.000), (s_2, 0.000), (s_2, 0.000)>
A_4	<(s_4, 0.268), (s_4, 0.472), (s_3, −0.172)>	<(s_5, −0.414), (s_3, −0.172), (s_2, 0.449)>
A_5	<(s_3, 0.000), (s_3, 0.000), (s_1, 0.414)>	<(s_4, −0.449), (s_3, −0.172), (s_3, 0.162)>
	G_3	G_4
A_1	<(s_4, −0.449), (s_2, 0.000), (s_2, 0.449)>	<(s_4, 0.000), (s_1, 0.000), (s_3, 0.000)>
A_2	<(s_3, 0.000), (s_3, −0.172), (s_2, 0.449)>	<(s_4, 0.000), (s_2, 0.000), (s_3, 0.000)>
A_3	<(s_3, 0.172), (s_4, −0.127), (s_3, 0.172)>	<(s_3, −0.464), (s_3, −0.172), (s_4, 0.000)>
A_4	<(s_3, 0.000), (s_4, 0.000), (s_5, 0.000)>	<(s_5, 0.000), (s_2, −0.268), (s_2, 0.000)>
A_5	<(s_4, 0.000), (s_3, 0.162), (s_2, −0.268)>	<(s_5, 0.000), (s_3, 0.000), (s_3, −0.172)>

Step 2. According to Table 4, we can aggregate all of the 2TLNNs r_{ij} by using the DG2TLNNWBM (DG2TLNNWGBM) operator to get the overall 2TLNNs A_i $(i = 1, 2, 3, 4, 5)$ of the green suppliers A_i. Suppose that $P = (1, 1, 1, 1)$, then the aggregating results are shown in Table 5.

Table 5. The aggregating results of the green suppliers by the DG2TLNNWBM (DG2TLNNWGBM) operator.

	DG2TLNNWBM	DG2TLNNWGBM
A_1	<(s_4, 0.209), (s_2, 0.299), (s_2, 0.251)>	<(s_4, 0.192), (s_2, 0.336), (s_2, 0.262)>
A_2	<(s_3, 0.418), (s_3, −0.454), (s_3, −0.286)>	<(s_3, 0.414), (s_3, −0.446), (s_3, −0.283)>
A_3	<(s_3, 0.259), (s_3, −0.390), (s_3, 0.316)>	<(s_3, 0.250), (s_3, −0.371), (s_3, 0.326)>
A_4	<(s_4, 0.226), (s_3, 0.354), (s_3, 0.067)>	<(s_4, 0.201), (s_3, 0.396), (s_3, 0.108)>
A_5	<(s_4, −0.073), (s_3, 0.023), (s_2, 0.079)>	<(s_4, −0.098), (s_3, 0.023), (s_2, 0.095)>

Step 3. According to the aggregating results shown in Table 5 and the score functions of the green suppliers are shown in Table 6.

Table 6. The score functions of the green suppliers.

	DG2TLNNWBM	DG2TLNNWGBM
A_1	(s_4, −0.114)	(s_4, −0.135)
A_2	(s_3, 0.386)	(s_3, 0.381)
A_3	(s_3, 0.111)	(s_3, 0.098)
A_4	(s_3, 0.268)	(s_3, 0.232)
A_5	(s_4, −0.392)	(s_4, −0.405)

Step 4. According to the score functions shown in Table 6 and the comparison formula of score functions, the ordering of the green suppliers is shown in Table 7. Note that ">" means "preferred to". As we can see, depending on the aggregation operators that were used, the best green supplier is A_1.

Table 7. Ordering of the green suppliers.

	Ordering
DG2TLNNWBM	$A_1 > A_5 > A_2 > A_4 > A_3$
DG2TLNNWGBM	$A_1 > A_5 > A_2 > A_4 > A_3$

6.2. Influence of the Parameter on the Final Result

In order to show the effects on the ranking results by changing parameters of P in the DG2TLNNWBM (DG2TLNNWGBM) operators, all of the results are shown in Tables 8 and 9.

Table 8. Ranking results for different operational parameters of the DG2TLNNWBM operator.

P	$s(A_1)$	$s(A_2)$	$s(A_3)$	$s(A_4)$	$s(A_5)$	Ordering
(1, 1, 1, 1)	$(s_4, -0.114)$	$(s_3, 0.386)$	$(s_3, 0.111)$	$(s_3, 0.268)$	$(s_4, -0.392)$	$A_1 > A_5 > A_2 > A_4 > A_3$
(2, 2, 2, 2)	$(s_5, 0.085)$	$(s_5, -0.267)$	$(s_5, -0.455)$	$(s_5, -0.263)$	$(s_5, -0.073)$	$A_1 > A_5 > A_4 > A_2 > A_3$
(3, 3, 3, 3)	$(s_5, 0.333)$	$(s_5, 0.032)$	$(s_5, -0.078)$	$(s_5, 0.163)$	$(s_5, 0.234)$	$A_1 > A_5 > A_4 > A_2 > A_3$
(4, 4, 4, 4)	$(s_5, 0.410)$	$(s_5, 0.121)$	$(s_5, 0.055)$	$(s_5, -0.326)$	$(s_5, 0.339)$	$A_1 > A_5 > A_4 > A_2 > A_3$
(5, 5, 5, 5)	$(s_5, 0.445)$	$(s_5, 0.156)$	$(s_5, 0.115)$	$(s_5, 0.404)$	$(s_5, 0.389)$	$A_1 > A_4 > A_5 > A_2 > A_3$
(6, 6, 6, 6)	$(s_5, 0.466)$	$(s_5, 0.174)$	$(s_5, 0.149)$	$(s_5, 0.448)$	$(s_5, 0.421)$	$A_1 > A_4 > A_5 > A_2 > A_3$
(7, 7, 7, 7)	$(s_5, 0.482)$	$(s_5, 0.185)$	$(s_5, 0.171)$	$(s_5, 0.476)$	$(s_5, 0.444)$	$A_1 > A_4 > A_5 > A_2 > A_3$
(8, 8, 8, 8)	$(s_5, 0.496)$	$(s_5, 0.195)$	$(s_5, 0.187)$	$(s_5, 0.495)$	$(s_5, 0.463)$	$A_1 > A_4 > A_5 > A_2 > A_3$
(9, 9, 9, 9)	$(s_6, -0.492)$	$(s_5, 0.203)$	$(s_5, 0.200)$	$(s_6, -0.490)$	$(s_5, 0.479)$	$A_4 > A_1 > A_5 > A_2 > A_3$
(10, 10, 10, 10)	$(s_5, -0.482)$	$(s_5, 0.210)$	$(s_5, 0.210)$	$(s_6, -0.479)$	$(s_5, 0.493)$	$A_4 > A_1 > A_5 > A_2 > A_3$

Table 9. Ranking results for different operational parameters of the DG2TLNNWGBM operator.

P	$s(A_1)$	$s(A_2)$	$s(A_3)$	$s(A_4)$	$s(A_5)$	Ordering
(1, 1, 1, 1)	$(s_4, -0.135)$	$(s_3, 0.381)$	$(s_3, 0.098)$	$(s_3, 0.232)$	$(s_4, -0.405)$	$A_1 > A_5 > A_2 > A_4 > A_3$
(2, 2, 2, 2)	$(s_3, -0.046)$	$(s_3, -0.422)$	$(s_2, 0.283)$	$(s_2, 0.300)$	$(s_3, -0.242)$	$A_1 > A_5 > A_2 > A_4 > A_3$
(3, 3, 3, 3)	$(s_3, -0.430)$	$(s_2, 0.326)$	$(s_2, 0.018)$	$(s_2, -0.122)$	$(s_2, 0.450)$	$A_1 > A_5 > A_2 > A_3 > A_4$
(4, 4, 4, 4)	$(s_2, 0.380)$	$(s_2, 0.218)$	$(s_2, -0.102)$	$(s_2, -0.352)$	$(s_2, 0.310)$	$A_1 > A_5 > A_2 > A_3 > A_4$
(5, 5, 5, 5)	$(s_2, 0.271)$	$(s_2, 0.156)$	$(s_2, -0.174)$	$(s_2, -0.498)$	$(s_2, 0.233)$	$A_1 > A_5 > A_2 > A_3 > A_4$
(6, 6, 6, 6)	$(s_2, 0.200)$	$(s_2, 0.111)$	$(s_2, -0.225)$	$(s_1, 0.400)$	$(s_2, 0.186)$	$A_1 > A_5 > A_2 > A_3 > A_4$
(7, 7, 7, 7)	$(s_2, 0.150)$	$(s_2, 0.073)$	$(s_2, -0.263)$	$(s_1, 0.326)$	$(s_2, 0.153)$	$A_5 > A_1 > A_2 > A_3 > A_4$
(8, 8, 8, 8)	$(s_2, 0.113)$	$(s_2, 0.040)$	$(s_2, -0.294)$	$(s_1, 0.269)$	$(s_2, 0.130)$	$A_5 > A_1 > A_2 > A_3 > A_4$
(9, 9, 9, 9)	$(s_2, 0.085)$	$(s_2, 0.011)$	$(s_2, -0.320)$	$(s_1, 0.225)$	$(s_2, 0.113)$	$A_5 > A_1 > A_2 > A_3 > A_4$
(10, 10, 10, 10)	$(s_2, 0.063)$	$(s_2, -0.016)$	$(s_2, -0.343)$	$(s_1, 0.189)$	$(s_2, 0.099)$	$A_5 > A_1 > A_2 > A_3 > A_4$

6.3. Comparative Analysis

Then, we compare our proposed method with other existing methods, including the LNNWAA operator and the LNNWGA operator proposed by Fang & Ye [53] and cosine measures of linguistic neutrosophic numbers [54]. The comparative results are shown in Table 10.

Table 10. Ordering of the green suppliers.

	Ordering
LNNWAA [53]	$A_1 > A_5 > A_4 > A_2 > A_3$
LNNWGA [53]	$A_1 > A_5 > A_4 > A_2 > A_3$
$C^{w_1}{}_{LNNs}$ [54]	$A_1 > A_5 > A_2 > A_3 > A_4$
$C^{w_2}{}_{LNNs}$ [54]	$A_1 > A_5 > A_2 > A_4 > A_3$

From above, we can that we get the same results to show the practicality and effectiveness of the proposed approaches. However, the existing aggregation operators, such as the LNNWAA operator and the LNNWGA operator, do not consider the information about the relationship between the arguments being aggregated, and thus cannot eliminate the influence of unfair arguments on the decision result. Our proposed DG2TLNNWBM and DG2TLNNWGBM operators consider the information about the relationship among arguments being aggregated.

7. Conclusions

In this paper, we investigate the MADM problems with 2TLNNs. Then, we utilize the Bonferroni mean (BM) operator, generalized Bonferroni mean (GBM) operator, and dual generalized Bonferroni mean (DGBM) operator to develop some Bonferroni mean aggregation operators with 2TLNNs: 2-tuple linguistic neutrosophic number weighted Bonferroni mean (2TLNNWBM) operator, 2-tuple linguistic neutrosophic number weighted geometric Bonferroni mean (2TLNNWGBM) operator, generalized 2-tuple linguistic neutrosophic number weighted Bonferroni mean (G2TLNNWBM) operator, generalized 2-tuple linguistic neutrosophic number weighted geometric Bonferroni mean (G2TLNNWGBM) operator, dual generalized 2-tuple linguistic neutrosophic number weighted Bonferroni mean (DG2TLNNWBM) operator, and dual generalized 2-tuple linguistic neutrosophic number weighted geometric Bonferroni mean (DG2TLNNWGBM) operator. The prominent characteristic of these proposed operators are studied. Then, we have utilized these operators to develop some approaches to solve the MADM problems with 2TLNNs. Finally, a practical example for green supplier selection in green supply chain management is given to verify the developed approach and to demonstrate its practicality and effectiveness. In the future, the application of the proposed aggregating operators of 2TLNNs needs to be explored in the decision making, risk analysis, and many other fields under uncertain environments [55–77].

Acknowledgments: The work was supported by the National Natural Science Foundation of China under Grant No. 71571128 and the Humanities and Social Sciences Foundation of Ministry of Education of the People's Republic of China (16XJA630005) and the Construction Plan of Scientific Research Innovation Team for Colleges and Universities in Sichuan Province (15TD0004).

Author Contributions: Jie Wang, Guiwu Wei and Yu Wei conceived and worked together to achieve this work, Jie Wang compiled the computing program by Matlab and analyzed the data, Jie Wang and Guiwu Wei wrote the paper. Finally, all the authors have read and approved the final manuscript.

Conflicts of Interest: The authors declare no conflicts of interest.

References

1. Zadeh, L.A. Fuzzy sets. *Inf. Control* **1965**, *8*, 338–353. [CrossRef]
2. Atanassov, K. Intuitionistic fuzzy sets. *Fuzzy Sets Syst.* **1986**, *20*, 87–96. [CrossRef]
3. Atanassov, K.; Gargov, G. Interval-valued intuitionistic fuzzy sets. *Fuzzy Sets Syst.* **1989**, *31*, 343–349. [CrossRef]
4. Smarandache, F. *Neutrosophy: Neutrosophic Probability, Set, and Logic: Analytic Synthesis & Synthetic Analysis;* American Research Press: Rehoboth, DE, USA, 1998.
5. Wang, H.; Smarandache, F.; Zhang, Y.Q.; Sunderraman, R. Single valued neutrosophic sets. *Multispace Multistructure* **2010**, *4*, 410–413.
6. Wang, H.; Smarandache, F.; Zhang, Y.Q.; Sunderraman, R. *Interval Neutrosophic Sets and Logic: Theory and Applications in Computing;* Hexis: Phoenix, AZ, USA, 2005.
7. Ye, J. A multicriteria decision-making method using aggregation operators for simplified neutrosophic sets. *J. Intell. Fuzzy Syst.* **2014**, *26*, 2459–2466.
8. Ye, J. Multicriteria decision-making method using the correlation coefficient under single-valued neutrosophic environment. *Int. J. Gen. Syst.* **2013**, *42*, 386–394. [CrossRef]
9. Broumi, S.; Smarandache, F. Correlation coefficient of interval neutrosophic set. *Appl. Mech. Mater.* **2013**, *436*, 511–517. [CrossRef]
10. Biswas, P.; Pramanik, S.; Giri, B.C. TOPSIS method for multi-attribute group decision-making under single-valued neutrosophic environment. *Neural Comput. Appl.* **2016**, *27*, 727–737. [CrossRef]
11. Liu, P.D.; Chu, Y.C.; Li, Y.W.; Chen, Y.B. Some generalized neutrosophic number Hamacher aggregation operators and their application to Group Decision Making. *Int. J. Fuzzy Syst.* **2014**, *16*, 242–255.
12. Şahin, R.; Liu, P.D. Maximizing deviation method for neutrosophic multiple attribute decision making with incomplete weight information. *Neural Comput. Appl.* **2016**, *27*, 2017–2029. [CrossRef]
13. Ye, J. Similarity measures between interval neutrosophic sets and their applications in multicriteria decision-making. *J. Intell. Fuzzy Syst.* **2014**, *26*, 165–172.
14. Zhang, H.Y.; Wang, J.Q.; Chen, X.H. Interval neutrosophic sets and their application in multicriteria decision making problems. *Sci. Word J.* **2014**, *2014*, 1–15. [CrossRef] [PubMed]

15. Peng, J.J.; Wang, J.Q.; Zhang, H.Y.; Chen, X.H. An outranking approach for multi-criteria decision-making problems with simplified neutrosophic sets. *Appl. Soft Comput.* **2014**, *25*, 336–346. [CrossRef]
16. Zhang, H.; Wang, J.Q.; Chen, X.H. An outranking approach for multi-criteria decision-making problems with interval-valued neutrosophic sets. *Neural Comput. Appl.* **2016**, *27*, 615–627. [CrossRef]
17. Liu, P.D.; Xi, L. The neutrosophic number generalized weighted power averaging operator and its application in multiple attribute group decision making. *Int. J. Mach. Learn. Cybernet.* **2016**. [CrossRef]
18. Peng, J.J.; Wang, J.Q.; Wu, X.H.; Wang, J.; Chen, X.H. Multi-valued neutrosophic sets and power aggregation operators with their applications in multi-criteria group decision-making problems. *Int. J. Comput. Intell. Syst.* **2015**, *8*, 345–363. [CrossRef]
19. Zhang, H.Y.; Ji, P.; Wang, J.Q.; Chen, X.H. An improved weighted correlation coefficient based on integrated weight for interval neutrosophic sets and its application in multi-criteria decision-making problems. *Int. J. Comput. Intell. Syst.* **2015**, *8*, 1027–1043. [CrossRef]
20. Chen, J.Q.; Ye, J. Some Single-Valued Neutrosophic Dombi Weighted Aggregation Operators for Multiple Attribute Decision-Making. *Symmetry* **2017**, *9*, 82. [CrossRef]
21. Liu, P.D.; Wang, Y.M. Multiple attribute decision making method based on single-valued neutrosophic normalized weighted Bonferroni mean. *Neural Comput. Appl.* **2014**, *25*, 2001–2010. [CrossRef]
22. Wu, X.H.; Wang, J.Q.; Peng, J.J.; Chen, X.H. Cross-entropy and prioritized aggregation operator with simplified neutrosophic sets and their application in multi-criteria decision-making problems. *J. Intell. Fuzzy Syst.* **2016**, *18*, 1104–1116. [CrossRef]
23. Li, Y.; Liu, P.; Chen, Y. Some Single Valued Neutrosophic Number Heronian Mean Operators and Their Application in Multiple Attribute Group Decision Making. *Informatica* **2016**, *27*, 85–110. [CrossRef]
24. Wang, J.; Tang, X.; Wei, G. Models for Multiple Attribute Decision-Making with Dual Generalized Single-Valued Neutrosophic Bonferroni Mean Operators. *Algorithms* **2018**, *11*, 2. [CrossRef]
25. Wei, G.W.; Zhang, Z.P. Some Single-Valued Neutrosophic Bonferroni Power Aggregation Operators in Multiple Attribute Decision Making. *J. Ambient Intell. Humaniz. Comput.* **2018**. [CrossRef]
26. Peng, X.D.; Dai, J.G. Approaches to single-valued neutrosophic MADM based on MABAC, TOPSIS and new similarity measure with score function. *Neural Comput. Appl.* **2018**, *29*, 939–954. [CrossRef]
27. Herrera, F.; Martinez, L. A 2-tuple fuzzy linguistic representation model for computing with words. *IEEE Trans. Fuzzy Syst.* **2000**, *8*, 746–752.
28. Herrera, F.; Martinez, L. An approach for combining linguistic and numerical information based on the 2-tuple fuzzy linguistic representation model in decision-making. *Int. J. Uncertain. Fuzz. Knowl.-Based Syst.* **2000**, *8*, 539–562. [CrossRef]
29. Merigó, J.M.; Casanovas, M.; Martínez, L. Linguistic aggregation operators for linguistic decision making based on the Dempster-Shafer theory of Evidence. *Int. J. Uncertain. Fuzz. Knowl.-Based Syst.* **2010**, *18*, 287–304. [CrossRef]
30. Wei, G.W. A method for multiple attribute group decision making based on the ET-WG and ET-OWG operators with 2-tuple linguistic information. *Expert Syst. Appl.* **2010**, *37*, 7895–7900. [CrossRef]
31. Wei, G.W. Extension of TOPSIS method for 2-tuple linguistic multiple attribute group decision making with incomplete weight information. *Knowl. Inf. Syst.* **2010**, *25*, 623–634. [CrossRef]
32. Merigo, J.M.; Casanovas, M. Decision Making with Distance Measures and Linguistic Aggregation Operators. *Int. J. Fuzzy Syst.* **2010**, *12*, 190–198.
33. Merigo, J.M.; Gil-Lafuente, A.M.; Zhou, L.G.; Chen, H.Y. Generalization of the linguistic aggregation operator and its application in decision making. *J. Syst. Eng. Electron.* **2011**, *22*, 593–603. [CrossRef]
34. Wei, G.W. Grey relational analysis method for 2-tuple linguistic multiple attribute group decision making with incomplete weight information. *Expert Syst. Appl.* **2011**, *38*, 4824–4828. [CrossRef]
35. Wei, G.W. Some harmonic averaging operators with 2-tuple linguistic assessment information and their application to multiple attribute group decision making. *Int. J. Uncertain. Fuzz. Knowl.-Based Syst.* **2011**, *19*, 977–998. [CrossRef]
36. Wei, G.W.; Zhao, X.F. Some dependent aggregation operators with 2-tuple linguistic information and their application to multiple attribute group decision making. *Expert Syst. Appl.* **2012**, *39*, 5881–5886. [CrossRef]
37. Xu, Y.J.; Wang, H.M. Approaches based on 2-tuple linguistic power aggregation operators for multiple attribute group decision making under linguistic environment. *Appl. Soft Comput.* **2011**, *11*, 3988–3997. [CrossRef]

38. Liao, X.W.; Li, Y.; Lu, B. A model for selecting an ERP system based on linguistic information processing. *Inf. Syst.* **2007**, *32*, 1005–1017. [CrossRef]
39. Wang, W.P. Evaluating new product development performance by fuzzy linguistic computing. *Expert Syst. Appl.* **2009**, *36*, 9759–9766. [CrossRef]
40. Tai, W.S.; Chen, C.T. A new evaluation model for intellectual capital based on computing with linguistic variable. *Expert Syst. Appl.* **2009**, *36*, 3483–3488. [CrossRef]
41. Zhao, X.F.; Li, Q.; Wei, G.W. Some prioritized aggregating operators with linguistic information and their application to multiple attribute group decision making. *J. Intell. Fuzzy Syst.* **2014**, *26*, 1619–1630.
42. Jiang, X.P.; Wei, G.W. Some Bonferroni mean operators with 2-tuple linguistic information and their application to multiple attribute decision making. *J. Intell. Fuzzy Syst.* **2014**, *27*, 2153–2162.
43. Beliakov, G.; James, S.; Mordelová, J.; Rückschlossová, T.; Yager, R.R. Generalized Bonferroni mean operators in multicriteria aggregation. *Fuzzy Sets Syst.* **2010**, *161*, 2227–2242. [CrossRef]
44. Bonferroni, C. Sulle medie multiple di potenze. *Boll. Unione Mat. Ital.* **1950**, *5*, 267–270.
45. Xia, M.M.; Xu, Z.; Zhu, B. Generalized intuitionistic fuzzy Bonferroni means. *Int. J. Intell. Syst.* **2012**, *27*, 23–47. [CrossRef]
46. Xu, Z.S.; Yager, R.R. Intuitionistic fuzzy Bonferroni means. *IEEE Trans. Syst. Man Cybern.* **2011**, *41*, 568–578.
47. Yager, R.R. On generalized Bonferroni mean operators for multi-criteria aggregation. *Int. J. Approx. Reason.* **2009**, *50*, 1279–1286. [CrossRef]
48. Wei, G.W. Picture 2-tuple linguistic Bonferroni mean operators and their application to multiple attribute decision making. *J. Intell. Fuzzy Syst.* **2017**, *19*, 997–1010. [CrossRef]
49. Tang, X.Y.; Wei, G.W. Models for green supplier selection in green supply chain management with Pythagorean 2-tuple linguistic information. *IEEE Access* **2018**, *6*, 18042–18060. [CrossRef]
50. Wei, G.W.; Zhao, X.F.; Lin, R.; Wang, H.J. Uncertain linguistic Bonferroni mean operators and their application to multiple attribute decision making. *Appl. Math. Model.* **2013**, *37*, 5277–5285. [CrossRef]
51. Zhu, B.; Xu, Z.S.; Xia, M.M. Hesitant fuzzy geometric Bonferroni means. *Inform. Sci.* **2012**, *205*, 72–85. [CrossRef]
52. Zhang, R.; Wang, J.; Zhu, X.; Xia, M.; Yu, M. Some Generalized Pythagorean Fuzzy Bonferroni Mean Aggregation Operators with Their Application to Multiattribute Group Decision-Making. *Complexity* **2017**, *2017*, 5937376. [CrossRef]
53. Fang, Z.; Ye, J. Multiple Attribute Group Decision-Making Method Based on Linguistic Neutrosophic Numbers. *Symmetry* **2017**, *9*, 111. [CrossRef]
54. Shi, L.; Ye, J. Cosine Measures of Linguistic Neutrosophic Numbers and Their Application in Multiple Attribute Group Decision-Making. *Information* **2017**, *8*, 117.
55. Chen, T. The inclusion-based TOPSIS method with interval-valued intuitionistic fuzzy sets for multiple criteria group decision making. *Appl. Soft Comput.* **2015**, *26*, 57–73. [CrossRef]
56. Wei, G.; Alsaadi, F.E.; Hayat, T.; Alsaedi, A. Projection models for multiple attribute decision making with picture fuzzy information. *Int. J. Mach. Learn. Cybern.* **2018**, *9*, 713–719. [CrossRef]
57. Merigo, J.M.; Gil-Lafuente, A.M. Fuzzy induced generalized aggregation operators and its application in multi-person decision making. *Expert Syst. Appl.* **2011**, *38*, 9761–9772. [CrossRef]
58. Wei, G.W.; Gao, H. The generalized Dice similarity measures for picture fuzzy sets and their applications. *Informatica* **2018**, *29*, 1–18. [CrossRef]
59. Gao, H.; Wei, G.W.; Huang, Y.H. Dual hesitant bipolar fuzzy Hamacher prioritized aggregation operators in multiple attribute decision making. *IEEE Access* **2018**, *6*, 11508–11522. [CrossRef]
60. Merigó, J.M.; Gil-Lafuente, A.M. Induced 2-tuple linguistic generalized aggregation operators and their application in decision-making. *Inform. Sci.* **2013**, *236*, 1–16. [CrossRef]
61. Yu, D.J.; Wu, Y.Y.; Lu, T. Interval-valued intuitionistic fuzzy prioritized operators and their application in group decision making. *Knowl.-Based Syst.* **2012**, *30*, 57–66. [CrossRef]
62. Gao, H.; Lu, M.; Wei, G.W.; Wei, Y. Some novel Pythagorean fuzzy interaction aggregation operators in multiple attribute decision making. *Fundam. Inform.* **2018**, *159*, 385–428. [CrossRef]
63. Wei, G.W.; Alsaadi, F.E.; Hayat, T.; Alsaedi, A. Bipolar fuzzy Hamacher aggregation operators in multiple attribute decision making. *J. Intell. Fuzzy Syst.* **2018**, *20*, 1–12. [CrossRef]
64. Wei, G.W. Some similarity measures for picture fuzzy sets and their applications. *Iran. J. Fuzzy Syst.* **2018**, *15*, 77–89.

65. Wei, G.W.; Wei, Y. Similarity measures of Pythagorean fuzzy sets based on cosine function and their applications. *Int. J. Intell. Syst.* **2018**, *33*, 634–652. [CrossRef]
66. Wei, G.W.; Lu, M. Pythagorean fuzzy power aggregation operators in multiple attribute decision making. *Int. J. Intell. Syst.* **2018**, *33*, 169–186. [CrossRef]
67. Ma, Z.M.; Xu, Z.S. Symmetric Pythagorean Fuzzy Weighted Geometric_Averaging Operators and Their Application in Multicriteria Decision-Making Problems. *Int. J. Intell. Syst.* **2016**, *31*, 1198–1219. [CrossRef]
68. Wei, G.W.; Lu, M. Pythagorean Fuzzy Maclaurin Symmetric Mean Operators in multiple attribute decision making. *Int. J. Intell. Syst.* **2018**, *33*, 1043–1070. [CrossRef]
69. Wei, G.W.; Alsaadi, F.E.; Hayat, T.; Alsaedi, A. Picture 2-tuple linguistic aggregation operators in multiple attribute decision making. *Soft Comput.* **2018**, *22*, 989–1002. [CrossRef]
70. Wei, G.W. Picture fuzzy Hamacher aggregation operators and their application to multiple attribute decision making. *Fundam. Inform.* **2018**, *157*, 271–320. [CrossRef]
71. Merigo, J.M.; Casanovas, M. Induced aggregation operators in decision making with the Dempster-Shafer belief structure. *Int. J. Intell. Syst.* **2009**, *24*, 934–954. [CrossRef]
72. Wei, G.W.; Lu, M.; Tang, X.Y.; Wei, Y. Pythagorean Hesitant Fuzzy Hamacher Aggregation Operators and Their Application to Multiple Attribute Decision Making. *Int. J. Intell. Syst.* **2018**, 1–37. [CrossRef]
73. Wei, G.W.; Gao, H.; Wei, Y. Some q-Rung Orthopair Fuzzy Heronian Mean Operators in Multiple Attribute Decision Making. *Int. J. Intell. Syst.* **2018**. [CrossRef]
74. Xu, Z.; Gou, X. An overview of interval-valued intuitionistic fuzzy information aggregations and applications. *Granul. Comput.* **2017**, *2*, 13–39. [CrossRef]
75. Meng, S.; Liu, N.; He, Y. GIFIHIA operator and its application to the selection of cold chain logistics enterprises. *Granul. Comput.* **2017**, *2*, 187–197. [CrossRef]
76. Wang, C.; Fu, X.; Meng, S.; He, Y. Multi-attribute decision making based on the SPIFGIA operators. *Granul. Comput.* **2017**, *2*, 321–331. [CrossRef]
77. Xu, Z.; Wang, H. Managing multi-granularity linguistic information in qualitative group decision making: An overview. *Granul. Comput.* **2016**, *1*, 21–35. [CrossRef]

symmetry

MDPI

Article

Multiple Criteria Group Decision-Making Considering Symmetry with Regards to the Positive and Negative Ideal Solutions via the Pythagorean Normal Cloud Model for Application to Economic Decisions

Jinming Zhou [1,2], Weihua Su [2], Tomas Baležentis [3,*] and Dalia Streimikiene [3]

1 School of Mathematics and Physics, Anhui Polytechnic University, Wuhu 241000, China; zjm@ahpu.edu.cn
2 School of Statistics and Mathematics, Zhejiang Gongshang University, Hangzhou 310018, China; zjsuweihua@163.com
3 Lithuanian Institute of Agrarian Economics, V. Kudirkos Str. 18, LT-03105 Vilnius, Lithuania; dalia@mail.lei.it
* Correspondence: tomas@laei.it; Tel.: +370-5-2622459

Received: 14 April 2018; Accepted: 26 April 2018; Published: 1 May 2018

Abstract: Pythagorean fuzzy sets are highly appealing in dealing with uncertainty as they allow for greater flexibility in regards to the membership and non-membership degrees by extending the set of possible values. In this paper, we propose a multi-criteria group decision-making approach based on the Pythagorean normal cloud. Some cloud aggregation operators are presented in this paper to facilitate the appraisal of the underlying utilities of the alternatives under consideration. The concept and properties of the Pythagorean normal cloud and its backward generation algorithm, aggregation operators and distance measurement are outlined. The proposed approach resembles the TOPSIS technique, which, indeed, considers the symmetry of the distances to the positive and negative ideal solutions. Furthermore, an example from e-commerce is presented to demonstrate and validate the proposed decision-making approach. Finally, the comparative analysis is implemented to check the robustness of the results when the aggregation rules are changed.

Keywords: Pythagorean fuzzy set; normal cloud; MCGDM; backward cloud transformation

1. Introduction

Decision-making is an important issue in the domain of economics and society in general [1,2], ass human input and interaction are often the decisive elements of the decision-making. Accordingly, expressing and handling cognitive information has been a focal topic related to the decision-making literature. What is more, it has been established that the use of exact ratings (e.g., exact numerical values) might not allow defining the preferences of the decision-makers to a substantial degree [3–5], which might reduce the effectiveness of the decision-making in general [6]. What is more, a decision-maker can fathom the limitations of his/her competences or possibilities to provide ratings in regards to certain alternatives and criteria in general. Realizing this, they might attach the corresponding information to their ratings [7,8], thus providing an additional dimension in the decision-making process. All in all, the information rendered by the decision-makers might be imprecise (in the case that no exact values are provided), incomplete (in the case that certain values are missing) and uncertain (in the case that the likelihood of observing different values can be specified). Under these circumstances, the theory of the fuzzy sets can be regarded as a possible means for handling the decision-making process and overcoming the limitations, which would have existed if conventional tools (e.g., crisp sets) had been applied. In order to account for different structures of uncertainty, the literature has proposed different strands of the fuzzy set theory since its initial definition by Zadeh [9,10]. In particular, the intuitionistic fuzzy set (IFS) theory [11] and interval-valued IFS theory [12] were put forward by Atanassov et al. Indeed, the application of such concepts allows one to account for the incompleteness and inconsistencies existing in information provided by the decision-makers. In principle, this implies that the underlying cognitive peculiarities of decision-makers can be accounted for. Yet, another example of the concepts for handling imprecise information is the hesitant fuzzy set proposed by Torra [13], which allows considering the hesitancy to provide certain ratings of the alternatives. The establishment of multiple theoretical concepts for imprecise information provides opportunities for more realistically handling multi-criteria decision-making problems (MCDM) in general.

However, the very existence of the multitude of the fuzzy set-based concepts does not warranty successful implementation of these in the area of MCDM. The practical implementation of the fuzzy sets requires certain restrictions to be satisfied. Turning to IFS, which is one of the most flexible tools for handling imprecise information, the decision-makers need to ensure that the sum of the degrees of membership and non-membership is not greater than unity. However, in some fuzzy MCDM problems, the decision-makers may fail to deliver their ratings in line with the requirements on the sum of the degrees of membership and non-membership as required by the theory of the IFS. In such instances, the application of the IFSs becomes rather complicated (e.g., decision-maker might be asked to reiterate the procedure of rating) and even impossible. Therefore, there have been certain attempts to rectify this shortcoming by modifying the underlying assumptions. The Pythagorean fuzzy sets (PFSs) proposed by Yager [14] can be identified as an option for modeling situations that cannot be defined in terms of IFSs due to difficulties associated with the restrictions on the degrees of membership and non-membership.

In order to ensure that the PFSs can be successfully applied in MCDM, dedicated techniques have been proposed. The aggregation operators for the PFSs can be considered as an important tool for the application of the PFSs in the MCDM problematique. The correspondence among membership degrees to the Pythagorean fuzzy numbers (PFNs) and the complex numbers was established by Yager [15]. More specifically, it was shown that the degrees of membership to PFS can be treated as a special subclass of complex numbers. The mathematical representations of the PFSs were further reviewed by Liang et al. [16]. They also defined the PFNs. The TOPSIS technique was then extended with the PFNs, providing the mathematical expressions of PFSs and presenting the concept of the PFN. Furthermore, the latter study put forward the Pythagorean fuzzy TOPSIS (technique for order preference by similarity to an ideal solution) for handling the MCDM problems with PFNs. The averaging functions for the PFSs were discussed by Beliakov and James [17]. In particular, they sought to ensure that the aggregation of the membership degrees of the PFSs led to consistent results. The goal of operationalizing the collaboration-based recommender system by using the PFSs was addressed by Reformat and Yager [18]. Zeng [19,20] developed a Pythagorean fuzzy multi-attribute group decision-making (MAGDM) method based on probabilistic information and the ordered weighted averaging (OWA) approach.

Even though the fuzzy sets can describe the degree of membership to a certain concept with regards to the attitude of the decision-maker, as well as their confidence, the spread of such ratings might not be fully represented. Accordingly, the need for a more flexible representation of the uncertain data triggered the development of the normal cloud (NC) concept. Li et al. [21] acknowledged the random nature of the membership functions and unified the probability theory and fuzzy set theory, thus devising the NC. Taking the normal distribution as a reference, one can employ the NC to model deviations from the theoretical distribution. By doing so, one is able to describe the random phenomena and use this information in MCDM [22]. There are three numerical characteristics that characterize the random phenomenon in terms of an NC: Ex (expectation), En (entropy) and He (hyper entropy). Ex is the expected value of the sample data; En is the spread of the sample values defining the uncertainty of the sample; He is the uncertainty of the degree of membership. The NC theory has been revised by introducing the additional types of NCs. Numerous extensions of the NC have been developed. For example, the integral cloud was put forward by Li et al. [23]. In addition, a multidimensional cloud was proposed [24]. Jiang et al. [25] developed a trapezoidal cloud (TC) model. The combination of intuitionistic fuzzy set theory and conventional NC theory was offered by Wang and Yang [26] and yielded the intuitionistic normal cloud (INC) model.

As NCs represent uncertain information, they can be applied to handle MCDM problems [26]. The distance measure, the similarity measure, the entropy and the inclusion measure for PFSs have been discussed by Peng et al. [27,28]. The cloud generator algorithm was applied by Yang et al. [29] in a linguistic hesitant fuzzy decision-making framework. Distance measures play an important role in constructing the NC-based procedures for MCDM [30]. The aggregation of NCs is yet another research avenue that deserves much attention. When proposing the linguistic MCDM technique, such aggregation operators as the cloud weighted arithmetic averaging operator, cloud weighted geometric averaging operator, cloud-ordered weighted arithmetic averaging operator and cloud hybrid aggregation operator were developed by Wang et al. [31]. Given the existing different types of the clouds, the aggregation operators have been revised accordingly. Therefore, the operators for TC and INC have been proposed. Wang et al. [32] presented a number of arithmetic aggregation operators for the TCs (including weighted arithmetic averaging operator, ordered weighted arithmetic averaging operator and hybrid arithmetic operator). Turning to the PFNs, there have also been advancements in the sense of the development of the aggregation operators and rules of comparison. The Pythagorean fuzzy uncertain linguistic Maclaurin symmetric mean aggregation (PFULMSMA) operator and the weighted PFULMSMA (WPFULMSMA) operator have been put forward by Liu et al. [33]. Furthermore, Garg presented an improved accuracy function for the ranking order of interval-valued Pythagorean fuzzy sets (IVPFSs) [34].

This paper combines the notion of the NC with the PFS and develops the Pythagorean normal cloud (PNC). Thereafter, the group decision-making procedure based on the PNCs is proposed. the proposed approach relies on the backward cloud generator, aggregation operators and distance measures to deal with the proposed PNCs. These concepts are presented in the paper. In the proposed framework, the ratings provided by the experts are treated as could drops of PNCs. The proposed approach is based on the TOPSIS (Technique for Order of Preference by Similarity to Ideal Solution) [35]. Application of the backward cloud generators allows considering the spread of the ratings provided by the experts rather than the average values only. The use of the aggregation operators allows constructing the aggregate indicators, which can further be used for decision-making. The paper concludes with an illustrative example where the proposed approach is tested by considering the case study of e-commerce.

2. Preliminaries

This section presents the focal concepts underpinning the proposed approach for cloud-based MCDM. More specifically, we describe the IFSs and a generalization thereof, namely the PFSs. The, we discuss the normal clouds and the generator algorithm, which allows transforming drops into a cloud. The concepts presented in this section will be further revised by incorporating the PFSs.

2.1. Intuitionistic Fuzzy Sets and Pythagorean Fuzzy Sets

The conventional fuzzy set has been generalized by Atanassov [11]. The resulting concept was termed intuitionistic fuzzy set. The IFS allows for a more detailed representation of vagueness and uncertainty, which makes it a promising tool for MCDM problems. The key feature of the IFS that makes it different from the conventional fuzzy sets is the different set of parameters describing membership to a certain fuzzy set. Therefore, an IFS can be defined as follows:

Definition 1. *Assuming there exists a certain fixed set $X = \{x_1, x_2, \cdots, x_n\}$, one can define an instance of IFS I in the following terms:*

$$I = \{< x, I(\mu_I(x), \nu_I(x)) > | x \in X\} \tag{1}$$

Values $\mu_I(x)$ and $\nu_I(x)$ are the degrees of membership and non-membership, respectively, and they define the extent to which a certain element x belongs to set I, under condition $0 \leq \mu_I(x) + \nu_I(x) \leq 1$, for all $x \in X$. Given the presence of inequality in the condition for the degrees of membership and non-membership to an IFS, their sum might be lower than unity, which would imply the presence of indeterminacy in the decision-making process. Formally, the degree of indeterminacy for x with regards to X is defined as $\pi_I(x) = 1 - \mu_I(x) - \nu_I(x)$. Indeed, the MCDM requires assessments of multiple alternatives against different criteria, which is associated with assessing the membership of the elements to sets multiple times. In order to define this process in a more concise manner, the notion of the intuitionistic fuzzy number (IFN) has been introduced [33,34]. Specifically, a certain IFN can be defined in terms of the two-tuple containing the degrees of membership and non-membership, i.e., pair $(\mu_I(x), \nu_I(x))$. Furthermore, one can introduce notation $\alpha = I(\mu_\alpha, \nu_\alpha)$ for IFN α, where $\mu_\alpha \in [0, 1], \nu_\alpha \in [0, 1], \mu_\alpha + \nu_\alpha \leq 1$.

For any three IFNs $\alpha = I(\mu_\alpha, \nu_\alpha), \alpha_1 = I(\mu_{\alpha_1}, \nu_{\alpha_1})$ and $\alpha_2 = I(\mu_{\alpha_2}, \nu_{\alpha_2})$, some operational laws of IFNs are introduced as follows [36]:

1. $\alpha_1 \oplus \alpha_2 = I(\mu_{\alpha_1} + \mu_{\alpha_2} - \mu_{\alpha_1}\mu_{\alpha_2}, \nu_{\alpha_1}\nu_{\alpha_2})$;
2. $\alpha_1 \otimes \alpha_2 = I(\mu_{\alpha_1}\mu_{\alpha_2}, \nu_{\alpha_1} + \nu_{\alpha_2} - \nu_{\alpha_1}\nu_{\alpha_2})$;
3. $\lambda\alpha = I(1 - (1 - \mu_\alpha)^\lambda, \nu_\alpha^\lambda), \lambda > 0$;
4. $\alpha^\lambda = I(\mu_\alpha^\lambda, 1 - (1 - \nu_\alpha)^\lambda), \lambda > 0$.

The experts' preferences might not be defined in terms of IFSs in case the experts cannot ensure that the constraint on the membership and non-membership degrees is maintained (i.e., their sum is not less than unity). Indeed, this might happen whenever the experts are not familiar with the IFS theory. In order to reduce the likelihood of such situations and improve the possibilities for the application of the fuzzy sets in the MCDM, a generalization of the IFS has been offered by Yager [14,15]. The resulting concept was termed PFS. The PFS can be defined as follows:

Definition 2. *Let us consider a fixed set $X = \{x_1, x_2, \cdots, x_n\}$, then a PFS P is defined in the following terms:*

$$P = \{< x, P(\mu_P(x), \nu_P(x)) > | x \in X\} \tag{2}$$

Similarly to the case of IFS, values $\mu_P(x)$ and $\nu_P(x)$ are the degrees of membership and non-membership, which define the extent to which a certain element x belongs to set P. However, the constraints on these two values are altered in the case of the PFS so that $0 \leq \mu_P^2(x) + \nu_P^2(x) \leq 1$, for all $x \in X$. These changes imply that the calculation of the degree of indeterminacy is changed, as well: for any PFS P and $x \in X$, the degree of indeterminacy is calculated as $\pi_P(x) = \sqrt{1 - \mu_P^2(x) - \nu_P^2(x)}$. For the ease of notation, let the pair $(\mu_P(x), \nu_P(x))$ be called the Pythagorean fuzzy number (PFN) [14]. Further on, a shorthand notation can be used to refer to a certain PFN, namely $\beta = P(\mu_\beta, \nu_\beta)$, where the usual conditions hold $\mu_\beta \in [0, 1], \nu_\beta \in [0, 1]$, and $\mu_\beta^2 + \nu_\beta^2 \leq 1$.

According to Definitions 1 and 2, one can note that the key delineation between PFN and IFN is the way degrees of membership and non-membership are restricted. More specifically, the case of IFN involves $0 \leq \mu_I(x) + \nu_I(x) \leq 1$, whereas the corresponding constraint in the case of PFN is $0 \leq \mu_P^2(x) + \nu_P^2(x) \leq 1$. The relationships among IFNs and PFNs can be established by considering simple mathematical facts. Note that for any given set of values $(a, b), (a, b \in [0, 1])$, if $a + b \leq 1$, then $a^2 + b^2 \leq 1$; thus, if a certain number is an IFN, then it is definitely a PFN, yet the opposite does not hold.

Given three PFNs $\beta = I(\mu_\beta, \nu_\beta), \beta_1 = I(\mu_{\beta_1}, \nu_{\beta_1})$ and $\beta_2 = I(\mu_{\beta_2}, \nu_{\beta_2})$, Zhang et al. [37] presented the main operations for them, shown as:

1. $\beta_1 \oplus \beta_2 = P(\sqrt{\mu_{\beta_1} + \mu_{\beta_2} - \mu_{\beta_1}\mu_{\beta_2}}, \nu_{\beta_1}\nu_{\beta_2})$;
2. $\beta_1 \otimes \beta_2 = P(\mu_{\beta_1}\mu_{\beta_2}, \sqrt{\nu_{\beta_1} + \nu_{\beta_2} - \nu_{\beta_1}\nu_{\beta_2}})$;

3. $\lambda\beta = P(1-(1-\mu_\beta)^\lambda, v_\beta^\lambda), \lambda > 0$;
4. $\beta^\lambda = I(\mu_\beta^\lambda, 1-(1-v_\beta)^\lambda), \lambda > 0$.

Definition 3. *Given any two PFNs, $\beta_j = P(\mu_{\beta_j}, v_{\beta_j}), j = 1,2$, there can be a natural quasi-ordering on the PFNs established in the following manner: $\beta_1 \geq \beta_2$ if and only if $\mu_{\beta_1} \geq \mu_{\beta_2}$ and $v_{\beta_1} \leq v_{\beta_2}$.*

In order to facilitate the comparison of PFNs, Zhang and Xu [38] defined the following principles:

Definition 4. *For a PFN $\beta = P(\mu_\beta, v_\beta)$, $s(\beta) = \mu_\beta^2 - v_\beta^2$ is referred to as the score function of β. The score function can then be exploited when comparing these PFNs. For two PFNs $\beta_1 = P(\mu_{\beta_1}, v_{\beta_1})$ and $\beta_2 = P(\mu_{\beta_2}, v_{\beta_2})$, if $s(\beta_1) > s(\beta_2)$, then $\beta_1 \geq \beta_2$; if $s(\beta_1) = s(\beta_2)$, then $\beta_1 = \beta_2$.*

Definition 5. *Let $\beta_1 = P(\mu_{\beta_1}, v_{\beta_1})$ and $\beta_2 = P(\mu_{\beta_2}, v_{\beta_2})$ be two PFNs, then:*

$$d_{PFD}(\beta_1, \beta_2) = \frac{1}{2}(|\mu_{\beta_1}^2 - \mu_{\beta_2}^2| + |v_{\beta_1}^2 - v_{\beta_2}^2| + |\pi_{\beta_1}^2 - \pi_{\beta_2}^2|) \tag{3}$$

is referred to as the Pythagorean fuzzy distance (PFD) between β_1 and β_2.

2.2. NC and the Backward Cloud Generator

The observed sample data can be used to recover the underlying data generation process (DGP). In the case of multi-criteria decision-making, this procedure can be used to describe uncertain phenomena. In this sub-section, we discuss the procedure for establishing an NC, which represents the underlying DGP.

Definition 6. *Let us assume there exists a universe of discourse denoted by U. Furthermore, let there be a qualitative concept in U that is denoted as T. Then, let $x \in U$ be a random realization of concept T, such that x follows $x \sim N(Ex, (En^*)^2)$, where $En^* \sim N(En, He^2)$. Given the conditions on the distribution of x, one can model the degree of certainty that x belongs to the concept T in the following way:*

$$y = \exp(-\frac{(x-Ex)^2}{2(En^*)^2}) \tag{4}$$

Thus, an NC defines the distribution of x in the universe U. In particular, a certain value of x is attached with a corresponding degree of certainty y, thus forming a cloud drop. The backward cloud generator of an NC allows aggregating separate drops into a cloud that defines the concept under analysis in a general manner. The backward generator proceeds as follows:

Step 1. Calculate the sample average $\bar{X} = \frac{1}{n}\sum_{i=1}^n x_i$ along with first-order sample absolute central moment $\frac{1}{n}\sum_{i=1}^n |x_i - \bar{X}|$, and sample variance $S^2 = \frac{1}{n-1}\sum_{i=1}^n (x_i - \bar{X})^2$;

Step 2. Obtain the estimates of Ex, He, En: $\hat{Ex} = \bar{X}$, $\hat{En} = \sqrt{\frac{\pi}{2}\frac{1}{n}\sum_{i=1}^n |x_i - \hat{Ex}|}$, $\hat{He} = \sqrt{S^2 - \frac{1}{3}\hat{En}^2}$;

Output: The estimate $(\hat{Ex}, \hat{En}, \hat{He})$ of (Ex, En, He).

3. Pythagorean Normal Cloud

The NC only defines the uncertainty surrounding the membership to a concept. However, the assessments might be associated with different degrees of confidence. This situation is present in IFSs and PFSs. Therefore, we update the concept of the NC with the PFSs in order to derive a more comprehensive means of the representations of uncertain information.

Definition 7. *For a given universe of discourse, U, one can characterize a PNC C in U in terms of Ex, En and He. Furthermore, Ex can be represented by a Pythagorean fuzzy number (PFN) $< Ex, \mu_\beta, v_\beta >$.*

Then, the PNC C is defined as: $C(<Ex, \mu_\beta, \nu_{\beta>}, En, He)$

Definition 8. *Let there be a set of PNCs* $C_i(<Ex_i, \mu_{\beta_i}, \nu_{\beta_i}>, En_i, He_i), i = 1, 2, \cdots, n$ *with associated weighting vector* $w = (w_1, w_2, \cdots, w_n)$ *of* $(C_1, C_2, \cdots, C_n)$*, such that* $w_i \in [0,1], (i = 1, 2, \cdots, n)$ *and* $\sum_{i=1}^{n} w_i = 1$*. The PNCWAA operator is:*

$$PNCWAA_w(C_1, C_2, \cdots, C_n) = (<\sum_{i=1}^{n} w_i Ex_i, \frac{\sum_{i=1}^{n} \mu_{\beta_i} w_i Ex_i}{\sum_{i=1}^{n} w_i Ex_i}, \frac{\sum_{i=1}^{n} \nu_{\beta_i} w_i Ex_i}{\sum_{i=1}^{n} w_i Ex_i}>, \sqrt{\sum_{i=1}^{n} w_i En_i^2}, \sqrt{\sum_{i=1}^{n} w_i He_i^2}) \quad (5)$$

In the earlier literature, Wang et al. discussed the main operations for NC [32], whereas other studies further developed those for INC [26]; besides, arithmetic operations for handling the TCs were outlined [25]. Based on the earlier literature, we present the following operational laws for the PNCs:

Definition 9. *Let* $C_1(<Ex_1, \mu_{\beta_1}, \nu_{\beta_1}>, En_1, He_1)$ *and* $C_2(<Ex_2, \mu_{\beta_2}, \nu_{\beta_2}>, En_2, He_2)$ *be two PNCs. Then, the following operational rules apply for* C_1 *and* C_2*:*

1. $C_1 + C_2 = (<Ex_1 + Ex_2, \frac{\mu_{\beta_1} Ex_1 + \mu_{\beta_2} Ex_2}{Ex_1 + Ex_2}, \frac{\nu_{\beta_1} Ex_1 + \nu_{\beta_2} Ex_2}{Ex_1 + Ex_2}>, \sqrt{En_1^2 + En_2^2}, \sqrt{He_1^2 + He_2^2})$;
2. $C_1 \times C_2 = (<Ex_1 Ex_2, \frac{\mu_{\beta_1} Ex_1 + \mu_{\beta_2} Ex_2}{Ex_1 Ex_2}, \frac{\nu_{\beta_1} Ex_1 + \nu_{\beta_2} Ex_2}{Ex_1 Ex_2}>, \sqrt{(Ex_2 En_1)^2 + (Ex_1 En_2)^2}, \sqrt{(Ex_2 He_1)^2 + (Ex_1 He_2)^2})$;
3. $\lambda C_1 = (<\lambda Ex_1, \mu_{\beta_1}, \nu_{\beta_1}>, \sqrt{\lambda} En_1, \sqrt{\lambda} He_1)$;
4. $C_1^\lambda = (<Ex_1^\lambda, \mu_{\beta_1}, \nu_{\beta_1}>, \sqrt{\lambda} Ex_1^{\lambda-1} En_1, \sqrt{\lambda} Ex_1^{\lambda-1} He_1)$;

Theorem 1. *Let there be any three PNCs* $C_1(<Ex_1, \mu_{\beta_1}, \nu_{\beta_1}>, En_1, He_1)$, $C_2(<Ex_2, \mu_{\beta_2}, \nu_{\beta_2}>, En_2, He_2)$ *and* $C_3(<Ex_3, \mu_{\beta_3}, \nu_{\beta_3}>, En_3, He_3)$*. For these PNCs, the following observations hold:*

1. $C_1 + C_2 = C_2 + C_1$;
2. $(C_1 + C_2) + C_3 = C_1 + (C_2 + C_3)$;
3. $\lambda(C_1 + C_2) = \lambda C_1 + \lambda C_2$;
4. $\lambda_1 C_1 + \lambda_2 C_1) = (\lambda_1 + \lambda_2) C_1$;
5. $C_1 \times C_2 = C_2 \times C_1$.

Proof. According to Definition 9, we can obtain

$$\begin{aligned} & C_1 + C_2 \\ & = (<Ex_1 + Ex_2, \frac{\mu_{\beta_1} Ex_1 + \mu_{\beta_2} Ex_2}{Ex_1 + Ex_2}, \frac{\nu_{\beta_1} Ex_1 + \nu_{\beta_2} Ex_2}{Ex_1 + Ex_2}>, \sqrt{En_1^2 + En_2^2}, \sqrt{He_1^2 + He_2^2}) \\ & = (<Ex_2 + Ex_1, \frac{\mu_{\beta_2} Ex_2 + \mu_{\beta_1} Ex_1}{Ex_2 + Ex_1}, \frac{\nu_{\beta_2} Ex_2 + \nu_{\beta_1} Ex_1}{Ex_2 + Ex_1}>, \sqrt{En_2^2 + En_1^2}, \sqrt{He_2^2 + He_1^2}) \\ & = C_2 + C_1; \end{aligned}$$

According to Definition 9, we can also obtain

$$\begin{aligned} & (C_1 + C_2) + C_3 \\ & = (<(Ex_1 + Ex_2) + Ex_3, \frac{(\mu_1 Ex_1 + \mu_2 Ex_2) + \mu_3 Ex_3}{(Ex_1 + Ex_2) + Ex_3}, \frac{(\nu_1 Ex_1 + \nu_2 Ex_2) + \nu_3 Ex_3}{(Ex_1 + Ex_2) + Ex_3}>, \sqrt{(En_1^2 + En_2^2) + En_3^2}, \\ & \sqrt{(He_1^2 + He_2^2) + He_3^2}) \\ & = (<Ex_1 + (Ex_2 + Ex_3), \frac{\mu_1 Ex_1 + (\mu_2 Ex_2 + \mu_3 Ex_3)}{Ex_1 + (Ex_2 + Ex_3)}, \frac{\nu_1 Ex_1 + (\nu_2 Ex_2 + \nu_3 Ex_3)}{Ex_1 + (Ex_2 + Ex_3)}>, \sqrt{En_1^2 + (En_2^2 + En_3^2)}, \\ & \sqrt{He_1^2 + (He_2^2 + He_3^2)}) \\ & = C_1 + (C_2 + C_3); \end{aligned}$$

According to Definition 9, we can obtain

$$\begin{aligned} & \lambda C_1 + \lambda C_2 \\ & = (<\lambda Ex_1 + \lambda Ex_2, \frac{\mu_1 \lambda Ex_1 + \mu_2 \lambda Ex_2}{\lambda Ex_1 + \lambda Ex_2}, \frac{\nu_1 \lambda Ex_1 + \nu_2 \lambda Ex_2}{\lambda Ex_1 + \lambda Ex_2}>, \sqrt{\lambda En_1^2 + \lambda En_2^2}, \sqrt{\lambda He_1^2 + \lambda He_2^2}) \\ & = (<\lambda(Ex_1 + Ex_2), \frac{\mu_1 Ex_1 + \mu_2 Ex_2}{Ex_1 + Ex_2}, \frac{\mu_1 Ex_1 + \mu_2 Ex_2}{Ex_1 + Ex_2}>, \sqrt{\lambda}\sqrt{En_1^2 + En_2^2}, \sqrt{\lambda}\sqrt{He_1^2 + He_2^2}) \end{aligned}$$

$= \lambda(C_1 + C_2);$

The proof for the fourth result of Theorem 1 is similar to that for the third result.

According to Definition 9, we can also note

$C_1 \times C_2$
$= (< Ex_1Ex_2, \frac{\mu_{\beta_1}Ex_1+\mu_{\beta_2}Ex_2}{Ex_1Ex_2}, \frac{\nu_{\beta_1}Ex_1+\nu_{\beta_2}Ex_2}{Ex_1Ex_2} >, \sqrt{(Ex_2En_1)^2 + (Ex_1En_2)^2}, \sqrt{(Ex_2He_1)^2 + (Ex_1He_2)^2})$
$= (< Ex_2Ex_1, \frac{\mu_{\beta_2}Ex_2+\mu_{\beta_1}Ex_1}{Ex_2Ex_1}, \frac{\nu_{\beta_2}Ex_2+\nu_{\beta_1}Ex_1}{Ex_2Ex_1} >, \sqrt{(Ex_1En_2)^2 + (Ex_2En_1)^2}, \sqrt{(Ex_1He_2)^2 + (Ex_2He_1)^2})$
$= C_2 \times C_1$ □

3.1. Backward Cloud Generator and Aggregation Operators for PNCs

The extensive form of the data describing a certain concept (i.e., cloud drops) can be aggregated into an intensive form describing the same concept (i.e., cloud) by means of the backward cloud generator [39]. In general, the sample data are used in the backward cloud generator algorithm of a PNC to recover the estimates $(< \hat{Ex}, \hat{\mu}_\beta, \hat{\nu}_\beta >, \hat{En}, \hat{He})$, which describe a PFN. In the context of group decision-making, the backward cloud generator can be used to aggregate the ratings provided by different experts into a single cloud (e.g., PNC), which considers not only the tendency, but also the spread of the assessments. Li et al. [21] proposed a backward cloud generator algorithm, which can be applied to generate the NCs. The backward cloud generator algorithm can be implemented by following these steps:

Step 1. Calculate the sample mean $\bar{X} =< \bar{Ex}, \bar{\mu}_\beta, \bar{\nu}_\beta >$, where $\bar{Ex} = \frac{1}{n}\sum_{i=1}^{n} Ex_i, \bar{\mu}_\beta = \frac{\sum_{i=1}^{n} \mu_\beta Ex_i}{\sum_{i=1}^{n} Ex_i}, \bar{\nu}_\beta = \frac{\sum_{i=1}^{n} \nu_\beta Ex_i}{\sum_{i=1}^{n} Ex_i}$, The first-order sample absolute central moment can be expressed as $\frac{1}{n}\sum_{i=1}^{n} |x_i - \bar{X}|$, and sample variance can be expressed as $S^2 = \frac{1}{n-1}\sum_{i=1}^{n}(x_i - \bar{X})^2$;

Step 2. Estimate the value of Ex, μ_β, ν_β $\hat{Ex} = \bar{Ex}, \hat{\mu}_\beta = \bar{\mu}_\beta, \hat{\nu}_\beta = \bar{\nu}_\beta$;

Step 3. Estimate the value of He, En, $\hat{En} = \sqrt{\frac{\pi}{2}}\frac{1}{n}\sum_{i=1}^{n} |x_i - \hat{Ex}|, \hat{He} = \sqrt{S^2 - \frac{1}{3}\hat{En}^2}$;

Output: The estimated value $(< \hat{Ex}, \hat{\mu}_\beta, \hat{\nu}_\beta >, \hat{En}, \hat{He})$ of $(< Ex, \mu_\beta, \nu_\beta >, En, He)$.

The clouds need to be aggregated in order to facilitate the decision-making process. To this end, the aggregation operators can be used. The cloud weighted arithmetic averaging operator and cloud weighted geometric averaging operator were brought forward by Wang et al. [30]. Further on, Wang and Yang [26] extended the weighted arithmetic averaging operator and presented an instance of aggregation operators for the intuitionistic normal clouds. In order to derive the utility of the alternatives considered in the MCDM problem when the PNCs are applied, one also needs the appropriate aggregation operators. Below, we present some aggregation operators for the PNCs, as well as discuss the properties thereof.

Definition 10. *Let $C_i(< Ex_i, \mu_{\beta_i}, \nu_{\beta_i} >, En_i, He_i), (i = 1, 2, \cdots, n)$ be a set of PNCs. The PNC weighted arithmetic averaging operator (PNCWAA) is defined as:*

$$PNCWAA(C_1, C_2, \cdots, C_n) = \sum_{i=1}^{n} w_i C_i \tag{6}$$

where w_i is the weight associated with $C_i, i = 1, 2, \cdots, n$,, such that $w_i \in [0, 1], i = 1, 2, \cdots, n$ and $\sum_{i=1}^{n} w_i = 1$. If $w_i = \frac{1}{n}$, the PNCWAA boils down to the PNC arithmetic average (PNCAA) operator, defined as:

$$PNCWAA(C_1, C_2, \cdots, C_n) = \frac{1}{n}\sum_{i=1}^{n} C_i \tag{7}$$

Theorem 2. *Let $C_i(<Ex_i, \mu_{\beta_i}, \nu_{\beta_i}>, En_i, He_i), (i = 1, 2, \cdots, n)$ be a set of PNCs. Then, the result of the aggregation based on the PNCWAA operator is also a PNC, and:*

$$PNCWAA(C_1, C_2, \cdots, C_n) = (<\sum_{i=1}^{n} w_i Ex_i, \frac{\sum_{i=1}^{n} w_i \mu_{\beta_i} Ex_i}{\sum_{i=1}^{n} w_i Ex_i}, \frac{\sum_{i=1}^{n} w_i \nu_{\beta_i} Ex_i}{\sum_{i=1}^{n} w_i Ex_i}>, \sqrt{\sum_{i=1}^{n} w_i (En_i)^2}, \sqrt{\sum_{i=1}^{n} w_i (He_i)^2}) \tag{8}$$

Proof. According to Theorem 1 and Definitions 9 and 10, we can obtain

$$\begin{aligned} &PNCWAA(C_1, C_2, \cdots, C_n) \\ &= \sum_{i=1}^{n} w_i C_i \\ &= \sum_{i=1}^{n} (<w_i Ex_i, \mu_{\beta_i}, \nu_{\beta_i}>, \sqrt{w_i} En_i, \sqrt{w_i} He_i) \\ &= (<\sum_{i=1}^{n} w_i Ex_i, \frac{\sum_{i=1}^{n} w_i \mu_{\beta_i} Ex_i}{\sum_{i=1}^{n} w_i Ex_i}, \frac{\sum_{i=1}^{n} w_i \nu_{\beta_i} Ex_i}{\sum_{i=1}^{n} w_i Ex_i}>, \sqrt{\sum_{i=1}^{n} w_i (En_i)^2}, \sqrt{\sum_{i=1}^{n} w_i (He_i)^2}) \quad \square \end{aligned}$$

Definition 11. *Let $C_i(<Ex_i, \mu_{\beta_i}, \nu_{\beta_i}>, En_i, He_i), (i = 1, 2, \cdots, n)$ be a set of PNCs. The PNC weighted geometric averaging operator (PNCWGA) is defined as:*

$$PNCWGA(C_1, C_2, \cdots, C_n) = \prod_{i=1}^{n} C_i^{w_i} \tag{9}$$

where w_i is the weight attached to $C_i, i = 1, 2, \cdots, n$,, such that $w_i \in [0, 1], i = 1, 2, \cdots, n$ and $\sum_{i=1}^{n} w_i = 1$. If $w_i = \frac{1}{n}$, the PNCWGA is reduced to an PNC geometric average (PNCGA) operator, defined as:

$$PNCGA(C_1, C_2, \cdots, C_n) = \sqrt[n]{\prod_{i=1}^{n} C_i} \tag{10}$$

Theorem 3. *Let $C_i(<Ex_i, \mu_{\beta_i}, \nu_{\beta_i}>, En_i, He_i), (i = 1, 2, \cdots, n)$ be a set of PNCs. Then, the result of the aggregation based on the PNCWGA operator is also a PNC, and:*

$$\begin{aligned} &PNCWGA(C_1, C_2, \cdots, C_n) \\ &= (<\prod_{i=1}^{n} Ex_i^{w_i}, \frac{\sum_{i=1}^{n} \mu_{\beta_i} Ex_i^{w_i}}{\prod_{i=1}^{n} Ex_i^{w_i}}, \frac{\sum_{i=1}^{n} \nu_{\beta_i} Ex_i^{w_i}}{\prod_{i=1}^{n} Ex_i^{w_i}}>, \prod_{i=1}^{n} Ex_i^{w_i} \sqrt{\sum_{i=1}^{n} w_i (\frac{En_i}{Ex_i})^2}, \prod_{i=1}^{n} Ex_i^{w_i} \sqrt{\sum_{i=1}^{n} w_i (\frac{He_i}{Ex_i})^2}) \end{aligned} \tag{11}$$

Proof. According to Theorem 1 and Definitions 9 and 11, we can obtain

$$\begin{aligned} &PNCWGA(C_1, C_2, \cdots, C_n) \\ &= \prod_{i=1}^{n} C_i^{w_i} \\ &= \prod_{i=1}^{n} (<Ex_i^{w_i}, \mu_{\beta_i}, \nu_{\beta_i}>, \sqrt{w_i} Ex_i^{w_i - 1} En_i, \sqrt{w_i} Ex_i^{w_i - 1} He_i) \\ &= (<\prod_{i=1}^{n} Ex_i^{w_i}, \frac{\sum_{i=1}^{n} \mu_{\beta_i} Ex_i^{w_i}}{\prod_{i=1}^{n} Ex_i^{w_i}}, \frac{\sum_{i=1}^{n} \nu_{\beta_i} Ex_i^{w_i}}{\prod_{i=1}^{n} Ex_i^{w_i}}>, \prod_{i=1}^{n} Ex_i^{w_i} \sqrt{\sum_{i=1}^{n} w_i (\frac{En_i}{Ex_i})^2}, \prod_{i=1}^{n} Ex_i^{w_i} \sqrt{\sum_{i=1}^{n} w_i (\frac{He_i}{Ex_i})^2}) \quad \square \end{aligned}$$

3.2. Distance Measures for PNCs

The distance measures are an important concept in MCDM. Indeed, they can be used to compare the alternatives considered against a reference point. In this subsection, the distance measures for PNCs alongside the properties of these measures are discussed. There have been distance measures for the integrated clouds developed by Wang and Liu [33]. Wang et al. [40] further considered the distance measures for interval integrated clouds. Following the principles outlined in the aforementioned papers, the distance measures for PNCs can be established as follows.

Definition 12. *Let there be any two PNCs $C_1(<Ex_1, \mu_{\beta_1}, \nu_{\beta_1}>, En_1, He_1)$ and $C_2(<Ex_2, \mu_{\beta_2}, \nu_{\beta_2}>, En_2, He_2)$. The distance measure for the PNCs can be defined as:*

$$d(C_1, C_2) = |(1 - \tau_1)\rho_1 Ex_1 - (1 - \tau_2)\rho_2 Ex_2| \tag{12}$$

where $\tau_1 = \frac{\sqrt{En_1^2+He_1^2}}{\sqrt{En_1^2+He_1^2}\sqrt{En_2^2+He_2^2}}$, $\tau_2 = \frac{\sqrt{En_2^2+He_2^2}}{\sqrt{En_1^2+He_1^2}\sqrt{En_2^2+He_2^2}}$, $\rho_1 = \min\{\mu_{\beta_1}, \sqrt{1-\nu_{\beta_1}^2}\}$, $\rho_2 = \min\{\mu_{\beta_2}, \sqrt{1-\nu_{\beta_2}^2}\}$. *In addition, when* $En_1 = En_2 = 0, He_1 = He_2 = 0$, *then* $\tau_1 = \tau_2 = 0$; *the distance measure between two PNCs can be expressed as* $d(C_1, C_2) = |\rho_1 Ex_1 - \rho_2 Ex_2|$. *Furthermore, when* $En_1 = En_2 = 0, He_1 = He_2 = 0$ *and* $\rho_1 = \rho_2 = 1$, *the distance between the two PNCs is the distance between two real numbers, and* $d(C_1, C_2) = |Ex_1 - Ex_2|$.

Property 1. *Let* C_1, C_2 *and* C_3 *be three PNCs,* $\Omega = \{1,2,3\}$. *Then, the distance measure given in Definition 12 satisfies the following properties:*

1. $d(C_i, C_j) \geq 0, i, j \in \Omega$;
2. $d(C_i, C_j) = d(C_j, C_i), i, j \in \Omega$;
3. $d(C_i, C_j) = 0$, *iff* $C_i = C_j, i, j \in \Omega$;
4. $d(C_i, C_k) \leq d(C_i, C_j) + d(C_j, C_k), i, j, k \in \Omega$.

4. PNC-Based MCGDM Method

An MCGDM approach for handling the problems with Pythagorean information is outlined in this section. The proposed approach is based on the TOPSIS [35]. We chose the TOPSIS approach due to the effectiveness and low computational burden associated with the computations underlying this approach. However, the MCDM framework based on the PNCs could be revised by applying such techniques as VIKOR (VIsekriterijumsko KOmpromisno Rangiranje) or Grey relational analysis, for instance.

Say we consider s alternatives $A = \{a_1, a_2, \cdots, a_s\}$, t decision-makers $M = \{m_1, m_2, \cdots, m_t\}$ and n criteria $c = \{c_1, c_2, \cdots, c_n\}$. The criteria may have different importance as defined by the weight vector $w = (w_1, w_2, \cdots, w_n)$, where $w_j > 0, (j = 1, 2, \cdots, n)$ and $\sum_{j=1}^{n} w_j = 1$. The ratings are provided by each decision-maker m_r for each alternative a_i against criterion c_j in terms of the PFN $x_{ijr} = < \mu_{\beta_{ijr}}, \nu_{\beta_{ijr}} >$.

The group MCDM proceeds by applying the backward cloud generator for the PNCs. The resulting data are then processed by applying the aggregation operators (PNCWAA or PNCWGA). The detailed procedure can be described in the following manner:

Step 1. The ratings provided by the experts are aggregated for each alternative and each criterion. The backward cloud generator algorithm described in Section 3.1 is applied to populate the PNC, which represents the aggregate rating $e_{ij} = (< Ex_{ij}, \mu_{\beta_{ij}}, \nu_{\beta_{ij}} >, En_{ij}, He_{ij})$ for alternative a_i against c_j.

Step 2. The ratings for each alternative are aggregated across the criteria. The resulting overall utility r_i of the alternative a_i can be obtained by using the PNCWAA (or PNCWGA):

$$\begin{aligned} r_i &= (< Ex_i, \mu_{\beta_i}, \nu_{\beta_i} >, En_i, He_i) \\ &= PNCWAA(e_{i1}, e_{i2}, \cdots, e_{in}) \\ &= (< \sum_{j=1}^{n} w_j Ex_{ij}, \frac{\sum_{j=1}^{n} w_j \mu_{\beta_{ij}} Ex_{ij}}{\sum_{j=1}^{n} w_j Ex_{ij}}, \frac{\sum_{j=1}^{n} w_j \nu_{\beta_{ij}} Ex_{ij}}{\sum_{j=1}^{n} w_j Ex_{ij}} >, \sqrt{\sum_{j=1}^{n} w_j (En_{ij})^2}, \sqrt{\sum_{j=1}^{n} w_j (He_{ij})^2}) \end{aligned} \tag{13}$$

Step 3. Calculate the coordinates of the positive and negative ideal solutions. The following equation defines the way the coordinates of the positive ideal solution can be obtained:

$$y^+ = (< Ex^+, \mu_\beta^+, \nu_\beta^+ >, En^+, He^+) = (< \max_{1 \leq i \leq n} Ex_i, \max_{1 \leq i \leq n} \mu_{\beta_i}, \min_{1 \leq i \leq n} \nu_{\beta_i} >, \min_{1 \leq i \leq n} En_i, \min_{1 \leq i \leq n} He_i) \tag{14}$$

The coordinates of the negative ideal solution can be obtained by considering the following equation:

$$y^- = (< Ex^-, \mu_\beta^-, \nu_\beta^- >, En^-, He^-) = (< \min_{1 \leq i \leq n} Ex_i, \min_{1 \leq i \leq n} \mu_{\beta_i}, \max_{1 \leq i \leq n} \nu_{\beta_i} >, \max_{1 \leq i \leq n} En_i, \max_{1 \leq i \leq n} He_i) \tag{15}$$

Step 4. Each alternative is positioned in between the positive and negative ideal solutions. The distances to the ideal solutions are obtained by considering the distance measure given by Equation (12). For the i-th alternative, its distance to the positive ideal solution defined by Equation (14) is obtained as:

$$\begin{aligned} d_i^+ &= d(r_i, y^+) \\ &= |(1 - \frac{\sqrt{En_i^2 + He_i^2}}{\sqrt{En_i^2 + He_i^2}\sqrt{(En^+)^2 + (He^+)^2}})\rho_i Ex_i - (1 - \frac{\sqrt{(En^+)^2 + (He^+)^2}}{\sqrt{En_i^2 + He_i^2}\sqrt{(En^+)^2 + (He^+)^2}})\rho^+ Ex^+| \end{aligned} \tag{16}$$

where $\rho_i = \min\{\mu_{\beta_i}, \sqrt{1 - v_{\beta_i}^2}\}, \rho^+ = \min\{\mu_{\beta^+}, \sqrt{1 - v_{\beta^+}^2}\}$.

Step 5. Similarly, Equation (14) is exploited to measure the distance between the i-th alternative and the negative ideal solution defined by Equation (15):

$$\begin{aligned} d_i^- &= d(r_i, y^-) \\ &= |(1 - \frac{\sqrt{En_i^2 + He_i^2}}{\sqrt{En_i^2 + He_i^2}\sqrt{(En^-)^2 + (He^-)^2}})\rho_i Ex_i - (1 - \frac{\sqrt{(En^-)^2 + (He^-)^2}}{\sqrt{En_i^2 + He_i^2}\sqrt{(En^-)^2 + (He^-)^2}})\rho^- Ex^-| \end{aligned} \tag{17}$$

where $\rho_i = \min\{\mu_{\beta_i}, \sqrt{1 - v_{\beta_i}^2}\}, \rho^- = \min\{\mu_{\beta^-}, \sqrt{1 - v_{\beta^-}^2}\}$.

Step 6. Rank the alternatives. The alternatives can be ordered on the basis of the normalized distance:

$$d_i^* = \frac{d_i^+}{d_i^+ + d_i^-} \tag{18}$$

where smaller values of d_i^* are associated with better alternatives a_i. Therefore, the asymmetry between the two ideal solutions and an alternative is used to rank the alternatives.

5. Results

We follow the application presented by [41] to show the possibilities for the application of the proposed approach for decision-making in e-commerce. More specifically, the case of an Internet shop is considered. The data from a business-to-consumer (B2C) website based in China, Tmall.com, are used to implement the proposed approach and rank the goods sold online against several criteria.

The website allows the customers to express their opinions (ratings) regarding the products they have already bought. These ratings can be used for new costumers when making decisions to buy. As several aspects of the goods purchased can be evaluated, the MCDM problem emerges. Indeed, the opinions of the existing customers might be aggregated for different articles (alternatives) and criteria, thus defining a decision matrix.

Let us consider the case of four cameras (x_1, x_2, x_3, x_4), which are compared against each other in order to identify the most appealing one. Therefore, we set $s = 4$. The cameras are compared in terms of the three criteria ($n = 3$). The criteria considered are: the quality of the logistics service provider (c_1), the level of service provided by the vendor (c_2) and the quality of each item (c_3). The ratings are expressed on a five-point scale. The criteria are assumed to have different importance as manifested by the associated weight vector $w = (0.3, 0.2, 0.5)$. The ratings provided by the previous consumers (experts) are aggregated into the PNCs (we omit the detailed description of this step for the sake of brevity).

5.1. Empirical Application

The MCDM procedure based on the resulting PNCs proceeds as follows:

Step 1. Aggregate the evaluations of each alternative under a certain criterion provided by all of the decision-makers by applying the backward generator. The criteria values of cameras x_1, x_2, x_3 and x_4 under the three criteria can be expressed as the following PNCs:

$$\begin{aligned} e_{11} &= (< 2.65, 0.81, 0.28 >, 0.94, 1.33), e_{12} = (< 2.56, 0.81, 0.18 >, 0.89, 1.26), \\ e_{13} &= (< 2.62, 0.81, 0.29 >, 0.89, 1.28), e_{21} = (< 3.57, 0.75, 0.33 >, 0.42, 0.67), \\ e_{22} &= (< 2.67, 0.73, 0.25 >, 0.90, 1.29), e_{23} = (< 2.73, 0.79, 0.21 >, 0.91, 1.30), \\ e_{31} &= (< 2.60, 0.75, 0.29 >, 0.92, 1.30), e_{32} = (< 3.56, 0.81, 0.25 >, 0.39, 0.65), \\ e_{33} &= (< 2.69, 0.85, 0.20 >, 0.92, 1.29), e_{41} = (< 2.98, 0.80, 0.32 >, 0.74, 0.95), \\ e_{42} &= (< 2.65, 0.90, 0.14 >, 0.92, 1.32), e_{43} = (< 3.57, 0.82, 0.16 >, 0.42, 0.68). \end{aligned}$$

Step 2. Obtain the overall utility scores for each camera. Based on Equation (13), the PNC utility scores of each camera can be expressed as:

$$\begin{aligned} r_1 &= (< 2.6170, 0.8100, 0.2654 >, 0.9053, 1.2913), r_2 = (< 2.9700, 0.7648, 0.2605 >, 0.7931, 1.1456), \\ r_3 &= (< 2.8370, 0.8125, 0.2373 >, 0.8412, 1.2931), r_4 = (< 3.2090, 0.8276, 0.2013 >, 0.6494, 0.9222). \end{aligned}$$

Step 3. Identify the coordinates describing the positive and negative ideal solutions. Following Equation (14), the positive ideal solution can be expressed as:

$$y^+ = (< 3.2090, 0.8276, 0.2013 >, 0.6494, 0.9222).$$

According to (15), the negative ideal solution can be expressed as:

$$y^- = (< 3.6170, 0.7648, 0.2654 >, 0.9053, 1.2913).$$

Step 4. Measure the distance between the vector defining a certain camera and the positive ideal solution. Following Equation (16), we obtain the following distances:

$$d_1^+ = 0.7314, d_2^+ = 0.4921, d_3^+ = 0.5751, d_4^+ = 0.0000.$$

Step 5. Measure the distance between the vector defining a certain camera and the negative ideal solution. Following Equation (17), we obtain the following distances:

$$d_1^- = 0.0433, d_2^- = 0.2661, d_3^- = 0.2130, d_4^- = 0.7448.$$

Step 6. Rank the cameras. According to (18), the normalized distance can be used to rank the alternatives:

$$d_1^* = 0.9441, d_2^* = 0.6491, d_3^* = 0.7298, d_4^* = 0.0000.$$

Clearly, $d_1^* > d_3^* > d_2^* > d_4^*$; thus, the cameras can be ranked as $x_4 \succ x_2 \succ x_3 \succ x_1$. Then, the best camera is x_4.

The robustness of the MCDM approach needs to be checked by means of the sensitivity analysis. Specifically, we look at the changes in the criterion weights and the resulting changes in the ranking of the alternatives. First, we define the design of variations in the weighting vector. The vector for positive weights of criteria is $w = (w_1, w_2, \cdots w_k)$ such that the weights are normalized, that is $\sum_{j=1}^{k} w_j = 1$. Then, if the weight of one criterion changes, the weight of other criteria must change accordingly in order to ensure they add up to unity. The resulting vector is then denoted as $w' = (w_1', w_2', \cdots w_k')$. Let us change the weight of criterion c_q, w_q, by a margin of $\triangle_q$. Then, the weights of the other criteria change by $\triangle_j, j = 1, 2, \cdots, k,$. Indeed, the following identity holds: $\triangle_j = \frac{\triangle_q w_j}{w_q - 1}, j = 1, 2, \cdots, k, j \neq q$. Observing that w and w' are related as $w_q' = w_q + \triangle_q$ and $w_j' = \frac{1 - w_q'}{1 - w_q} w_j, j = 1, 2, \cdots q-1, q+1, \cdots, k$, we can get $-w_q < \triangle_q < 1 - w_q$. Since $0 < w_q' < 1$, it is obvious that $\triangle_q \in (-w_q, 1 - w_q)$.

In order to proceed with the sensitivity analysis, we manipulate w_3 and set $\triangle_3 = -0.499, -0.3, -0.1, 0.2, 0.4, 0.499$. The resulting weighting vectors and the corresponding orders of ranking are then summarized in Table 1. The key message is that the rating is stable in terms of the best- and worst-performing alternatives for $\triangle_3 \in [-0.1, 0.4]$. Figure 1 presents the differences in the normalized distances d^* due to changes in the weighting vector induced by different values of $\triangle_3$.

Table 1. Sensitivity of the ranking to the weights of criteria.

Case No.	$\triangle_3$	w'	The Final Ranking
1	−0.499	(0.5994, 0.3996, 0.0010)	$x_2 \succ x_3 \succ x_4 \succ x_1$
2	−0.3	(0.4800, 0.3200, 0.2000)	$x_2 \succ x_4 \succ x_3 \succ x_1$
3	−0.1	(0.3600, 0.2400, 0.4000)	$x_4 \succ x_2 \succ x_3 \succ x_1$
4	0.2	(0.2400, 0.1600, 0.6000)	$x_4 \succ x_2 \succ x_3 \succ x_1$
5	0.4	(0.1200, 0.0800, 0.8000)	$x_4 \succ x_3 \succ x_2 \succ x_1$
6	0.499	(0.0006, 0.0004, 0.9990)	$x_4 \succ x_3 \succ x_1 \succ x_2$

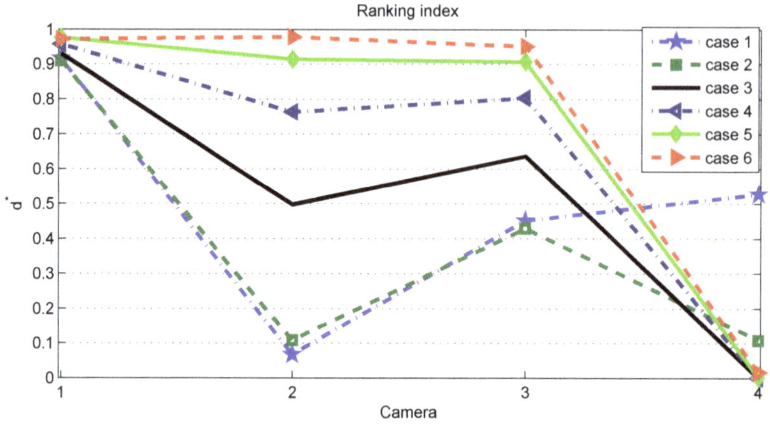

Figure 1. Changes in the normalized distances d_i^* due to changes in the weighting vector. Note: the cases represent those given in Table 2.

5.2. Comparative Analysis

The proposed PFNs approach is also compared to those based on the INC and neutrosophic normal cloud (NNC) approach developed by Wang and Yang [26] and Zhang et al. [41], respectively. Since the approach [26] was based on IFNs, the expert assessments used in the MCDM need to be transformed from the PFNs into IFNs. This step involves alterations in the membership and non-membership degrees: as the sum of these values and the degree of indeterminacy needs to be equal to unity, we normalize these values for each PFN. The resulting IFNs and the three-sigma principle are then applied when constructing the INCs for each alternative and criterion. The clouds are aggregated by considering the scores (ratings) and membership degrees of each drop within a corresponding cloud. The cloud drops are then generated from the aggregate clouds by Monte Carlo simulations. Then, the items are ranked by considering the average INC drop values for each alternative.

Aiming to increase the robustness of the analysis, we implement the Monte Carlo simulations so as to achieve 10,000 cloud drops. As we can see above, the proposed approach and the frameworks outlined by [26,41] all identify x_1 as the worst item. Anyways, the results regarding the most preferable item diverge. Specifically, the framework by [41] suggests x_2 as the best item, while the proposed approach and the method in [26] both identify x_4 as the best item. The results based on the methodology developed by Zhang et al. [41] are, therefore, the most divergent from the other two.

We further employ the uncertain pure linguistic information cloud (UPLC) model based on [42]. Utilizing UPLC to convert the uncertain linguistic values into an integrated cloud renders the following ranking of all alternatives: $x_4 \succ x_2 \succ x_3 \succ x_1$; and the best camera is x_4. Then, the generalized interval aggregation operator [43] is applied to aggregate the uncertain linguistic variables in the initial decision matrices to derive the individual utilities for the alternatives in the first stage and the uncertain pure linguistic hybrid harmonic averaging (UPLHAA) operator to aggregate the individual utilities in the second stage. The resulting ranking order of the alternatives is $x_4 \succ x_3 \succ x_1 \succ x_2$, and the best camera is x_4.

Therefore, we apply different aggregation principles, which differ in terms of the order of aggregations (across decision-makers, across criteria) and the underlying aggregation operators. The results of the comparative analysis are summarized in Table 2. The proposed method appears to be valid in identifying the best- and worst-performing alternatives.

Table 2. The ranking orders rendered by different methods. UPLC, pure linguistic information cloud; UPLHAA, pure linguistic hybrid harmonic averaging.

Method	The Final Ranking	The Best Camera	The Worst Camera
The method in [26]	$x_4 \succ x_3 \succ x_2 \succ x_1$	x_4	x_1
The method in [41]	$x_2 \succ x_4 \succ x_3 \succ x_1$	x_2	x_1
UPLC [42]	$x_4 \succ x_2 \succ x_3 \succ x_1$	x_4	x_1
UPLHAA [43]	$x_4 \succ x_3 \succ x_1 \succ x_2$	x_4	x_2
The proposed method	$x_4 \succ x_2 \succ x_3 \succ x_1$	x_4	x_1

6. Conclusions

The concept of the Pythagorean normal cloud was proposed in this study. It allows expressing the expected value of the normal cloud as a Pythagorean fuzzy number (note that the Pythagorean fuzzy numbers offer more possibilities for constructing fuzzy ratings if opposed to the conventional fuzzy numbers). The proposed approach, therefore, allows for a greater flexibility in accounting for the confidence of the decision-makers and the distribution of their ratings.

The group MCDM procedure based on the Pythagorean normal clouds was developed. The backward cloud generator was applied to aggregate the expert assessments into the Pythagorean normal clouds. The concept of the PNC was then incorporated into aggregation operators. As a result, the Pythagorean normal cloud weighted arithmetic averaging operator and the Pythagorean normal cloud weighted geometric averaging operator were developed. Application of these operators allowed calculating the Pythagorean fuzzy utility of the alternatives considered. Based on the symmetry among the alternatives and the ideal solutions, the alternatives were ranked according to the values of the normalized distances.

An empirical application to e-commerce was presented in order to demonstrate the operationality of the proposed approach. The existing customers expressed opinions on the goods purchased, as well as their confidence in the form of the Pythagorean fuzzy numbers. These were further aggregated into Pythagorean normal clouds and processed in line with the suggested approach. The comparative analysis was carried out in order to demonstrate the validity of the proposed approach. Future research could aim to improve the weighting schemes used in the aggregation approach. For instance, the deviation of experts from the sample mean (i.e., their competence) could be taken into account when constructing the clouds.

Author Contributions: Conceptualization, J.Z. and T.B. Methodology, J.Z. and W.S. Software, J.Z. Writing, original draft preparation, J.Z. Writing, review and editing, T.B. and D.S. Supervision, W.S.

Funding: This research was funded by the National Natural Science Foundation of China (Nos. 71671165, 2017yyrw07), Natural Science in Universities of the General Project of Anhui Province (No. TSKJ2017B22) and the Financial engineering research and development center of Anhui Polytechnic University (No. JRGCKF201506).

Conflicts of Interest: The authors declare no conflict of interest.

References

1. Rostamzadeh, R.; Esmaeili, A.; Nia, A.S.; Saparauskas, J.; Ghorabaee, M.K. A fuzzy ARAS method for supply chain management performance measurement in SMEs under uncertainty. *Transform. Bus. Econ.* **2017**, *16*, 319–348.
2. Hu, K.H.; Jianguo, W.; Tzeng, G.H. Improving China's regional financial center modernization development using a new hybrid MADM model. *Technol. Econ. Dev. Econ.* **2018**, *24*, 429–466. [CrossRef]
3. Czubenko, M.; Kowalczuk, Z.; Ordys, A. Autonomous driver based on an intelligent system of decision-making. *Cogn. Comput.* **2015**, *7*, 1–13.
4. Akusok, A.; Miche, Y.; Hegedus, J.; Nian, R.; Lendasse, A. A two stage methodology using K-NN and false-positive minimizing ELM for nominal data classification. *Cogn. Comput.* **2014**, *6*, 432–445.
5. Mak, D.K. A fuzzy probabilistic method for medical diagnosis. *J. Med. Syst.* **2015**, *39*, 26–35.
6. Meng, F.Y.; Chen, X.H. Correlation coefficients of hesitant fuzzy sets and their application based on fuzzy measures. *Cogn. Comput.* **2015**, *7*, 445–463.
7. Zenebe, A.; Zhou, L.N.; Norcio, A.F. User preferences discovery using fuzzy models. *Fuzzy Sets Syst.* **2010**, *161*, 3044–3063.
8. Vahidov, R.; Ji, F. A diversity-based method for infrequent purchase decision support in e-commerce. *Electron. Commer. Res. Appl.* **2015**, *4*, 143–158.
9. Zadeh, L.A. Fuzzy sets. *Inform. Control* **1965**, *8*, 338–353.
10. Zadeh, L.A. Probability measures of fuzzy events. *J. Math. Anal. Appl.* **1968**, *23*, 421–427.
11. Atanassov, K.T. Intuitionistic fuzzy sets. *Fuzzy Sets Syst.* **1986**, *20*, 87–96.
12. Atanassov, K.T.; Gargov, G. Interval valued intuitionistic fuzzy sets. *Fuzzy Sets Syst.* **1998**, *31*, 343–349.
13. Torra, V. Hesitant fuzzy sets. *Int. J. Intell. Syst.* **2010**, *25*, 529–539.
14. Yager, R.R. Pythagorean membership grades in multi-criteria decision-making. *IEEE Trans. Fuzzy Syst.* **2014**, *22*, 958–965.
15. Yager, R.R. Properties and applications of Pythagorean fuzzy sets. In *Imprecision and Uncertainty in Information Representation and Processing*; Angelov, P., Sotirov, S., Eds.; Springer: Cham, Switzerland, 2016; pp. 119–136, ISBN 978-3-319-26301-4.
16. Liang, D.C.; Xu, Z.S.; Darko, A.P. Projection model for fusing the information of Pythagorean fuzzy multicriteria group decision-making based on geometric Bonferroni mean. *Int. J. Intell. Syst.* **2017**, *32*, 966–987. [CrossRef]
17. Beliakov, G.; James, S. Averaging aggregation functions for preferences expressed as Pythagorean membership grades and fuzzy orthopairs. In Proceedings of the IEEE International Conference on Fuzzy Systems, Beijing, China, 6–11 July 2014; pp. 298–305.
18. Reformat, M.Z.; Yager, R.R. Suggesting recommendations using Pythagorean fuzzy sets illustrated using Netflix movie data. In *Information Processing and Management of Uncertainty in Knowledge-Based Systems*; Laurent, A., Ed.; Springer: Berlin, Germany, 2014; pp. 546–556, ISBN 978-3-319-08794-8.
19. Zeng, S.Z. Pythagorean fuzzy multiattribute group decision-making with probabilistic information and OWA approach. *Int. J. Intell. Syst.* **2017**, *32*, 1136–1150. [CrossRef]
20. Zeng, S.Z.; Chen, J.P.; Li, X.S. A hybrid method for Pythagorean fuzzy multiple-criteria decision-making. *Int. J. Inf. Technol. Decis.* **2016**, *15*, 403–422.
21. Li, D.Y.; Liu, C.Y.; Du, Y.; Han, X. Artificial intelligence with uncertainty. *J. Softw.* **2004**, *15*, 1583–1594.
22. Wang, G.Y.; Xu, C.L.; Li, D.Y. Generic normal cloud model. *Inf. Sci.* **2014**, *280*, 1–15.
23. Li, D.Y.; Han, J.W.; Shi, X.M.; Chan, M.C. Knowledge representation and discovery based on linguistic atoms. *Knowl.-Based Syst.* **1998**, *10*, 431–440.
24. Wang, D.; Zeng, D.B.; Singh, V.P.; Xu, P.C.; Liu, D.F.; Wang, Y.K.; Zeng, X.K.; Wu, J.C.; Wang, L.C. A multidimension cloud model-based approach for water quality assessment. *Environ. Res.* **2016**, *149*, 113–121.
25. Jiang, J.B.; Liang, J.R.; Jiang, W.; Gu, Z.P. Application of trapezium cloud model in conception division and conception exaltation. *Comput. Eng. Des.* **2008**, *29*, 1235–1240.
26. Wang, J.Q.; Yang, W.E. Multiple criteria group decision-making method based on intuitionistic normal cloud by Monte Carlo simulation. *Syst. Eng. Theory Pract.* **2013**, *33*, 2859–2865.

27. Peng, X.D.; Yuan, H.Y.; Yang, Y. Pythagorean fuzzy measures and their applications. *Int. J. Intell. Syst.* **2017**, *32*, 991–1029. [CrossRef]
28. Peng, X.D.; Yang, Y. Fundamental properties of interval-valued Pythagorean fuzzy aggregation operators. *Int. J. Intell. Syst.* **2016**, *31*, 444–487.
29. Yang, W.E.; Wang, J.Q.; Ma, C.Q.; Wang, X.F. Hesitant linguistic multiple criteria decision-making method based on cloud generating algorithm. *Control Decis.* **2015**, *30*, 371–374.
30. Wang, J.Q.; Liu, T. Uncertain linguistic multi-criteria group decision-making approach based on integrated cloud. *Control Decis.* **2012**, *27*, 1185–1190.
31. Wang, J.Q.; Peng, L.; Zhang, H.Y.; Chen, X.H. Method of multi-criteria group decision-making based on cloud aggregation operators with linguistic information. *Inf. Sci.* **2014**, *274*, 177–191.
32. Wang, J.Q.; Wang, P.; Wang, J.; Zhang, H.Y.; Chen, X.H. Atanassov's interval-valued intuitionistic linguistic multi-criteria group decision-making method based on trapezium cloud model. *IEEE Trans. Fuzzy Syst.* **2014**, *23*, 542–554.
33. Liu, C.; Tang, G.L.; Liu, P.D. An approach to multicriteria group decision-making with unknown weight information based on Pythagorean fuzzy uncertain linguistic aggregation operators. *Math. Probl. Eng.* **2017**. [CrossRef]
34. Garg, H. A novel improved accuracy function for interval valued Pythagorean fuzzy sets and its applications in the decision-making process. *Int. J. Intell. Syst.* **2017**, *32*, 1247–1260. [CrossRef]
35. Hwang, C.L.; Yoon, K. *Multiple Attribute Decision Making: Methods and Applications*; Springer-Verlag: New York, NY, USA, 1981; ISBN 978-3-540-10558-9.
36. Xu, Z.S. Intuitionistic fuzzy aggregation operators. *IEEE Trans. Fuzzy Syst.* **2008**, *14*, 1179–1187.
37. Zhang, X.L.; Zhao, L.; Zang, J.Y.; Fan, H.M.; Cheng, L. Flatness intelligent control based on T-S cloud inference neural network. *Trans. ISIJ* **2014**, *54*, 2608–2617.
38. Zhang, X.L.; Xu, Z.S. Extension of TOPSIS to multiple criteria decision making with Pythagorean fuzzy sets. *Int. J. Intell. Syst.* **2014**, *29*, 1061–1078.
39. Luo, Z.Q.; Zhang, G.W. A new algorithm of backward normal onevariate cloud. *J. Front. Comput. Sci. Technol.* **2007**, *1*, 234–240.
40. Wang, J.Q.; Peng, J.J.; Zhang, H.Y.; Liu, T.; Chen, X.H. An uncertain linguistic multi-criteria group decision-making method based on a cloud model. *Group Decis. Negot.* **2015**, *24*, 171–192.
41. Zhang, H.Y.; Ji, P.; Wang, J.Q.; Chen, X.H. A Neutrosophic normal cloud and its application in decision-making. *Cogn. Comput.* **2016**, *8*, 649–669.
42. Peng, B.; Zhou, J.M.; Peng, D.H. Cloud model based approach to group decision-making with uncertain pure linguistic information. *J. Intell. Fuzzy Syst.* **2017**, *32*, 1959–1968.
43. Peng, B.; Ye, C.M.; Zeng, S.Z. Uncertain pure linguistic hybrid harmonic averaging operator and generalized interval aggregation operator based approach to group decision-making. *Knowl.-Based Syst.* **2012**, *36*, 175–181.

symmetry

MDPI

Article

Optimal Dividend and Capital Injection Problem with Transaction Cost and Salvage Value: The Case of Excess-of-Loss Reinsurance Based on the Symmetry of Risk Information

Qingyou Yan [1,2], Le Yang [1,*], Tomas Baležentis [3,*], Dalia Streimikiene [3] and Chao Qin [1]

1 School of Economics and Management, North China Electric Power University, Beijing 102206, China; qingyouyan@126.com (Q.Y.); qinchao08@163.com (C.Q.)
2 Beijing Key Laboratory of New Energy and Low-Carbon Development, North China Electric Power University, Beijing 102206, China
3 Lithuanian Institute of Agrarian Economics, V. Kudirkos Str. 18, LT-03105 Vilnius, Lithuania; dalia@mail.lei.lt
* Correspondence: ncepu_yl@126.com (L.Y.); tomas@laei.lt (T.B.)

Received: 8 June 2018; Accepted: 9 July 2018; Published: 12 July 2018

Abstract: This paper considers the optimal dividend and capital injection problem for an insurance company, which controls the risk exposure by both the excess-of-loss reinsurance and capital injection based on the symmetry of risk information. Besides the proportional transaction cost, we also incorporate the fixed transaction cost incurred by capital injection and the salvage value of a company at the ruin time in order to make the surplus process more realistic. The main goal is to maximize the expected sum of the discounted salvage value and the discounted cumulative dividends except for the discounted cost of capital injection until the ruin time. By considering whether there is capital injection in the surplus process, we construct two instances of suboptimal models and then solve for the corresponding solution in each model. Lastly, we consider the optimal control strategy for the general model without any restriction on the capital injection or the surplus process.

Keywords: optimal dividend; capital injection; salvage value; transaction cost; excess-of-loss reinsurance

1. Introduction

The expansion of the economic activities in the sense of time and space triggered the need for managing the exposure to risk for different types of businesses (Aniunas et al. [1], Lakstutiene et al. [2], and Kurach [3]). However, the increasing scale and scope of the activities of the insurance companies require identification of effective strategies for risk management in the insurance business itself. Therefore, there have been attempts to model the optimal strategies allowing for stable operation of the insurance companies.

Since the stochastic control theory is a primal approach towards handling the risk issues in recent years, more and more scholars pay attention to the aspect of the optimal dividend for an insurance company. A plethora of mathematical models for managing these issues have been developed. For instance, Assmussen and Taksar [4] applied a controlled diffusion approach in order to address the issue of the optimal dividend in a more advanced framework. They showed that a singular type of control indicating pay out when the sum to be paid out (the accumulated surplus) exceeds a certain level (and no payment in case the threshold is not reached) that can be used as the optimal strategy. Gerber and Shiu [5] pointed out that the barrier strategies solve the mathematical problems. However, the dividend stream corresponding to this solution is not acceptable in reality. Belhaj [6] considered a Brownian risk and a Poisson risk as the two kinds of liquidity risk within the unified framework.

The resulting model implied that the barrier strategy is still the optimal one. Taking an insurance company into consideration, Azue and Muler [7] examined the optimal dividend exercise in case the uncontrolled reserve process follows a classical Cramér-Lundberg model, which involves claim-size distribution of an unknown type. The closed-form solutions for different cases were obtained by Meng et al. [8] who studied an optimal dividend problem taking nonlinear insurance risk processes into consideration. The nonlinearity was related to internal competition factors.

Although these papers argued that the optimal dividend strategy is a barrier strategy where the expected cumulative discounted value of the dividend flow is maximized within the time horizon until the ruin event. By assuming different conditions, ruin happens for the insurance company following the risk process with probability 1. Apparently, it is actually unrealistic. Taking some pharmaceutical or petroleum companies, for example, the shareholders focus on the economic returns and the social benefits as well. Therefore, once their company is on the edge of bankruptcy, they will prevent that from happening by raising sufficient funds. Therefore, in the real financial market, a company always raises funds and, subsequently, reduces exposure to risk by the virtue of capital injection. Therefore, when capital injection is taken into account, the company is assumed to survive forever. Afterward, the expected cumulative discounted dividends are less than the expected discounted cost of capital injection in the infinite time horizon, which is regarded as a critical value and should be maximized when deciding on the strategies for dividend and capital injection management for a certain company. There have also been a number of papers studying this aspect. The conventional risk model was taken into consideration by Kulenko and Schmidli [9] to streamline the dividend payments and capital injections. The issue of the dividend payments and capital injections was also tackled by Yao et al. [10] in terms of the dual risk model involving fixed transaction costs. The latter study identified the bond strategy with an upper and lower barriers as the optimal one. In the context of the random time horizon and a ruin penalty, Zhao and Yao [11] investigated the optimal dividend and capital injection strategy. Yin and Yuen [12] studied the issue of optimal control at the company level when there is a surplus process characterized by an upward jump diffusion and random return on investment.

Besides capital injection, reinsurance is also considered an effective method for a company to control its risk exposure. This is because an appropriate reinsurance strategy can protect a company against the potentially large loss and, therefore, reduce the earning volatility. In practice, there are many different types of insurance policies adopted by companies. Due to its great value both in theory and practice, the issue of the combined dividend and reinsurance has attracted substantial attention and now there is plenty of research on this issue, which includes Høgaard and Taksar [13], Peng et al. [14], Yao et al. [15,16], Yao and Fan [17], and other references. Among the possible strategies, options such as the proportional reinsurance and the excess-of-loss reinsurance have also been investigated extensively (see, e.g., Candenillas et al. [18], Meng and Siu [19], Xu and Zhou [20], Yao et al. [21], A et al. [22], Yao et al. [23]). In these papers, the excess-of-loss reinsurance and dividend strategies are explored and the corresponding solutions to the value function are obtained as well. However, the fixed transaction cost incurred by capital injection has not been discussed in-depth, which is crucial.

Accordingly, we focus our research on the optimal dividend and capital injection policies for an insurance company that manages its risk by the virtue of both the excess-of-loss reinsurance and capital injection based on the symmetric of risk information. In this paper, the symmetry of risk information requires the reinsurance and insurance companies to have complete information on each other. In other words, the possible loss is the common information of both sides in order for there to be no moral hazard caused by asymmetric information of risk. We also add the fixed and proportional transaction cost incurred by capital injection and the salvage value of the company at the ruin time into the surplus process. In reality, transaction cost is unavoidable when the managers run the business especially since the fixed transaction cost is always generated by advisories and consultants when capital injection happens, which makes the impulse control problems more difficult (see, e.g., Paulsen [24], Bai et al. [25], Peng et al. [14], Liu and Hu [26]). In addition, the salvage value of the company can be interpreted as a company's liquidation value at the time of bankruptcy such as the company's brand name or agency network. There has also been more research that examines the optimal dividend policy for an insurance company in the presence of the salvage value for bankruptcy (see, e.g., Loeffen and Renaud [27], Liang and Young [28], Yao et al. [15]). Therefore, introducing the fixed transaction cost and salvage value make our model closer to reality. In order to find out a strategy that maximizes the expected sum of the discounted salvage value and the discounted cumulative dividends minus the expected discounted cost of capital injection until the ruin time, we construct two auxiliary suboptimal models in which one never goes bankrupt by capital injection and the other is a classical model without capital injection. After identifying the corresponding solutions to these two auxiliary models and the corresponding optimal strategy, we solve the general control problem without any restrictions on capital injection or the surplus process.

The outline of the paper is as follows. Section 2 presents the optimal control problem and then gives the definition of the value function by using a diffusion approximation to the compound Poisson model with excess-of-loss reinsurance. Sections 3 and 4 consider two auxiliary suboptimal models, respectively. Section 5 explores the solution to the general control problem. The last Section concludes the study.

2. Model Formulation and the Control Problem

Let $(\Omega, F, \mathbb{P})$ be a probability space with the filtration $\{F_t\}_{t\geq 0}$ satisfying the usual conditions. In the classical risk theory, without reinsurance and dividend payments, an insurance company's surplus following the compound Poisson risk process on this filtered probability space is given by the equation below.

$$X_t = x + pt - \sum_{i=1}^{N_t} Y_i, \tag{1}$$

where x is the initial surplus, $p > 0$ is the premium rate, $\{N_t\}$ is a Poisson process with intensity λ, and the individual claim size $Y_1, Y_2, \dots$, independent of $\{N_t\}$ are independent identically distributed positive random variables with a common continuous distribution $F(y) = 1 - \overline{F}(y) = P(Y_i < y)$ where the corresponding finite first and second moments are $\mu^{(1)} = E[Y_1] > 0$ and $\mu^{(2)} = E\left[Y^2\right] > 0$. Define $M = \sup\{y : F(y) < 1\}$, in this paper we only consider the case where the claim distribution has an upper bound, which means $M < \infty$.

If the excess-of-loss reinsurance is taken by an insurance company to cede the potential risk (denote by $m \in [0, M]$ the excess-of-loss retention level), then, for each claim Y_i, the retained risk level is $Y_i^{(m)} = Y_i \wedge m$, and its first and second moments are shown below:

$$\mu^{(1)}(m) = E\left[Y_i^{(m)}\right] = \int_0^m \overline{F}(y)dy, \tag{2}$$

$$\mu^{(2)}(m) = E\left[(Y_i^{(m)})^2\right] = \int_0^m 2y\overline{F}(y)dy. \tag{3}$$

Assume that the excess-of-loss reinsurance premium rate is calculated based on the expected value principle with the safe loading $\theta > 0$. Then the company's surplus process with reinsurance can be rewritten as the equation below:

$$X_t^m = x + (p - p^{(m)})t - \sum_{i=1}^{N_t} Y_i^{(m)}, \tag{4}$$

where $p^{(m)} = (1+\theta)E\left[\sum_{i=1}^{N_t}\left(Y_i - Y_i^{(m)}\right)\right] = (1+\theta)\lambda\left(\mu^{(1)} - \mu^{(1)}(m)\right)$. According to many previous studies, the diffusion approximation can be described by the formula below:

$$X_t^m = x + \left[\theta\lambda\mu^{(1)}(m) + p - (1+\theta)\lambda\mu^{(1)}\right]t + \sqrt{\lambda\mu^{(2)}(m)}B_t, \tag{5}$$

where $\{B_t\}$ is a standard Brownian motion, which is adapted to the filtration $\mathcal{F}_t^B = \sigma\{B_s : 0 \leq s \leq t\}$. In this paper, we consider the cheap reinsurance, which is shown as $p = (1+\theta)\lambda\mu^{(1)}$ and the insurer can dynamically control the retention level m to expose the risk, which means the surplus process becomes the equation below:

$$X_t^m = x + \theta\lambda\mu^{(1)}(m_t)t + \sqrt{\lambda\mu^{(2)}(m_t)}B_t.$$

Now, we incorporate dividend payments and capital injection into the model. Let $\{L_t\}$ denote the cumulative amount of dividend pay until time t and $\{G\}$ denote the capital injection described by a sequence of increasing stopping times $\{\tau_n | n = 1, 2, 3, \ldots\}$ and the corresponding amount $\{\eta_n | n = 1, 2, 3, \ldots\}$. With a control strategy $\pi = \{m_t^\pi; L_t^\pi; G^\pi\} = \{m_t^\pi; L_t^\pi; \tau_1^\pi, \ldots, \tau_n^\pi; \eta_1^\pi, \ldots, \eta_n^\pi\}$, at time t, the surplus process becomes the equation below:

$$X_t^\pi = x + \theta\lambda\mu^{(1)}(m_t^\pi)t + \sqrt{\lambda\mu^{(2)}(m_t^\pi)}B_t - L_t^\pi + \sum_{n=1}^{\infty} I_{\{\tau_n^\pi \leq t\}}\eta_n^\pi. \tag{6}$$

Definition 1. *A control strategy π is admissible if it meets the following conditions.*

(i) $\{m_t^\pi\}$ is an $\{\mathcal{F}_t\}_{t\geq 0}$ adapted process with $m_t^\pi \in [0, M]$ for all $t \geq 0$.
(ii) $\{L_t^\pi\}$ is an increasing, $\{\mathcal{F}_t\}_{t\geq 0}$-adapted cádlág process and $\Delta L_t^\pi \leq X_{t-}^\pi$.
(iii) $\{\tau_n^\pi\}$ is a sequence stopping times with respect to $\{\mathcal{F}_t\}_{t\geq 0}$ and $0 \leq \tau_1^\pi \leq \cdots \leq \tau_n^\pi \leq \cdots$, a.s.
(iv) $\eta_n^\pi (n = 1, 2, 3, \ldots)$ is measurable and non-negative with respect to $\{\mathcal{F}_{\tau_n^\pi}\}$.
(v) $\forall T > 0$, it has $P(\lim_{n\to\infty} \tau_n^\pi < T) = 0$.

For each admissible strategy π, we establish respective ruin time as $\tau^\pi := \inf\{t \geq 0; X_t^\pi < 0\}$, which is a $\{\mathcal{F}_t\}_{t\geq 0}$ stopping time. If capital injection occurs, this stopping time could be infinite.

Therefore, we estimate the value of an insurance company by exploiting the performance index function. The performance index function is defined as the expected sum of the discounted salvage value and the discounted cumulative dividends except for the expected discounted costs of capital injection until the ruin time:

$$V(x, \pi) = E_x\left[\int_0^{\tau^\pi} \beta_1 e^{-\delta s} dL_s^\pi - \sum_{n=1}^{\infty} e^{-\delta\tau_n^\pi}(\beta_2\eta_n^\pi + K)I_{\{\tau_n^\pi \leq \tau^\pi\}} + Pe^{-\delta\tau^\pi}\right], \tag{7}$$

where $\delta > 0$ is the interest force, $P \geq 0$ is the salvage value of an insurance company at the ruin time, $\beta_1 < 1$ means the proportional transaction cost in the dividend payout process, and $\beta_2 > 1$ and $K > 0$ are the proportional and fixed transaction costs associated with the capital injection, respectively.

With the initial surplus x, the objective is to obtain the value function

$$V(x) = \sup_{\pi\in\Pi_x} V(x, \pi), \tag{8}$$

and the corresponding optimal control strategy $\pi^* = \left\{m^{\pi^*}; L^{\pi^*}; G^{\pi^*}\right\}$ such that $V(x) = V(x, \pi^*)$.

Remark 1. *The compound Poisson risk model is applied to descript the surplus process of an insurance company in this research. In fact, the compound Poisson risk process also known as the Cramér–Lundberg process is a commonly used jump process. Lots of work related to the jump process has been done in various contexts and the literature includes as Nguyen et al. [29], Nguyen and Vuong [30], and Hoang and Vuong [31].*

Remark 2. *In this paper, we only focus on the instance in which $P \geq 0$. As for the case of $P < 0$, it is not included in our study because the surplus will be always non-negative if the insurer can cede all the risk to* *the reinsurer. Following the optimality of the value function, we have $V(0) \geq 0$. Therefore, the case where $V(0) = P < 0$ is impossible.*

Aiming to proceed with the results, we first present some useful operators. For a function $v \in C^2$, the operator of maximum capital injection $\mathcal{M}$ is defined below:

$$\mathcal{M}v(x) := \sup_{y \geq 0}\{v(x+y) - \beta_2 y - K\},$$

and the operator $\mathcal{L}^m$ is represented by the equation below:

$$\mathcal{L}^m v(x) = \frac{1}{2}\lambda\mu^{(2)}(m)v''(x) + \theta\lambda\mu^{(1)}(m)v'(x) - \delta v(x).$$

3. The Solution to the Problem Where Bankruptcy is Not Allowed

In this section, we consider one suboptimal control model that the insurance company will not go bankrupt in finite time horizon due to capital injection. Then, the objective is to maximize the expected present value of the discounted cumulative dividend payout except for the discounted cost of capital injection in the infinite time horizon.

Defined by $\pi_c = \{m^{\pi_c}; L^{\pi_c}; G^{\pi_c}\} \in \Pi_x$, the control strategy of the insurance company won't ruin. Therefore, for each admissible strategy, the performance index function becomes the equation below

$$V(x, \pi_c) = E_x\left[\int_0^\infty \beta_1 e^{-\delta s} dL_s^{\pi_c} - \sum_{n=1}^{\infty} e^{-\delta \tau_n^{\pi_c}}(\beta_2\eta_n^{\pi_c} + K) I_{\{\tau_n^{\pi_c} \leq \infty\}}\right]. \tag{9}$$

The objective is to obtain the value function shown below

$$V_c(x) = \sup_{\pi_c \in \Pi_x} V(x, \pi_c), \tag{10}$$

and the associated optimal strategy π_c^* with $V_c(x) = V(x, \pi_c^*)$.

Assume that the value function defined by Equation (10) is sufficiently smooth. Then, based on the stochastic control theorem, one can derive that the HJB Equation along with the boundary condition of this suboptimal control problem are given below:

$$\max\left\{\max_{0 \leq m \leq M} \mathcal{L}^m V_c(x),\ \beta_1 - V_c'(x),\ \mathcal{M}V_c(x) - V_c(x)\right\} = 0, \tag{11}$$

$$\max\{\mathcal{M}V_c(0) - V_c(0),\ -V_c(0)\} = 0. \tag{12}$$

The operator $\mathcal{M}V_c(x)$ denotes the value of a strategy to choose the optimal immediate capital injection. If x is the starting point for the surplus process and the process is governed in line with the optimal strategy, then the performance index function associated with this strategy is $V_c(x)$. However, suppose that the surplus process still starts at x. If we choose an appropriate time to inject the capital and, after that, the surplus process is also governed in line with the optimal strategy, then the performance index function becomes $\mathcal{M}V_c(x)$. It is easy to show that the performance index function with the first strategy is greater than the second one. Moreover, the two functions can be equal if and only if the time of capital injection is optimal. Therefore, it follows that $\mathcal{M}V_c(x) \leq V_c(x)$. Furthermore, the time value of money implies that the optimal time to inject the capital only comes at the moment when the surplus becomes zero. Mathematically, it has stated that $\mathcal{M}V_c(0) = V_c(0)$ and $\mathcal{M}V_c(x) < V_c(x)$ for all $x > 0$.

Furthermore, when an insurance company is on the edge of bankruptcy, it usually has two ways to tackle the risk. The first one is to inject new capital and its surplus immediately jumps to some level $\eta^* > 0$. If the time for this capital injection is optimal, by the definition of the operator M, the optimal amount of capital injection should be $\eta^* := \inf\{x : V'_c(x) = \beta_2\}$ and the boundary condition satisfies $\mathcal{M}V_c(0) = V_c(0) = V_c(\eta^*) - \beta_2\eta^* - K > 0$. Therefore, the optimal strategy of capital injection $G^{\pi_c^*}$ is constructed in the equations below:

$$\tau_1^{\pi_c^*} = \inf\{t \geq 0 : X_{t-}^{\pi_c^*} = 0\}, \tag{13}$$

$$\tau_n^{\pi_c^*} = \inf\{t \geq \tau_{n-1}^{\pi_c^*} : X_{t-}^{\pi_c^*} = 0\}, n = 2,\ 3,\ \ldots, \tag{14}$$

$$\eta_n^{\pi_c^*} \equiv \eta^*, n = 1,\ 2,\ 3,\ \ldots \tag{15}$$

The second one is to cede all the potential risks to a reinsurance company and keep the insurance company's surplus at the barrier 0 forever. Since the insurance company never goes bankrupt, it does not need any capital injection. If this choice is optimal, we can deduce that the boundary condition should satisfy $V_c(0) = 0$ and $\mathcal{M}V_c(0) < V_c(0)$. Correspondingly, the optimal strategy of capital injection $G^{\pi_c^*}$ is shown below

$$G^{\pi_c^*} \equiv 0. \tag{16}$$

In addition, if there is some value $x_{1c}^* := \inf\{x : V'(x) = \beta_1\}$ such that $x_{1c} \geq \eta^*$, the optimal dividend strategy is a linear barrier strategy with the barrier x_{1c}^*. That is, $L^{\pi_c^*}$ satisfies the equation below

$$L^{\pi_c^*} = (x - x_{1c}^*)^+ + \int_0^t I_{(X_s^{\pi_c^*} = x_{1c}^*)} dL_s^{\pi_c^*}, \text{ for all } t \geq 0. \tag{17}$$

Therefore, the optimal excess-of-loss reinsurance retention level $m^{\pi_c^*}(x)$ should satisfy the equation below

$$\mathcal{L}^{m^{\pi_c^*}(x)} V_c(x) = \max_{0 \leq m \leq M} \mathcal{L}^m V_c(x) = 0, \text{ for } 0 \leq x \leq x_{1c}^*. \tag{18}$$

Theorem 1. *Let $g(x)$ be an increasing concave and twice continuously differentiable solution to the Equations (11) and (12). In this case, one arrives at the following outcomes.*

(i) For each admissible strategy π_c, there exists $g(x) \geq V(x, \pi_c)$ and, therefore, $g(x) \geq V_c(x)$ for all $x \geq 0$.

(ii) In case of the strategy, $\pi_c^* = \left\{m^{\pi_c^*}; L^{\pi_c^*}; G^{\pi_c^*}\right\}$ is constructed by Equations (13)–(18) with $g(x) = V(x, \pi_c^*)$. Then $g(x) = V_c(x)$ and π_c^* is the optimal control strategy.

Proof. (i) Fixing a strategy $\pi_c \in \Pi_x$, define the sets $\Lambda = \{s : L_{s-}^{\pi_c} \neq L_s^{\pi_c}\}$ and $\Lambda' = \{s : G_{s-}^{\pi_c} \neq G_s^{\pi_c}\} = \{\tau_1^{\pi}, \ldots, \tau_n^{\pi}, \ldots\}$, then let $\hat{L}_t^{\pi_c} = \sum_{s \in \Lambda, s \leq t} (L_s^{\pi_c} - L_{s-}^{\pi_c})$ and $\tilde{L}_t^{\pi_c} = L_t^{\pi_c} - \hat{L}_t^{\pi_c}$ be the discontinuous and continuous parts of $L_t^{\pi_c}$, respectively. By the virtue of Itô formula, it can be shown by the equation below.

$$\begin{aligned} e^{-\delta(t \wedge \tau^{\pi_c})} g(X_t^{\pi_c}) = {} & g(x) + \int_0^{t \wedge \tau^{\pi_c}} e^{-\delta s} \mathcal{L}^{m^{\pi_c}} g(X_s^{\pi_c})\, ds + \int_0^{t \wedge \tau^{\pi_c}} e^{-\delta s} \sqrt{\lambda \mu^{(2)}(m_s^{\pi_c})} g'(X_s^{\pi_c})\, dB_s \\ & - \int_0^{t \wedge \tau^{\pi_c}} e^{-\delta s} g'(X_s^{\pi_c})\, d\tilde{L}_s^{\pi_c} + \sum_{s \in \Lambda \cup \Lambda', s \leq t \wedge \tau^{\pi_c}} e^{-\delta s} [g(X_s^{\pi_c}) - g(X_{s-}^{\pi_c})]. \end{aligned} \tag{19}$$

The sum of discontinuous parts of $e^{-\delta t} g(X_t^{\pi_c})$ can be rearranged in the following manner:

$$
\begin{aligned}
&\sum_{s\in\Lambda\cup\Lambda', s\le t\wedge\tau^{\pi_c}} e^{-\delta s}\left[g(X_s^{\pi_c})-g(X_{s-}^{\pi_c})\right] \\
&= \sum_{s\in\Lambda, s\le t\wedge\tau^{\pi_c}} e^{-\delta s}\left[g(X_s^{\pi_c})-g(X_{s-}^{\pi_c})\right] + \sum_{\tau_n^{\pi_c}\le t\wedge\tau^{\pi_c}} e^{-\delta s}\left[g(\eta_n^{\pi_c})-g(0)\right] \\
&\le -\sum_{s\in\Lambda, s\le t\wedge\tau^{\pi_c}} e^{-\delta s}\beta_1(L_s^{\pi_c}-L_{s-}^{\pi_c}) + \sum_{i=1}^{\infty} e^{-\delta\tau_n^{\pi_c}}(\beta_2\eta_n^{\pi_c}+K)I_{\{\tau_n^{\pi_c}\le t\wedge\tau^{\pi_c}\}}.
\end{aligned} \tag{20}
$$

Since $g(x)$ satisfies the HJB Equation with $g'(x)\ge\beta_1$ and $\mathcal{M}g(0)\le g(0)$, we can see that the above inequality holds. Moreover, the second term on the right side of Equation (19) is non-positive. Then inserting Equation (20) into Equation (19) yields the equation below

$$
\begin{aligned}
e^{-\delta(t\wedge\tau^{\pi_c})}g(X_t^{\pi_c}) \;\le\; & g(x)+\int_0^{t\wedge\tau^{\pi_c}} e^{-\delta s}\sqrt{\lambda\mu^{(2)}(m_s^{\pi_c})}g'(X_s^{\pi_c})\,dB_s \\
& -\int_0^{t\wedge\tau^{\pi_c}} e^{-\delta s}\beta_1 dL_s^{\pi_c} + \sum_{i=1}^{\infty} e^{-\delta\tau_n^{\pi_c}}(\beta_2\eta_n^{\pi_c}+K)I_{\{\tau_n^{\pi_c}\le t\wedge\tau^{\pi_c}\}}.
\end{aligned} \tag{21}
$$

Owing to capital injection or ceding all the risk to the reinsurer, the bankruptcy never happens, which means $\tau^{\pi_c}=\infty$. $X_t^{\pi_c}$ has a "continuous" path and $g(x)$ is increasing, which is defined by the equation below

$$
\lim_{t\to\infty}\inf e^{-\delta t}g(X_t^{\pi_c})\ge\lim_{t\to\infty}e^{-\delta t}g(0)=0.
$$

Note that the stochastic integral with respect to Brownian motion is a uniformly integral martingale. Therefore, applying the expectations on both sides of Equation (21) and setting $t\to\infty$, one arrives at the formula below

$$
g(x)\ge V(x,\pi_c).
$$

From Equation (10), it follows that $g(x)\ge V_c(x)$.

(ii) If the strategy π_c^* is constructed according to Equations (13)–(18), by replacing π_c with π_c^* in Equation (19) and taking some simple calculations, we have outlined the equation below for $x\le x_{1c}^*$.

$$
\begin{aligned}
&\int_0^{t\wedge\tau^{\pi_c}} e^{-\delta s}\beta_1 dL_s^{\pi_c^*} - \sum_{n=1}^{\infty} e^{-\delta\tau_n^{\pi_c^*}}(\beta_2\eta_n^{\pi_c^*}+K)\, I_{\{\tau_n^{\pi_c^*}\le t\wedge\tau^{\pi_c^*}\}} \\
&= -e^{-\delta(t\wedge\tau^{\pi_c^*})}g(X_t^{\pi_c^*})+g(x)+\int_0^{t\wedge\tau^{\pi_c^*}} e^{-\delta s}\sqrt{\lambda\mu^{(2)}(m_s^{\pi_c^*})}g'(X_s^{\pi_c^*})\,dB_s.
\end{aligned} \tag{22}
$$

Applying the expectations and the limits on both sides of Equation (22) and noting that the controlled process $X_t^{\pi_c^*}$ is a double barrier reflecting process, the theorem can be proven. □

Next, before obtaining the closed-form solution of the value function and the retention level $m^{\pi_c^*}(x)$, we present the following lemma, which plays a key role in the solution procedure.

Lemma 1. *Let $\hat{m}_0\in(0,M)$ be the unique solution to the equation given below*

$$
C\exp\left[\int_0^{G(M)-G(\hat{m}_0)}\frac{\theta}{G^{-1}[x+G(\hat{m}_0)]}ds\right]=\beta_2, \tag{23}
$$

where $C>0$ is a constant and $G(x)=\int_0^x\frac{\mu^{(2)}(y)}{\frac{2\delta}{\theta\lambda}y^2+2\theta y\mu^{(1)}(y)-\theta\mu^{(2)}(y)}dy$. For each $m_0\in[0,\hat{m}_0]$, define a function

$$
F(x,m_0)=C\exp\left[\int_x^{x_0}\frac{\theta}{G^{-1}[x+G(m_0)]}ds\right],\ 0\le x\le x_0, \tag{24}
$$

where $x_0 = G(M) - G(m_0)$. As such, there is a unique $\eta \in (0, x_0)$ such that $F(\eta, m_0) = \beta_2$. Furthermore, the function below:

$$\Phi(m_0) = \int_0^{\eta(m_0)} [F(x, m_0) - \beta_2]dx \tag{25}$$

is decreasing in m_0 in the range $[0,\ \Phi(0)]$.

Proof. From the analytic form of $F(x, m_0)$, differentiating the function with respect to x yields the formula below

$$\frac{\partial F}{\partial x} = -\frac{C}{G^{-1}[x + G(m_0)]} \exp\left[\int_x^{x_0} \frac{\theta}{G^{-1}[x + G(m_0)]} ds\right] < 0,\ 0 < x < x_0.$$

Therefore, $F(x, m_0)$ is a decreasing function. Then, we let $m(x) = G^{-1}[x + G(m_0)]$ and we have the equation below:

$$m(x_0) = M,$$

$$m'(x) = \frac{1}{\mu^{(2)}(m)}\left[\frac{2\delta}{\theta\lambda}m^2 + 2\theta m\mu^{(1)}(m) - \theta\mu^{(2)}(m)\right].$$

Doing a variable change of $y = m(s)$ for $F(x, m_0) = C\exp\left[\int_x^{x_0} \frac{\theta}{m(s)} ds\right]$ and combining with the above two Equations, we have the equation below

$$F(x, m_0) = C\exp\left[\int_{m(x)}^{M} \frac{\mu^{(2)}(y)}{2y^2\left(\frac{\delta}{\theta^2\lambda}y + \mu^{(1)}(y) - \frac{\mu^{(2)}(y)}{2y}\right)} ds\right],\ 0 \le x \le x_0. \tag{26}$$

Clearly, $F(x, m_0)$ can be also viewed as a decreasing function of m_0. Then, Equation (23) can be rewritten as $F(0, \hat{m}_0) = \beta_2$. Therefore, we can deduce that if $m_0 \in [0, \hat{m}_0]$, there is a unique solution $\eta(m_0) \in [0, x_0]$ to the Equation $F(\eta, m_0) = \beta_2$. Furthermore, $\eta(m_0)$ is a decreasing function of m_0. Therefore, the minimum $\eta_{\min} = \eta(\hat{m}_0) = 0$ and the maximum $\eta_{\max} = \eta(0) < x_0$, which is uniquely determined by $F(\eta(0), 0) = C\exp\left[\int_{\eta(0)}^{G(M)} \frac{\theta}{G^{-1}(s)} ds\right] = \beta_2$. It's easy to see that $\Phi(m_0)$ is non-negative and decreasing in $[0, \hat{m}_0]$, which satisfies $\Phi(\hat{m}_0) = 0$ and $\Phi(0) = \int_0^{\eta(0)} [F(\eta(0), 0) - \beta_2]\, dx$. Therefore, it holds $\Phi(m_0) \in [0, \Phi(0)]$ for $m_0 \in [0, \hat{m}_0]$. □

In the following part, we will solve the explicit solution to the HJB Equation with the boundary condition. From Theorem 1, we know that Equation (11) can be rewritten using the formula below.

$$\max_{0 \le m \le M}\left[\frac{1}{2}\lambda\mu^{(2)}(m)g''(x) + \theta\lambda\mu^{(1)}(m)g'(x) - \delta g(x)\right] = 0, \text{ for } 0 \le x \le x_{1c}^* \tag{27}$$

and

$$g'(x) = \beta_1, \text{ for } x \ge x_{1c}^*.$$

For $0 \le x \le x_{1c}^*$, by differentiating on both sides of Equation (27) with respect to m and setting the derivative equal to zero, we have the equation below:

$$m(x) = -\theta\frac{g'(x)}{g''(x)}. \tag{28}$$

Substituting Equation (28) back into Equation (27) leads to the formula below:

$$\theta\lambda\left[\mu^{(1)}(m)-\frac{\mu^{(2)}(m)}{2m}\right]g'(x)-\delta g(x)=0. \tag{29}$$

Differentiating the above Equation with respect to x and using Equation (28) again, we obtain the formula below:

$$m'(x)=\frac{1}{\mu^{(2)}(m)}\left[\frac{2\delta}{\theta\lambda}m^2+2\theta m\mu^{(1)}(m)-\theta\mu^{(2)}(m)\right]. \tag{30}$$

Let:

$$G(x)=\int_0^x\frac{\mu^{(2)}(y)}{\frac{2\delta}{\theta\lambda}y^2+2\theta y\mu^{(1)}(y)-\theta\mu^{(2)}(y)}dy$$

It is not hard to verify that $G'(x)>0$ and then the inverse function of $G(x)$ exists. Therefore, the equation is shown below:

$$m(x)=G^{-1}[x+G(m_0)]. \tag{31}$$

From Equation (28), we can see that $m(x)$ is strictly increasing, which implies that there exists $x^*_{0c}<x^*_{1c}$ such that the insurance company will keep all the claims and not cede any risk to the insurer if the surplus exceeds x^*_{0c}. In the view of (28), we have the equation below:

$$g(x)=k_1\int_0^x\exp\left[\int_z^{x_0}\frac{\theta}{m(s)}ds\right]dz+k_2,\ 0\le x\le x^*_{0c}. \tag{32}$$

In addition, for $x^*_{0c}<x\le x^*_{1c}$, taking $m(x)\equiv M$, we have $g(x)$ satisfying the following ODE

$$\frac{1}{2}\lambda\mu^{(2)}g''(x)+\theta\lambda\mu^{(1)}g'(x)-\delta g(x)=0.$$

It has the solution shown below:

$$g(x)=k_3e^{r_+(x-x^*_{1c})}+k_4e^{r_-(x-x^*_{1c})},\ x^*_{0c}<x\le x^*_{1c}.$$

where $r_\pm$ are the two roots of the equation $\frac{1}{2}\lambda\mu^{(2)}r^2+\theta\lambda\mu^{(1)}r-\delta=0$, and:

$$r_\pm=\frac{-\theta\mu^{(1)}\pm\sqrt{\left(\theta\mu^{(1)}\right)^2+2\mu^{(2)}\delta}}{\mu^{(2)}}. \tag{33}$$

Lastly, for $x>x^*_{1c}$, $g'(x)\equiv\beta_1$ and $g(x^*_{1c})=k_3+k_4$ yields the following formula:

$$g(x)=\beta_1(x-x^*_{1c})+k_3+k_4.$$

The constants k_1, k_2, k_3, k_4, and the critical values x^*_{0c}, x^*_{1c} are determined by the principle of smooth fit. From the first and second derivatives of $g(x)$ at the points x^*_{0c} and x^*_{1c}, we can have the following equalities:

$$k_3r_++k_4r_-=\beta_1,$$

$$k_3(r_+)^2+k_4(r_-)^2=0,$$

$$k_3r_+e^{r_+(x^*_{0c}-x^*_{1c})}+k_4r_-e^{r_-(x^*_{0c}-x^*_{1c})}=k_1,$$

$$k_3(r_+)^2e^{r_+(x^*_{0c}-x^*_{1c})}+k_4(r_-)^2e^{r_-(x^*_{0c}-x^*_{1c})}=-\frac{\theta}{M}k_1.$$

Solving the above equations and doing some calculations, we have found the following formula:

$$k_3 = -\frac{r_-\beta_1}{r_+(r_+ - r_-)} > 0, \tag{34}$$

$$k_4 = -\frac{r_+\beta_1}{r_-(r_+ - r_-)} > 0, \tag{35}$$

$$k_1 = \frac{M\beta_1}{M+\theta/r_+}\left[\frac{M+\theta/r_-}{M+\theta/r_+}\right]^{\frac{r_-}{r_+-r_-}}. \tag{36}$$

It's obvious that $k_3 + k_4 = \frac{\theta\lambda\mu^{(1)}\beta_1}{\delta}$. Therefore, substituting the constants back into $g'(x_{0c}^*) = k_1$ yields the formula below:

$$x_{1c}^* = x_{0c}^* + \frac{1}{r_+ - r_-}\ln\left[\frac{M+\theta/r_+}{M+\theta/r_-}\right], \tag{37}$$

where x_{0c}^* satisfies the following equation:

$$x_{0c}^* = G(M) - G(m^{\pi_c^*}(0)). \tag{38}$$

In the view of Equation (29), Let $x = 0$ and performing the same variable change as in Lemma 1, we obtain the following formula:

$$k_2 = \tfrac{\theta\lambda}{\delta}k_1\left(\mu^{(1)}(m^{\pi_c^*}(0)) - \tfrac{\mu^{(2)}(m^{\pi_c^*}(0))}{2m^{\pi_c^*}(0)}\right)\exp\left(\int_{m^{\pi_c^*}(0)}^{M}\frac{\mu^{(2)}(y)}{2y^2\left(\frac{\delta}{\theta^2\lambda}y + \mu^{(1)}(y) - \frac{\mu^{(2)}(y)}{2y}\right)}dy\right). \tag{39}$$

Therefore, the unknown constants x_{0c}^*, x_{1c}^*, and k_2 are clear from Equations (37)–(39) once $m^{\pi_c^*}(0)$ is determined. Considering the analysis before and the boundary condition $\mathcal{M}g(0) = g(0)$, we can obtain the value of $m^{\pi_c^*}(0)$ in the following two cases.

(i) If $0 < K \leq \Phi(0)$, it's conjectured that $\mathcal{M}g(0) = g(0)$ holds under the condition that there exist some $m^{\pi_c^*}(0) \in [0, M]$ and $\eta^*(m^{\pi_c^*}(0)) > 0$ in which the following equations are true:

$$g'(\eta^*) = \beta_2, \tag{40}$$

$$g(0) = g(\eta^*) - \beta_2\eta^* - K = \mathcal{M}g(0). \tag{41}$$

which can be rewritten as:

$$K = \int_0^{\eta^*}[g'(x) - \beta_2]dx = \int_0^{\eta^*(m^{\pi_c^*}(0))}[F(x, m^{\pi_c^*}(0)) - \beta_2]\,dx = \Phi(m^{\pi_c^*}(0)). \tag{42}$$

By Lemma 1, it follows that $m^{\pi_c^*}(0) \in [0, \hat{m}_0)$ and η^* exist if and only if $0 < K \leq \Phi(0)$.

(ii) If $K > \Phi(0)$, the value $\eta^*(m^{\pi_c^*}(0)) > 0$ satisfying Equations (40) and (41) doesn't exist and $\mathcal{M}g(0) < g(0)$, which implies that $G^{\pi_c^*} \equiv 0$. Therefore, in order to meet the boundary condition in Equation (12), we have $g(0) = 0$ and $m^{\pi_c^*}(0) \equiv 0$. Now, summarizing the above discussions, we have the following result.

Theorem 2. *If the insurance company doesn't allow for bankruptcy, the value function $V_c(x)$ coincides with the formulas below.*

$$g(x) = \begin{cases} k_1\int_0^x \exp\left[\int_z^{x_0}\frac{\theta}{m(s)}ds\right]dz + k_2, & 0 \leq x \leq x_{0c}^*, \\ k_3 e^{r_+(x - x_{1c}^*)} + k_4 e^{r_-(x - x_{1c}^*)}, & x_{0c}^* \leq x \leq x_{1c}^*, \\ \beta_1(x - x_{1c}^*) + \frac{\theta\lambda\mu^{(1)}\beta_1}{\delta}, & x > x_{1c}^*, \end{cases} \tag{43}$$

where the constants k_1, k_2, k_3, and k_4 are given by Equations (36), (39), (34), and (35), respectively. In addition, the critical values x^*_{0c} and x^*_{1c} are shown in Equations (37) and (38). Correspondingly, the optimal dividends strategy $L^{\pi^*_c}$ satisfies the equation below:

$$L^{\pi^*_c} = (x - x^*_{1c})^+ + \int_0^t I_{\{X_s^{\pi^*_c} = x^*_{1c}\}} dL_s^{\pi^*_c}, \text{ for all } t \geq 0. \tag{44}$$

In addition, the optimal retention level of excess-of-loss reinsurance is shown below:

$$m^{\pi^*_c}(x) = \begin{cases} G^{-1}(x + G(m^{\pi^*_c}(0))), & 0 \leq x \leq x^*_{0c}, \\ M, & x \geq x^*_{0c}. \end{cases} \tag{45}$$

The value $m^{\pi^*_c}(0)$ and the optimal injection strategy $G_t^{\pi^*_c}$ are determined in the following cases.

(i) If $0 < K \leq \Phi(0)$, $m^{\pi^*_c}(0) = m_0 \in [0, \hat{m}_0]$ is the unique solution to the equation $\Phi(m_0) = K$. The optimal injection strategy $G_t^{\pi^*_c}$ is given by Equations (13)–(15) and the optimal amount of capital injection η^* is obtained by Equations (40) and (41). This means that, by injecting the capital, the insurance company's surplus immediately jumps to η^* when it hits the barrier 0. In this case, the boundary condition are $\mathcal{M}g(0) = g(0)$ and $g(0) \geq 0$.

(ii) If $K > \Phi(0)$, then $m^{\pi^*_c}(0) = 0$. The optimal strategy of capital injection satisfies $G_t^{\pi^*_c} \equiv 0$, which means the capital injection never happens. It suggests that, if the insurance company's surplus attains zero, it will cede all the potential risk to the reinsurance company and keep the surplus stay at 0. Therefore, the bankruptcy will never happen.

Remark 3. *We can easily verify that $g(x)$ is concave by checking its second derivative symbol and prove that $g(x)$ given in the three cases is the solution to the HJB Equation by substituting all the forms of $g(x)$ back into Equation (11) and applying the analysis before the Theorem. Therefore, we omit all the details here.*

Remark 4. *From Theorem 2, $\Phi(0)$ could be viewed as the maximum fixed transaction cost that the insurance company is willing to pay when the capital injection happens. With the increase of the fixed cost K, the company should reduce the retention level and raise the dividend barrier to increase the size of capital injection in order to reduce its amount. When K is larger than $\Phi(0)$, the best way to avoid raising new funds is to keep the company away from bankruptcy, which coincides with the real market.*

4. The Solution to the Problem without Capital Injection

In this section, we consider the other suboptimal control model for an insurance company seeking to maximize the expected discounted value of the cumulative dividend payout until the ruin time plus the salvage value at the ruin time.

Defined by $\pi_d = \{m^{\pi_d}; L^{\pi_d}; 0\} \in \Pi_x$ the control strategy without capital injection, the performance index function associated with this strategy becomes the equation below:

$$V(x, \pi_d) = E_x\left[\int_0^{\tau^{\pi_d}} \beta_1 e^{-\delta s} dL_s^{\pi_d} + Pe^{-\delta \tau^{\pi_d}}\right]. \tag{46}$$

The objective is to find the value function

$$V_d(x) = \sup_{\pi_d \in \Pi_x} V(x, \pi_d), \tag{47}$$

and the associated optimal control strategy π^*_d so that $V_d(x) = V(x, \pi^*_d)$.

Let the value function defined by Equation (47) be sufficiently smooth. Based on the stochastic control theorem, we obtain the corresponding HJB Equation below:

$$\max\left\{\max_{0\leq m\leq M}\mathcal{L}^{m}V_d(x),\ \beta_1 - V'_d(x)\right\} = 0 \tag{48}$$

with the boundary condition $V_d(0) = P$.

It's easy to see that it is a typical optimal dividend control problem. Inspired by some research on this issue, we conjecture that the optimal dividend strategy $L_t^{\pi_d^*}$ is still a barrier strategy with some critical values $x_{1d}^* = \inf\{x : V_d'(x) = \beta_1\}$. Mathematically, we define the formula below:

$$L_t^{\pi_d^*} = (x - x_{1d}^*)^+ + \int_0^t I_{\{X_s^{\pi_d^*} = x_{1d}^*\}} dL_s^{\pi_d^*}, \text{ for all } t \geq 0. \tag{49}$$

Furthermore, the optimal excess-of-loss reinsurance retention level $m^{\pi_d^*}(x)$ satisfies the equation below.

$$\mathcal{L}^{m^{\pi_d^*}(x)}V_d(x) = \max_{0\leq m\leq M}\mathcal{L}^{m}V_d(x) = 0, \text{ for } 0 \leq x \leq x_{1d}^*. \tag{50}$$

Therefore, based on the above analysis, we have the following result.

Theorem 3. *Let $f(x)$ be an increasing, concave, and twice continuously differential solution to the HJB Equation (48) with the boundary condition. In this case, one can obtain the following results.*

(i) For each $\pi_d \in \Pi_x$, it shows that $f(x) \geq V(x, \pi_d)$. Therefore, $f(x) \geq V_d(x)$ for all $x \geq 0$.
(ii) If the strategy $\pi_d^* = \{m^{\pi_d^*}; L^{\pi_d^*}; 0\}$ is constructed by Equations (49) and (50) so that $f(x) = V(x, \pi_d^*)$ is constructed by Equations (49) and (50) so that , then $f(x) = V_d(x)$ and π_d^* is optimal.

Proof. Fixing a strategy $\pi_d \in \Pi_x$, define the sets $\Lambda = \{s : L_{s-}^{\pi_d} \neq L_s^{\pi_d}\}$ and then let $\hat{L}_t^{\pi_d} = \sum_{s\in\Lambda, s\leq t}(L_s^{\pi_d} - L_{s-}^{\pi_d})$ and $\tilde{L}_t^{\pi_d} = L_t^{\pi_d} - \hat{L}_t^{\pi_d}$ be the discontinuous and continuous parts of $L_t^{\pi_d}$, respectively. Then, by the general Itô formula, we obtain the equation below.

$$\begin{aligned} e^{-\delta(t\wedge\tau^{\pi_d})}f(X_t^{\pi_d}) = {} & f(x) + \int_0^{t\wedge\tau^{\pi_d}} e^{-\delta s}\mathcal{L}^{m^{\pi_d}}f(X_s^{\pi_d})\,ds + \int_0^{t\wedge\tau^{\pi_d}} e^{-\delta s}\sqrt{\lambda\mu^{(2)}(m_s^{\pi_d})}f'(X_s^{\pi_d})\,dB_s \\ & - \int_0^{t\wedge\tau^{\pi_d}} e^{-\delta s}f'(X_s^{\pi_d})\,d\tilde{L}_s^{\pi_d} + \sum_{s\in\Lambda, s\leq t\wedge\tau^{\pi_d}} e^{-\delta s}\left[f(X_s^{\pi_d}) - f(X_{s-}^{\pi_d})\right] \end{aligned} \tag{51}$$

due to $f(x)$ satisfying the HJB Equation with $f'(x) = \beta_1$, the equation below shows:

$$\sum_{s\in\Lambda, s\leq t\wedge\tau^{\pi_d}} e^{-\delta s}\left[f(X_s^{\pi_d}) - f(X_{s-}^{\pi_d})\right] \leq -\sum_{s\in\Lambda, s\leq t\wedge\tau^{\pi_d}} e^{-\delta s}\beta_1(L_s^{\pi_d} - L_{s-}^{\pi_d}). \tag{52}$$

Moreover, the second term on the right side of Equation (51) is non-positive. Therefore, substituting Equation (52) into Equation (51) leads to the findings below:

$$e^{-\delta(t\wedge\tau^{\pi_d})}f(X_t^{\pi_d}) \leq f(x) + \int_0^{t\wedge\tau^{\pi_d}} e^{-\delta s}\sqrt{\lambda\mu^{(2)}(m_s^{\pi_d})}f'(X_s^{\pi_d})\,dB_s - \int_0^{t\wedge\tau^{\pi_d}} e^{-\delta s}\beta_1 dL_s^{\pi_d}. \tag{53}$$

since $f(x)$ is an increasing function and $f(0) = P$, the following equation was found.

$$\liminf_{t\to\infty} e^{-\delta(t\wedge\tau^{\pi_d})}f(X_t^{\pi_d}) \geq \lim_{t\to\infty} e^{-\delta(t\wedge\tau^{\pi_d})}f(0) = Pe^{-\delta\tau^{\pi_d}}.$$

The stochastic integral with respect to the Brownian motion is a uniformly integral martingale. Taking expectations on both sides of Equation (53) and letting $t \to \infty$, we have $f(x) \geq V(x, \pi_d)$ and, therefore, $f(x) \geq V_d(x)$.

(ii) If the strategy π_d^* is constructed according to Equations (49) and (50), replacing π_d by π_d^* in Equation (51) and taking some simple calculations, we found the equation below:

$$\int_0^{t\wedge\tau^{\pi_d}} e^{-\delta s}\beta_1 dL_s^{\pi_d} = -e^{-\delta(t\wedge\tau^{\pi_d})} f(X_t^{\pi_d^*}) + f(x) + \int_0^{t\wedge\tau^{\pi_d}} e^{-\delta s}\sqrt{\lambda\mu^{(2)}(m_s^{\pi_d})} f'(X_s^{\pi_d})\, dB_s. \quad (54)$$

Taking the expectation and the limits on both sides of Equation (54), we can obtain the result. □

We can see that both $V_d(x)$ and $V_c(x)$ satisfy the same HJB Equation but meet different boundary conditions. Therefore, we can get the expression of $V_d(x)$ and the retention level $m^{\pi_d^*}(x)$ by the same method where the value $m^{\pi_d^*}(0) \in [0, M]$ is determined by the boundary condition $V_d(0) = P$. In order to save space, we avoid the repeated calculations and give the result as follows.

Theorem 4. *If the insurance company does not allow for capital injection, according to the salvage value $P \geq 0$, the value function $V_d(x)$ coincides with $f(x)$ in the following three cases.*

(i) If $0 \leq P \leq \frac{\theta\lambda}{\delta} k_1\left(\mu^{(1)} - \frac{\mu^{(2)}}{2M}\right)$, then $f(x)$ has the form:

$$f(x) = \begin{cases} k_1\int_0^x \exp\left[\int_z^{x_0} \frac{\theta}{m(s)} ds\right] dz + P, & 0 \leq x < x_{0d}^*, \\ k_3 e^{r_+(x-x_{1d}^*)} + k_4 e^{r_-(x-x_{1d}^*)}, & x_{0d}^* \leq x < x_{1d}^*, \\ \beta_1(x - x_{1d}^*) + \frac{\theta\lambda\mu^{(1)}\beta_1}{\delta}, & x \geq x_{1d}^*, \end{cases} \quad (55)$$

where the constants k_1, k_3, and k_4 are given by Equations (36), (34), and (35), respectively. The critical level x_{0d}^* and x_{1d}^* satisfy the equation below:

$$x_{1d}^* = x_{0d}^* + \frac{1}{r_+ - r_-}\ln\left[\frac{M + \theta/r_+}{M + \theta/r_-}\right], \quad (56)$$

$$x_{0d}^* = G(M) - G(m^{\pi_d^*}(0)). \quad (57)$$

In addition, $m^{\pi_d^*}(0)$ is the solution to the following equation

$$\frac{\theta\lambda}{\delta}k_1\left(\mu^{(1)}(m^{\pi_d^*}(0)) - \frac{\mu^{(2)}(m^{\pi_d^*}(0))}{2m^{\pi_d^*}(0)}\right)\exp\left(\int_{m^{\pi_d^*}(0)}^M \frac{\mu^{(2)}(y)}{2y^2\left(\frac{\delta}{\theta^2\lambda}y + \mu^{(1)}(y) - \frac{\mu^{(2)}(y)}{2y}\right)}dy\right) = P. \quad (58)$$

Accordingly, the optimal dividend strategy $L_t^{\pi_d^*}$ should satisfy the equation below:

$$L_t^{\pi_d^*} = (x - x_{1d}^*)^+ + \int_0^t I_{\{X_s^{\pi_d^*} = x_{1d}^*\}} dL_s^{\pi_d^*}, \text{ for all } t \geq 0. \quad (59)$$

In addition, the optimal excess-of-loss reinsurance retention level $m^{\pi_d^*}(x)$ is shown below:

$$m^{\pi_d^*}(x) = \begin{cases} G^{-1}(x + G(m^{\pi_d^*}(0))), & 0 \leq x \leq x_{0d}^*, \\ M, & x \geq x_{0d}^*. \end{cases} \quad (60)$$

(ii) If $\frac{\theta\lambda}{\delta}k_1\left(\mu^{(1)}-\frac{\mu^{(2)}}{2M}\right)<P\leq\frac{\theta\lambda\mu^{(1)}}{\delta}$, then $f(x)$ has the form below:

$$f(x)=\begin{cases} k_3e^{r_+(x-x_{1d}^*)}+k_4e^{r_-(x-x_{1d}^*)}, & 0\leq x<x_{1d}^*, \\ \beta_1(x-x_{1d}^*)+\frac{\theta\lambda\mu^{(1)}\beta_1}{\delta}, & x\geq x_{1d}^*, \end{cases} \tag{61}$$

where the constants k_3 and k_4 are given by Equations (34) and (35). The critical level x_{1d}^* is determined by the equation below:

$$k_3e^{-r_+x_{1d}^*}+k_4e^{-r_-x_{1d}^*}=P. \tag{62}$$

Correspondingly, the optimal dividend strategy $L_t^{\pi_d^*}$ should satisfy the formula below:

$$L_t^{\pi_d^*}=(x-x_{1d}^*)^++\int_0^t I_{\{X_s^{\pi_d^*}=x_{1d}^*\}}dL_s^{\pi_d^*},\ \text{for all } t\geq 0. \tag{63}$$

In addition, the optimal excess-of-loss reinsurance retention level is $m^{\pi_d^*}(x)\equiv M$.

(iii) If $P>\frac{\theta\lambda\mu^{(1)}}{\delta}$, then $f(x)=\beta_1x+P$. The optimal dividend strategy is to pay the whole initial surplus x as the dividends and declare bankruptcy at once. Then, the salvage P is realized.

Remark 5. *As shown in Theorem 4, the determination of the value function and the optimal control strategy depends on P and the retained risk level of the insurance company is increasing with P. In the case of $P=0$, the optimal retention is zero. This means that the insurer will cede all the risk to the reinsurance company and keep the surplus at zero. Therefore, the ruin will never happen. When the salvage value is great enough, it's optimal to announce the bankruptcy and realize the salvage value at once.*

5. The Solution to the General Control Problem

If there are no restrictions on the capital injection or the surplus process, the general control problem seeks to maximize the expected sum of discounted salvage value and the discounted dividends except for the expected discounted cost of capital injection over all the admissible strategies. Therefore, the corresponding HJB Equation takes the following form:

$$\max\left\{\max_{0\leq m\leq M}\mathcal{L}^mV(x),\ \beta_1-V'(x),\ \mathcal{M}V(x)-V(x)\right\}=0. \tag{64}$$

The boundary condition is shown below:

$$\max\{\mathcal{M}V(0)-V(0),\ P-V(0)\}=0. \tag{65}$$

Theorem 5. *Let $v(x)$ be a concave, increasing, and twice continuously differentiable solution to Equations (64) and (65). We have the following result:*

(i) For each $\pi\in\Pi_x$, it shows that $v(x)\geq V(x,\pi)$. So $v(x)\geq V(x)$ for all $x\geq 0$.
(ii) If there is a strategy of $\pi^*=\left\{m^{\pi^*};L^{\pi^*};G^{\pi^*}\right\}$ so that $v(x)=V(x,\pi^*)$, then $v(x)=V(x)$ and π^* is optimal.

Proof. The proof of (i) is similar to the first statement's proof of Theorems 1 and 3 and the result (ii) can be obtained by considering the optimality of $v(x)$. We omit this here. □

Before deriving the optimal strategy by solving the general control problem, we give the two Lemmas to show that the sign of some important properties are determined by the relationships of parameters. This shows the equation below:

$$\hat{P} = \frac{\theta\lambda k_1}{\delta}\left[\mu^{(1)}(\hat{m}_0) - \frac{\mu^{(2)}(\hat{m}_0)}{2\hat{m}_0}\right]\exp\left(\int_{\hat{m}_0}^{M}\frac{\mu^{(2)}(y)}{2y^2\left(\frac{\delta}{\theta^2\lambda}y + \mu^{(1)}(y) - \frac{\mu^{(2)}(y)}{2y}\right)}dy\right).$$

One can note that $\hat{P} \in [0, \frac{\theta\lambda k_1}{\delta}(\mu^{(1)} - \frac{\mu^{(2)}}{2M})]$ with a unique root $\hat{m}_0 \in [0, M]$.

Lemma 2. *The sign of $\mathcal{M}V_c(0) - V_c(0)$ and $P - V_c(0)$ are determined in the following manner.*

(i) If $0 < K \leq \Phi(0)$, $P \leq \hat{P}$ and $m^{\pi_c^*}(0) \leq m^{\pi_d^*}(0)$, it has $\mathcal{M}V_c(0) - V_c(0) = 0$ and $P - V_c(0) \geq 0$.
(ii) If $0 < K \leq \Phi(0)$, $P \leq \hat{P}$ and $m^{\pi_c^*}(0) > m^{\pi_d^*}(0)$, it has $\mathcal{M}V_c(0) - V_c(0) = 0$ and $P - V_c(0) < 0$.
(iii) If $0 < K \leq \Phi(0)$ and $P > \hat{P}$, it has $\mathcal{M}V_c(0) - V_c(0) < 0$.
(iv) If $K > \Phi(0)$, it has $\mathcal{M}V_c(0) - V_c(0) < 0$.

Proof. If the function $h(x) = \mu^{(1)}(x) - \frac{\mu^{(2)}(x)}{2x}$, in Section 3, it follows that:

$$\begin{aligned} V_c(0) &= \frac{\theta\lambda k_1}{\delta}\left[\mu^{(1)}(m_0) - \frac{\mu^{(2)}(m_0)}{2m_0}\right]\exp\left(\int_{m_0}^{M}\frac{\mu^{(2)}(y)}{2y^2\left(\frac{\delta}{\theta^2\lambda}y+\mu^{(1)}(y)-\frac{\mu^{(2)}(y)}{2y}\right)}dy\right) \\ &= \frac{\theta\lambda k_1}{\delta}g(m_0), \end{aligned}$$

where the function $g(x)$ is given below:

$$g(x) = h(x)\exp\left(\int_x^M \frac{h'(y)}{\frac{\delta}{\theta^2\lambda}y + h(y)}dy\right), \; x > 0.$$

By checking $g(0+) = 0, g(M) = \mu^{(1)} - \frac{\mu^{(2)}}{2M}$ and the derivative below are discovered:

$$g'(x) = h'(x)\exp\left(\int_x^M \frac{h'(y)}{\frac{\delta}{\theta^2\lambda}y + h(y)}dy\right)\left(1 - \frac{h(x)}{\frac{\delta}{\theta^2\lambda}x + h(x)}\right) > 0.$$

We can deduce that $g(x)$ is increasing in $[0, M]$ and $\hat{P} = \theta\lambda k_1 g(\hat{m}_0)/\delta$ and the case of $P \leq \hat{P}$ leads to $m^{\pi_d^*}(0) \leq \hat{m}_0$. From Theorem 2, it follows that $\mathcal{M}V_c(0) - V_c(0) = 0$ holds with some $m^{\pi_c^*}(0) \in [0, \hat{m}_0)$. When $0 < K \leq \Phi(0)$, it's clear that $V_c(0) = \theta\lambda k_1 g(m^{\pi_c^*}(0))/\delta < \hat{P}$ since $m^{\pi_c^*}(0) < \hat{m}_0$. Therefore, if $m^{\pi_c^*}(0) \leq m^{\pi_d^*}(0)$, then $P - V_c(0) \geq 0$. On the other hand, if $m^{\pi_c^*}(0) > m^{\pi_d^*}(0)$, then $P - V_c(0) < 0$. In this paper, we have the statements (i) and (ii). As for $P > \hat{P}$, it shows that $V_c(0) < \hat{P} < P$ and $\mathcal{M}V_c(0) - V_c(0) = 0$, which holds with $m^{\pi_c^*}(0) \in [0, \hat{m}_0)$. The statement (iv) is a direct result of Theorem 2.

The proof is completed. □

Lemma 3. *If the equality $V_d(0) - P = 0$ holds, the sign of $\mathcal{M}V_d(0) - V_d(0)$ is determined by the different cases, which is shown below.*

(i) If $0 < K \leq \Phi(0)$, $P \leq \hat{P}$, and $m^{\pi_c^*}(0) < m^{\pi_d^*}(0)$, we find that $\mathcal{M}V_d(0) - V_d(0) < 0$.
(ii) If $0 < K \leq \Phi(0)$, $P \leq \hat{P}$, and $m^{\pi_c^*}(0) \geq m^{\pi_d^*}(0)$, we find that $\mathcal{M}V_d(0) - V_d(0) \geq 0$.
(iii) If $0 < K \leq \Phi(0)$ and $P > \hat{P}$, we find that $\mathcal{M}V_d(0) - V_d(0) < 0$.
(iv) If $K > \Phi(0)$, we find that $\mathcal{M}V_d(0) - V_d(0) < 0$.

Proof. In the case of $P \le \hat{P}$, the equality $V_d(0) - P = 0$ has a unique solution of $m^{\pi_d^*}(0) \le \hat{m}_0$. By Lemma 1, when $0 < K \le \Phi(0)$, it suggests that there exists some $\eta(m^{\pi_d^*}(0)) \in [0, x_0]$ such that $V'_d(\eta(m^{\pi_d^*}(0))) = F(\eta, m^{\pi_d^*}(0)) = \beta_2$. Since $\Phi(x)$ is a decreasing function, the following equation was found:

$$\begin{aligned}\mathcal{M}V_d(0) - V_d(0) &= \max_{y\ge 0}\{V_d(y) - \beta_2 y - K - V_d(0)\} = \max_{y\ge 0}\{\textstyle\int_0^y (V'_d(x) - \beta_2)dx\} - K \\ &= \textstyle\int_0^{\eta(m^{\pi_d^*}(0))} (V'_d(x) - \beta_2)dx - K \le \int_0^{\eta(m^{\pi_c^*}(0))} (V'_d(x) - \beta_2)dx - K = 0,\end{aligned}$$

where the inequality follows from $m^{\pi_d^*}(0) \ge m^{\pi_c^*}(0)$. Clearly, the following equation holds:

$$\mathcal{M}V_d(0) - V_d(0) = \Phi(\eta(m^{\pi_d^*}(0))) - K > 0,$$

then the condition satisfies $m^{\pi_c^*}(0) > m^{\pi_d^*}(0)$. As for the case of $P > \hat{P}$, the solution $m^{\pi_d^*}(0) \in [0, \hat{m}_0]$ doesn't exist. Therefore, there isn't some value $\eta\left(m^{\pi_d^*}(0)\right) > 0$ such that $V'_d(\eta(m^{\pi_d^*}(0))) = F(\eta, m^{\pi_d^*}(0)) = \beta_2$. It implies that $V'_d(x) < \beta_2$ holds for all $x \ge 0$. Since $V_d(x)$ is concave, then the following equation is found:

$$\mathcal{M}V_d(0) - V_d(0) = \max_{y\ge 0}\left\{\int_0^y (V'_d(x) - \beta_2)dx\right\} - K < 0.$$

From Lemma 1, we know that the maximum of $\mathcal{M}V_d(0) - V_d(0)$ is $\Phi(0) - K$. Therefore, it follows that $\mathcal{M}V_d(0) - V_d(0) < 0$ when $\Phi(0) < K$.

The proof is completed. □

Comparing the two different suboptimal models in Sections 3 and 4 and using the above two Lemmas, we obtain the following Theorem.

Theorem 6. *For any given initial surplus $x > 0$, if the general control problem seeks to maximize the performance index function over all admissible strategies, $g(x)$ and $f(x)$ are the solution to the HJB Equation in Theorems 2 and 4, respectively. Then the solution is given in the following two cases:*

Case 1. If $\mathcal{M}g(0) - g(0) = 0$ and $P - g(0) \le 0$, the following equivalent condition is valid

$$0 < K \le \Phi(0), P \le \hat{P} \text{ and } m^{\pi_c^*}(0) > m^{\pi_d^*}(0),$$

then $V(x) = V_c(x) = g(x)$ and the optimal strategy π^* is the same as the corresponding strategy $\pi_c^* = \left\{m^{\pi_c^*}; L^{\pi_c^*}; G^{\pi_c^*}\right\}$ in Theorem 2.

Case 2. If $\mathcal{M}f(0) - f(0) < 0$ and $P - f(0) = 0$, one of the following equivalent conditions holds.

(i) $0 < K \le \Phi(0), P \le \hat{P}$ and $m^{\pi_c^*}(0) < m^{\pi_d^*}(0)$;
(ii) $0 < K \le \Phi(0)$ and $P > \hat{P}$;
(iii) $K > \Phi(0)$.

then $V(x) = V_d(x) = f(x)$ and the optimal strategy π^* is the same as the corresponding strategy $\pi_d^* = \left\{m^{\pi_d^*}; L^{\pi_d^*}; 0\right\}$ in Theorem 4.

Proof. The proofs of (i) and (ii) resemble the second statement in Theorems 1 and 3, which we omit here. □

Remark 6. *Xu and Zhou [20] explored the optimal dividend policies with the terminal value and excess-of-loss reinsurance. It is mainly the general control problem with Case 2 in this paper. Liu and Hu [26] studied the* *optimal financing and dividend policies with excess-of-loss reinsurance in the case where* $P = K = 0$. *Those results can be perceived as the limiting form of our results when* $P \to 0$, $K \to 0$. *Since there exists the fixed transaction cost and the salvage value, whether the insurance company decides to inject new capital or declare bankruptcy relies on the underlying cost of injections and also the potential profits in the future.*

6. Conclusions

This paper investigated the optimal control problem for an insurance company with transaction costs and salvage value where the company controls the risk exposure by the excess-of-loss reinsurance and capital injection based on the symmetry of risk information. Besides the proportional cost, the fixed cost incurred by capital injection is also incorporated. The insurance company's objective is to maximize the expected discounted sum of the salvage value and the cumulative dividends minus the expected discounted cost of capital injection until the ruin time. By considering whether there is capital injection in the surplus process, we construct two categories of suboptimal models and then solve for the corresponding solution in each model. Lastly, we consider the optimal control strategy for the general model without any restriction on the capital injection or the surplus process.

The result shows that, with the excess-of-loss reinsurance, if the insurance company does not intend to inject the capital and allows for the possibility of bankruptcy, the determinations of both the value function and the optimal dividend strategy depend on the salvage value of the company at the ruin time. Furthermore, the retained risk level of excess-of-loss reinsurance also increases with this salvage value. In particular, if the salvage value is zero, the optimal retention is zero and, therefore, the insurance company should cede all the risk to the reinsurance company. However, if the salvage value is great enough, it's optimal to announce the bankruptcy and realize the salvage value at once. If the insurance company is willing to prevent itself from going bankrupt by injecting the capital, it should give more attention to the maximum fixed cost that can be paid when the capital injection occurs. In addition, with the increase of the fixed cost, the company should reduce the retention risk level and raise the dividend barrier at the same time. By doing this, the insurance company can increase the size of capital injection and, therefore, reduce the amount of the fixed cost. However, if the cost is large enough, the best way forward for the insurance company is not to collect new money in order to keep itself from going bankrupt. This coincides with the real-world market situation.

As it is widely known, dividends and capital injection are two important economic activities in an insurance company's operations. Therefore, how to decide the corresponding control strategies is always an imperative problem that remains to be solved. From the main result, we can see that, when an insurance company takes the excess-of-loss reinsurance as the main insurance strategy to manage and control the exposure to risk, the choice of the optimal strategy for the general control problem is determined by some key parameters in the surplus process. Furthermore, due to the existence of the fixed transaction cost and the salvage value, the insurance company should consider the cost of injections and the potential profits in the future when deciding to inject new capital or declare bankruptcy.

Author Contributions: Conceptualization, Q.Y.; Formal analysis, L.Y.; Funding acquisition, Q.Y., T.B., and L.Y.; Investigation, C.Q.; Resources, T.B. and D.S.; Supervision, Q.Y.; Writing—original draft, L.Y.; Writing—review & editing, T.B. and D.S.

Funding: This research was funded by the Fundamental Research Funds for the Central Universities [Grant No. 2016XS78], the Research Base Project of Beijing Social Science Foundation [Grant No. 16JDGLB023], and the 111 Project [Grant No. B18021].

Acknowledgments: We thank the two anonymous referees and the editor for their thoughtful comments and suggestions.

Conflicts of Interest: The authors declare that there are no conflicts of interests regarding the publication of this paper.

References

1. Aniunas, P.; Gipiene, G.; Valukonis, M.; Vijunas, M. Liquidity Risk Management Model for Local Banks. *Transform. Bus. Econ.* **2017**, *16*, 153–173.
2. Lakstutiene, A.; Witkowska, J.; Leskauskiene, E. Transformation of the EU Deposit Insurance System: Evaluation. *Transform. Bus. Econ.* **2017**, *16*, 147–170.
3. Kurach, R. International Diversification of Pension Funds Using the Cointegration Approach: The Case of Poland. *Transform. Bus. Econ.* **2017**, *16*, 42–55.
4. Asmussen, S.; Taksar, M. Controlled diffusion models for optimal dividend pay-out. *Insur. Math. Econ.* **1997**, *20*, 1–15. [CrossRef]
5. Gerber, H.U.; Shiu, E.S.W. On optimal dividends strategies in the compound Poisson model. *N. Am. Actuar. J.* **2006**, *10*, 76–93. [CrossRef]
6. Belhaj, M. Optimal dividend payments when cash reserves follow a jump-diffusion process. *Math. Financ.* **2010**, *20*, 313–325. [CrossRef]
7. Azcue, P.; Muler, N. Optimal dividend policies for compound Poisson processes: The case of bounded dividend rates. *Insur. Math. Econ.* **2012**, *51*, 26–42. [CrossRef]
8. Meng, H.; Siu, T.K.; Yang, H. Optimal dividends with debts and nonlinear insurance risk processes. *Insur. Math. Econ.* **2013**, *53*, 110–121. [CrossRef]
9. Kulenko, N.; Schmidli, H. Optimal dividend strategies in a Cramer–Lundberg model with capital injections. *Insur. Math. Econ.* **2008**, *43*, 270–278. [CrossRef]
10. Yao, D.; Yang, H.; Wang, R. Optimal dividend and capital injection problem in the dual model with proportional and fixed transaction costs. *Eur. J. Oper. Res.* **2011**, *211*, 568–576. [CrossRef]
11. Zhao, Y.X.; Yao, D.J. Optimal dividend and capital injection problem with a random time horizon and a ruin penalty in the dual model. *Appl. Math. A J. Chin. Univ.* **2015**, *30*, 325–339. [CrossRef]
12. Yin, C.; Yuen, K.C. Optimal dividend problems for a jump-diffusion model with capital injections and proportional transaction costs. *J. Ind. Manag. Optim.* **2015**, *11*, 1247–1262. [CrossRef]
13. Højgaard, B.; Taksar, M. Optimal dynamic portfolio selection for a corporation with controllable risk and dividend distribution policy. *Quant. Financ.* **2004**, *4*, 315–327. [CrossRef]
14. Peng, X.; Chen, M.; Guo, J. Optimal dividend and equity issuance problem with proportional and fixed transaction costs. *Insur. Math. Econ.* **2012**, *51*, 576–585. [CrossRef]
15. Yao, D.; Yang, H.; Wang, R. Optimal risk and dividend control problem with fixed costs and salvage value: Variance premium principle. *Econ. Model.* **2014**, *37*, 53–64. [CrossRef]
16. Yao, D.; Yang, H.; Wang, R. Optimal dividend and reinsurance strategies with financing and liquidation value. *ASTIN Bull. J. IAA* **2016**, *46*, 365–399. [CrossRef]
17. Yao, D.; Fan, K. Optimal risk control and dividend strategies in the presence of two reinsurers: Variance premium principle. *J. Ind. Manag. Optim.* **2018**, *14*, 1–15. [CrossRef]
18. Cadenillas, A.; Choulli, T.; Taksar, M.; Zhang, L. Classical and impulse stochastic control for the optimization of the dividend and risk policies of an insurance firm. *Math. Financ.* **2006**, *16*, 181–202. [CrossRef]
19. Meng, H.; Siu, T. On optimal reinsurance, dividend and reinvestment strategies. *Econ. Model.* **2011**, *28*, 211–218. [CrossRef]
20. Xu, J.; Zhou, M. Optimal risk control and dividend distribution policies for a diffusion model with terminal value. *Math. Comput. Model.* **2012**, *56*, 180–190. [CrossRef]
21. Yao, D.; Wang, R.; Cheng, G. Optimal dividend and capital injection strategy with excess-of-loss reinsurance and transaction costs. *J. Ind. Manag. Optim.* **2017**, *13*, 51. [CrossRef]
22. Chunxiang, A.; Lai, Y.; Shao, Y. Optimal excess-of-loss reinsurance and investment problem with delay and jump-diffusion risk process under the CEV model. *J. Comput. Appl. Math.* **2018**, *342*, 317–336.
23. Yao, D.; Wang, R.; Lin, X. Optimal dividend, capital injection and excess-of-loss reinsurance strategies for insurer with a terminal value of the bankruptcy. *Sci. Sin. Math.* **2017**, *47*, 969–994.
24. Paulsen, J. Optimal dividend payments and reinvestments of diffusion processes with both fixed and proportional costs. *SIAM J. Control Optim.* **2008**, *47*, 2201–2226. [CrossRef]
25. Bai, L.; Guo, J.; Zhang, H. Optimal excess-of-loss reinsurance and dividend payments with both transaction costs and taxes. *Quant. Financ.* **2010**, *10*, 1163–1172. [CrossRef]

26. Liu, W.; Hu, Y. Optimal financing and dividend control of the insurance company with excess-of-loss reinsurance policy. *Stat. Probab. Lett.* **2014**, *84*, 121–130. [CrossRef]
27. Loeffen, R.L.; Renaud, J.F. De Finetti's optimal dividends problem with an affine penalty function at ruin. *Insur. Math. Econ.* **2010**, *46*, 98–108. [CrossRef]
28. Liang, Z.; Young, V.R. Dividends and reinsurance under a penalty for ruin. *Insur. Math. Econ.* **2012**, *50*, 437–445. [CrossRef]
29. Nguyen, V.H.; Vuong, Q.H.; Tran, M.N. Central limit theorem for functional of jump Markov processes. *Vietnam J. Math.* **2005**, *33*, 443–461.
30. Nguyen, V.H.; Vuong, Q.H. On the martingale representation theorem and approximate hedging a contingent claim in the minimum mean square deviation criterion. *VNU J. Sci. Math. Phys.* **2007**, *23*, 143–154.
31. Hoang, T.P.T.; Vuong, Q.H. A Merton Model of Credit Risk with Jumps. *J. Stat. Appl. Probab. Lett.* **2015**, *2*, 97–103.

symmetry

MDPI

Article

Flight Stability Analysis of a Symmetrically-Structured Quadcopter Based on Thrust Data Logger Information

Endrowednes Kuantama [1,†], Ioan Tarca [2,†], Simona Dzitac [3,†], Ioan Dzitac [4,5,*,†,‡] and Radu Tarca [6,†]

1 Engineering Doctoral School, University of Oradea, St. Universitatii, 1, 410087 Oradea, Romania; endrowednes@gmail.com
2 Department of Mechanical Engineering and Automotives, University of Oradea, St. Universitatii, 1, 410087 Oradea, Romania; nelut@uoradea.ro
3 Department of Energy Engineering, University of Oradea, St. Universitatii, 1, 410087 Oradea, Romania; simona@dzitac.ro
4 Department of Mathematics, Computer Science, Aurel Vlaicu University of Arad, St. Elena Dragoi, 2, 310330 Arad, Romania
5 R & D Center: "Cercetare Dezvoltare Agora", Agora University of Oradea, St. Piata Tineretului, 8, 410526 Oradea, Romania
6 Department of Mechatronics, University of Oradea, St. Universitatii, 1, 410087 Oradea, Romania; rtarca@uoradea.ro
* Correspondence: professor.ioan.dzitac@ieee.org or ioan.dzitac@uav.ro; Tel.: +40-722-562-053
† These authors contributed equally to this work.
‡ Current address: Department of Mathematics—Computer Science, Aurel Vlaicu University of Arad, St Elena Dragoi, 2, 310025 Arad, Romania.

Received: 23 June 2018; Accepted: 17 July 2018; Published: 19 July 2018

Abstract: Quadcopter flight stability is achieved when all of the rotors–propellers generate equal thrust in hover and throttle mode. It requires a control system algorithm for rotor speed adjustment, which is related with the translational vector and rotational angle. Even with an identical propeller and speed, the thrusts generated are not necessarily equal on all rotors–propellers. Therefore, this study focuses on developing a data logger to measure thrust and to assist in flight control on a symmetrically-structured quadcopter. It is developed with a four load cells sensor with two-axis characterizations and is able to perform real-time signal processing. The process includes speed adjustment for each rotor, trim calibration, and a proportional integral derivative (PID) control tuning system. In the data retrieval process, a quadcopter was attached with data logger system in a parallel axis position. Various speeds between 1200 rpm to 4080 rpm in throttle mode were analyzed to determine the stability of the resulting thrust. Adjustment result showed that the thrust differences between the rotors were less than 0.5 N. The data logger showed the consistency of the thrust value and was proved by repeated experiments with 118 s of sampling time for the same quadcopter control condition. Finally, the quadcopter flight stability as the result of tuning process by the thrust data logger was validated by the flight controller data.

Keywords: thrust; data logger; sensor; quadcopter; measurement; control system; stability

1. Introduction

The control system of quadcopter movement has six degrees of freedom, consisting of three translational and three rotational movement, which are crucial for the maneuverings stability [1]. By controlling the rotational speed of each rotor, one can also manage the thrust. Orientation sensors such as gyroscope, accelerometer, magnetometer, and GPS (Global Positioning System) are used to continuously read the quadcopter's position, so each movement error can be assessed and corrected using sensor fusion algorithms [2,3]. The control system has the most important role in achieving a high stability [4–6]. Quadcopter maneuver stability can be discerned with naked eyes, and can be read through the log of sensor data used for each acceleration and angular change in the three-axis coordinates. The control system and orientation sensors will adjust continuously to maintain the desired flight position [7]. To improve the performance of the orientation sensors and to reduce the position error, monitoring of the speed rotors is needed, especially during hovering and throttling. Monitoring and control adjustment can be done in various ways, for example, using cascade iterative [8], test bench system [9], and spherical model for UAV (Unmanned Aerial Vehicle) [10]. In this study, however, the development of a thrust data logger was adapted, and it was specially designed to read the thrust generated on each rotor–propeller simultaneously.

A common problem with quadcopters is unequal thrust, despite using an identical propeller and angular velocity. The propeller's imbalances and vibrations, the values of the pulse width modulation (PWM) or pulse position modulation (PPM) on the electronic speed controller, as well as the remote-control rate, may cause uneven thrust. A data logger is needed to overcome this problem. It is used to read the vertical thrust from rotor–propeller and to assist in the pre-flight calibration process, which will result in equal thrust for each rotor–propeller in hover or throttle mode.

The thrust data logger was developed using four pieces of a load cell sensor. A load cell is one of the devices capable of measuring sheer force and bending movement [11,12]. In this study, the mechanic and electronic parts were modeled. The load cell position formed a rectangular coordinate system, which is parallel with the inertial frame of the quadcopter. Positions along the x-axis and y-axis of the coordinate system are used to discover the angular change in the roll and pitch position, respectively. The data obtained from the two-axis load cell allows a flight calibration to be performed on the control system to achieve flight stability. When the quadcopter has throttle with various speeds, all of the rotors must generate equal thrust. This system design covers the overall quadcopter movement through the generated thrust. Because the test was done on a quadcopter assembled entirely from spare parts, any propeller imbalance or differences in rotor speeds can be observed in detail. In this study, the specific mechanical design embedded with the load cell and electronic board will allow real-time signal processing, in which the serial data can be read every second via the USB (Universal Serial Bus) port as a serial monitor with a transmission rate of 9600 bits per second. An adjustment of the thrust value can be done easily by knowing the thrust error rate of each rotor. The thrust data logger is also used to perform trim calibration and a proportional integral derivative (PID) coefficient adjustment on the quadcopter control system. Thus, the thrust data must be simultaneously collected, this way allowing for real-time settings. The result will be validated by the orientation sensors data when the quadcopter flies under windless conditions.

2. Thrust Data Logger Model

The thrust data logger was designed to continuously measure the thrust value generated by the quadcopter's rotors and was used in the pre-flight calibration process. There are three components required to design this system, that is, a load cell sensor, an analog-front-end (AFE), and a microcontroller unit (MCU). A detailed schematic of this thrust data logger can be seen in Figure 1. The load cell sensor is usually used as a tool to measure weight and force [13,14]. In this study, it was used to measure thrust based on the correlation between weight and the magnitude of the gravity acceleration. The output of the sensor is a voltage signal, which needs to be translated into a digital form using an AFE containing a 24-bit analog digital converter (ADC) driver before being processed by the MCU. The MCU calculates the analog input voltage between 0 to 5 volts, which yields a resolution between a reading of 4.9 mV per unit. With a 16 MHz MCU clock, the ADC must be set to a 125 kHz sampling rate. The thrust calibration was performed on the MCU, and this system was designed to retrieve data once a second. The thrust data logger was calibrated using professional equipment. An automatic zero calibration was designed so that the weight of the quadcopter is eliminated; thus, only the thrust value on each axis has been obtained. The manual calibration using buttons was provided to ensure the initial values before starting the data retrieval process.

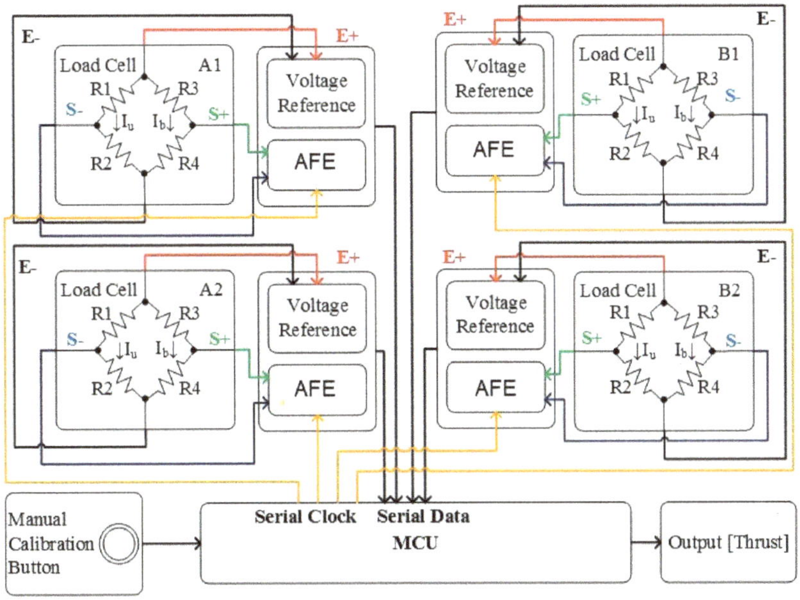

Figure 1. Data logger schematic. MCU—microcontroller unit; AFE—analog-front-end.

The magnitude of the reference voltage is denoted by (E+) and (E−), the voltage output is denoted by (S+) and (S−), and the load cell sensors are denoted by (A1), (A2), (B1), and (B2). Load cell sensors with a maximum load of 1 kg were used to measure the magnitude of the upward and downward force on the z-axis on each rotor [15]. The symmetrical position of the quadcopter rotor made the load sensors on the data logger be in symmetrical position as well. The quadcopter was placed together with the data logger sensor on the ground station, where the rotor and load cell sensor must be in parallel position.

The concept of measuring using a load cell sensor is to change the stimulus received by the sensor's mechanical part into an electrical signal. The changes manifest in the form of resistance modification. The system is based on the concept of the Wheatstone bridge circuit, as seen in Equations (1)–(3) [16].

$$V_{R2} = I_U R_2 = \frac{V_{ref}}{R_2 + R_2} R_2 \tag{1}$$

$$V_{R4} = I_B R_4 = \frac{V_{ref}}{R_3 + R_4} R_4 \tag{2}$$

$$V_{out} = V_{R1} - V_{R2} = \frac{R_2 R_3 - R_1 R_4}{(R_2 + R_1)(R_3 + R_4)} V_{ref} \tag{3}$$

The sensitivity of the sensors (S) can be calculated using Equation (4), where E_v is the excitation voltage, R_{out} is the rated output, and C is the maximum capacity of the load cell sensor. With a maximum capacity of 1 kg, the sensitivity of the load sensor is 0.7 mV/kg. The sensitivity of the load sensor still depends on the signal processing time [17].

$$S = \frac{E_v R_{out}}{C} \tag{4}$$

In this design, a 5 volt power supply was given as a reference voltage $\left(V_{ref}\right)$ or an excitation voltage. A detailed specification of the load cell sensor used can be seen in Table 1.

Table 1. Specifications of load cell sensor [18].

Load Cell Sensor—CZL635	
Load-cell material	Aluminum
Rated output	0.7 ± 0.15 mV/V
Repeatability	0.05% Full scale
Creep	0.05% Full scale for 10 min
Zero balance	±1.0% Full scale
Maximum excitation	18 Volt
Maximum capacity	1 kg
Input impedance	1000 ± 10 Ω
Output impedance	1000 ± 3 Ω

The design of the mechanical data logger and the details of the mechanical size can be seen in Figure 2. The load cell sensor was attached with two types of aluminum ring with thickness of 4 mm. At the bottom, there was an outer ring connecting the load cell with a flat surface, where, in this case, there was a table with a weight significantly greater than the total force generated by the quadcopter. The inner ring on the top of the load cell sensor was used as the foundation for connecting the quadcopter with the sensor. Four M6x100 screws were used as connectors and each screw was in a symmetrical position.

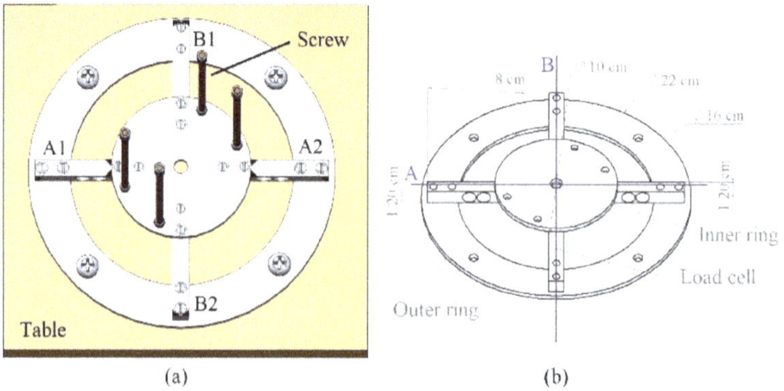

Figure 2. Model of thrust data logger (**a**) Mechanical layout (**b**) Detailed sketch

This model shows the two axis load cells (i.e., the *A*-axis and *B*-axis) used to measure thrust, which correlates with the quadcopter's three degrees of freedom. Knowing the changes in the thrust values for each of the four sensors points, the errors in the angular changes of the yaw (Ψ), pitch (ϕ), and roll (θ) angles in a 3-dimensional (3D) coordinate system (x, y, z) can be calculated and further reduced by adjusting the control system. One of the simplest concepts of quadcopter maneuvering is that, if one of the rotors produces a smaller thrust than the other, the quadcopter will move toward the smaller thrust. The prototype of the thrust data logger can be seen in Figure 3. The quadcopter used in this test has a symmetrical length and width of 46 cm and a height of 8 cm, having 33 cm long propellers.

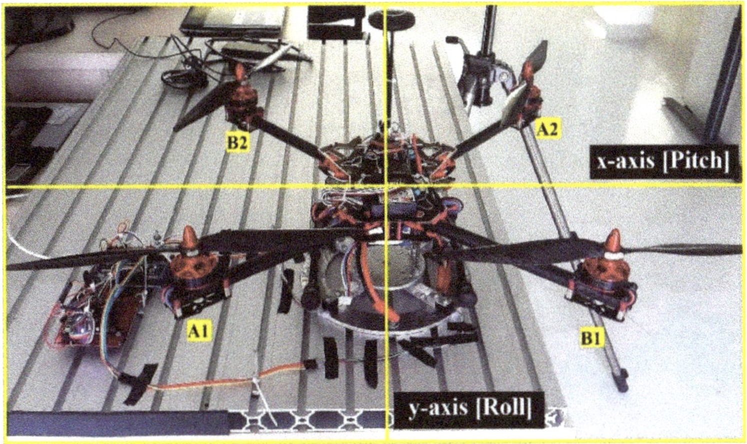

Figure 3. Data logger prototype and quadcopter thrust movement.

The experiment started with thrust data retrieval during the hovering state to see if each rotor produces the same value. Afterwards, when throttling, the speed of the quadcopter was increased from the lowest speed of 1200 rpm to the highest speed of around 4080 rpm. The angular velocity of the rotor was measured with a laser optical sensor using the time of arrival of the phase marker method, at a sampling frequency of 48 kHz and a resolution of 1.5625 Hz. The data obtained was then used to perform the calibration process on the quadcopter.

3. Data Processing

The MCU as the central processing unit on the data logger will process the serial data from all of the load sensors into thrust values. A dynamic compensation algorithm was needed to read the analog signal for each load cell [19,20]. The MCU was designed to be able to read the analog signal with a 125 kHz sampling rate, and a full-scale differential input voltage of 40 mV was used as a low noise programmable gain amplifier. In relation to the quadcopter movement stability calibration, signal processing occurs not only in the data logger, but also on the MCU on the quadcopter. It can be said that the data logger functions to continuously monitor the output during the quadcopter's calibration. A manually controlled quadcopter with a remote control can be calibrated through the trim process. Data processing on the quadcopter commonly uses a PID control system. The exact output value of the coefficient gain adjustment process can be read through the data logger, which measures the thrust simultaneously. The designed data logger can work steadily when the rotor is in a steady speed.

3.1. Thrust Data Logger

The data processing on the data logger was made by reading the output signal of the load cell used in the weighing system, in which the measured weight had been converted into the thrust value [21,22]. A certain period is needed in order to avoid the oscillation of the response signal from the load cell, for getting more accurate data. The load cell sensor has one degree of freedom (along the z-axis) with dynamic behavior, and generally is used in weighting the devices. The mechanical design of this sensor produces elastic behavior with a linear output signal to facilitate the calibration process [23]. The data processing speed for the load cell sensor can be calculated using Equation (5) [24].

$$\varepsilon(t) = \gamma f_n^2 T(t) \quad (5)$$

$$T_{(1,2,3,4)} = \left(\frac{M_{sample} - M_{ref}}{61.9}\right) g \quad (6)$$

where $\varepsilon(t)$ is the load cell response time, f_n is the natural frequency depending on MCU's clock frequency, γ is the load cell's static gain, and T is the thrust value. From Equation (3), it can be seen that the sensor output is a voltage, and it needs to be calculated using Equation (6). The 24-bit ADC data is translated into units of weight by performing a data comparison between the measured values $\left(M_{sample}\right)$, with the reference value at no load $\left(M_{ref}\right)$. These results will be converted into a mass value by being divided by 61.9 as a weight calibration value. The expression for the thrust value can be defined in terms of the normal gravity acceleration $\left(g = 9.80665ms^{-2}\right)$. This equation applies to all of the load cells' sensors that are used in this system, and the output data can be read continuously through serial monitor.

3.2. Dynamic Movement of Quadcopter

The quadcopter control system algorithm can be analyzed using Newton Euler theorem, and there are various control systems that can be used to improve the stability of the quadcopter [25–27]. The quadcopter's dynamic behavior on each axis can affect the angular change [28,29]. The amount of thrust generated depends on the rotor's angular velocity (ω_i). The value of the lifting constant depends on the air density generated by the propeller (B), as seen in Equation (7). From the other perspective, the amount of torque (τ) generated in each axis can be calculated based on the resulting thrust for the same axis (T) as well as the distance between the rotor and the center of mass (L), as seen in Equation (8).

$$T = B\omega_i^2 \quad (7)$$

$$\begin{bmatrix} \tau_x \\ \tau_y \\ \tau_z \\ T_z \end{bmatrix} = \begin{bmatrix} L(T_{A1} - T_{A2}) \\ L(T_{B1} - T_{B2}) \\ (T_{A1} - T_{B1} + T_{A2} - T_{B2}) \\ (T_{A1} + T_{B1} + T_{A2} + T_{B2}) \end{bmatrix} = \begin{bmatrix} LB(\omega_{A1}^2 - \omega_{A2}^2) \\ LB(\omega_{B1}^2 - \omega_{B2}^2) \\ B(\omega_{A1}^2 - \omega_{B1}^2 + \omega_{A2}^2 - \omega_{B2}^2) \\ B(\omega_{A1}^2 + \omega_{B1}^2 + \omega_{A2}^2 + \omega_{B2}^2) \end{bmatrix} \quad (8)$$

The torque generated on each axis affects the rotational angle alteration. The magnitude of the rotational acceleration for the resulting airframe will affect the quadcopter dynamic movement, and it can be calculated using Equation (9).

$$\begin{bmatrix} T \\ \tau \end{bmatrix} = \begin{bmatrix} (m.I_m) & 0 \\ 0 & I \end{bmatrix} \begin{bmatrix} a_m \\ \dot{\omega} \end{bmatrix} + \begin{bmatrix} \omega \ x \ v_B \\ \omega \ x \ I.\omega \end{bmatrix} \quad (9)$$

where m represents mass, I_m represents the 3×3 identity matrix, a_m represents the linear acceleration, v_B represents the linear velocity, ω is the angular velocity, τ represents the torque applied to the bodyframe, $\dot{\omega}$ represents the angular acceleration, and I represents the moment of inertia. The inertia matrix for the quadcopter can be written diagonally in a matrix transform because of the symmetry of the quadcopter's geometry, as expressed in Equation (10).

$$I = \begin{bmatrix} I_{xx} & -I_{xy} & -I_{xz} \\ -I_{yx} & I_{yy} & -I_{yz} \\ -I_{zx} & -I_{zy} & I_{zz} \end{bmatrix} = \begin{bmatrix} I_{xx} & 0 & 0 \\ 0 & I_{yy} & 0 \\ 0 & 0 & I_{zz} \end{bmatrix} \quad (10)$$

where I_{xx}, I_{yy}, and I_{zz} are presented in Equation (11), considering that the rotors are assimilated as material points. The inertia moments are calculated based on angular momentum in the body-frame coordinate, with the mass of a quadcopter (M_T) concentrated in its center of mass, and the radius (R_C). The weight of the rotor (m_r) and the distance between the frame and the rotor (L) will affect the inertia moments for each axis. Because of the quadcopter's symmetry, the value of the inertia moments for the x- and y-axis are equal. The motion of the quadcopter, especially the rotor's thrust, affects its attitude.

$$\begin{bmatrix} I_{xx} \\ I_{yy} \\ I_{zz} \end{bmatrix} = \begin{bmatrix} \frac{2}{5}M_T R_C^2 + (I_{A1} + I_{A2}) \\ \frac{2}{5}M_T R_C^2 + (I_{B1} + I_{B2}) \\ \frac{2}{5}M_T R_C^2 + (I_{A1} + I_{A2} + I_{B1} + I_{B2}) \end{bmatrix} = \begin{bmatrix} \frac{2}{5}M_T R_C^2 + 2L^2 m_r \\ \frac{2}{5}M_T R_C^2 + 2L^2 m_r \\ \frac{2}{5}M_T R_C^2 + 4L^2 m_r \end{bmatrix} \quad (11)$$

The Newton–Euler equation for the quadcopter's dynamic (Equation (9)) shows two components, one for the rotation movement and one for translation. The translational component is presented in Equation (12), in which the disturbances are highlighted in the d vector. The total vertical thrust is represented with (T_z) and is related to Equation (8). On the z-axis, the gravitational force (g) and the total mass of the quadcopter (m) affect all of the rotational angles (i.e., pitch, roll, and yaw angles). The vector (d_x, d_y, d_z) affects the magnitude of the translational vector.

$$m\begin{bmatrix} \ddot{x} \\ \ddot{y} \\ \ddot{z} \end{bmatrix} = T_z \begin{bmatrix} (\cos\phi \sin\theta \cos\psi + \sin\phi \sin\psi) \\ (\sin\theta \cos\phi \sin\psi - \sin\phi \cos\psi) \\ (\cos\theta \cos\phi) \end{bmatrix} + \begin{bmatrix} 0 \\ 0 \\ -mg \end{bmatrix} + \begin{bmatrix} d_x \\ d_y \\ d_z \end{bmatrix} \quad (12)$$

The correlation between the designed thrust data logger and the quadcopter dynamic movement can be seen in Equations (8), (9) and (12). On hover state, the value of the pitch, roll, and yaw angles must be equal to zero. If the thrust generated on each rotor–propeller are unequal, angle alteration and rotational acceleration on the airframe may occur. The rotational acceleration affects the translational vector on each axis on the quadcopter. In such a case, the quadcopter position can be maintained by adjusting the error. If the thrust values are known, the adjustment of the rotational speed on each rotor will be easier. It is important to note that the thrust generated are affected by the thrust factor, that is, the rotor rotation and air density generated by the propeller. The thrust factor is closely related with the propeller's aspects (i.e., size, angle of attack, vibration, and balance).

3.3. PID Control System

One of the thrust data logger system applications in a quadcopter is to set the coefficient values of the PID control system. This control system allows for minimizing errors during manoeuvers, by performing a manual adjustment to the coefficient of the PID in correlation with the quadcopter's angle and position on each axis. Details of the PID algorithm and simulation analysis are presented in previous research [30]. A translation or rotational error can occur with different magnitudes on the pitch, roll, and yaw motion. Based on this condition, the PID control system is divided into three sub-systems, namely pitch, roll, and yaw. By adjusting the coefficient value of each motion using the Ziegler–Nichols rules, movement stability can be achieved [31,32]. After the data is processed, the thrust generated at the two-axis coordinates system (illustrated on Figure 3) will affect the angular change on the pitch (x-axis) and roll (y-axis).

The dynamic value of the thrust on the x- and y-axis will affect the torque and thrust on the z-axis. From Equation (9), the pitch movement with a positive torque value will move the quadcopter backwards, while the opposite will move the quadcopter forwards. In the roll movement, a positive torque value will cause the quadcopter to move leftwards, and vice versa. The yaw movement or rotation on the z-axis can occur if a pair of rotors rotating in the same direction has a lower speed than the other. On the z-axis, the throttle level can occur if all of the rotors produce an equal thrust while increasing or decreasing the rotors speeds, thus affecting the altitude of the quadcopter.

$$\begin{bmatrix} \tau_\phi \\ \tau_\theta \\ \tau_\psi \end{bmatrix} = \begin{bmatrix} K_{p,x}\left(\phi'_p - \phi\right) + K_{i,x}(\phi'_i - \phi) + K_{d,x}(\phi'_d - \phi) \\ K_{p,y}\left(\theta'_p - \theta\right) + K_{i,y}(\theta'_i - \theta) + K_{d,y}(\theta'_d - \theta) \\ K_{p,z}\left(\psi'_p - \psi\right) + K_{i,z}(\psi'_i - \psi) + K_{d,z}(\psi'_d - \psi) \end{bmatrix} \quad (13)$$

All of the dynamic movement trajectories are read through the orientation sensor, so that each acceleration and angular change can be compared with the desired position. The magnitude of this error value can be minimized with a PID control system and can indirectly set the amount of thrust generated by each propeller. By using the Euler equation, the angular velocity vector of dynamic movement can be taken into account, and the influence between the resulting angle and the thrust can be seen in Equation (13). This equation shows the PID algorithm used to perform the adjustment for each of the trajectories, in which the desired values of the pitch, roll, and yaw angles are represented by (ϕ', θ', ψ'), and (ϕ, θ, ψ) is the output read through the orientation sensor.

4. Result and Discussion

A thrust data logger with a maximum load of 4 kg was tested to measure the thrust value and perform a stability adjustment for a 1.8 kg quadcopter. A laser optical sensor was used to measure the angular velocity of the rotor–propeller. At the time of the measurement, all of the electronic and mechanical parts of the quadcopter should be in the same position under conditions similar to flight. The quadcopter's center of the mass must be aligned with the thrust data logger's center, otherwise the total reaction of the torque on the airframe quadcopter may be affected. The first test was performed to determine the thrust value generated in throttle mode, with the rotor speed increasing gradually from 1200 rpm to 4080 rpm. As much as 24 sample points have been considered, each sample being acquired during a period of 1 s. The trust obtained from the data logger can be seen in Figure 4.

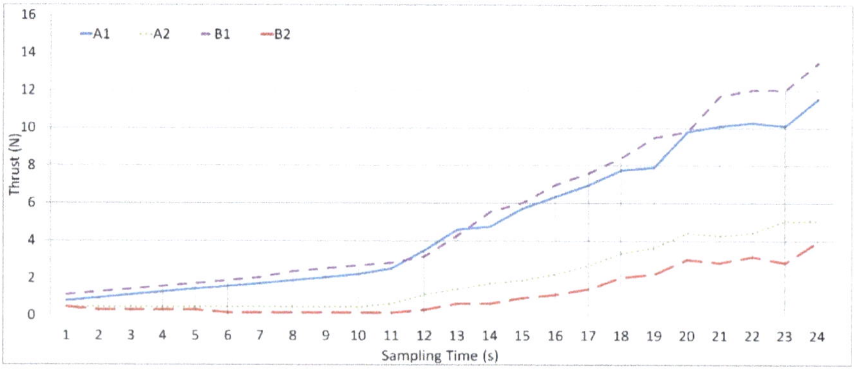

Figure 4. Thrust value without control system adjustment.

To analyze this data, it should be considered that the thrust produced by the propellers moving in the same direction (in *A*-axis or *B*-axis) will interfere with each other. Each rotor produces a different thrust with a significant error. In the *A*-axis, at 1200 rpm, the difference in the thrust value is 0.3 N while in the *B*-axis it is 0.5 N. With high speed, the thrust difference is approximately 5 N. The higher the speed, the greater the error, so stability cannot be obtained. If one rotor has a higher thrust than the other in the z-axis, then the position of said rotor will have a greater positive value than the other. Figure 5 shows the position and direction of the quadcopter for various thrust values. The acquired data has been split into three equal parts, for each of them the thrust value being the average of the acquired data, to find out the position error. The results show that the quadcopter has a tendency to move backward at any speed, and the field tests also yield the same result. In the hovering state, the quadcopter cannot maintain its position because of the imbalanced thrust.

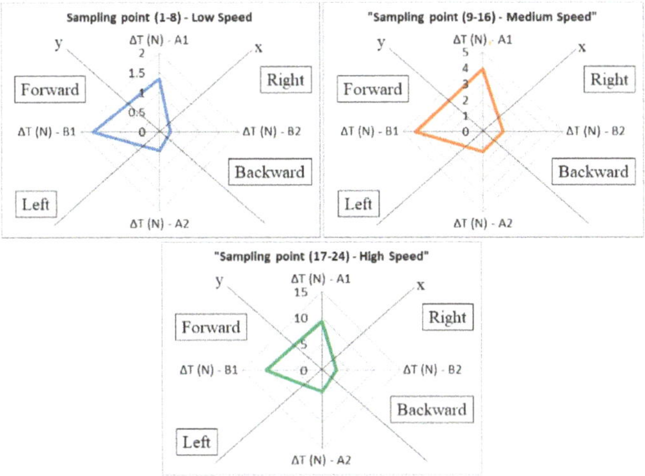

Figure 5. Thrust value related with quadcopter movement.

To overcome this problem, one can perform a trim calibration process and PID coefficient adjustment. Trim calibration is the adjustment of the zero value on both the remote control and the quadcopter, used only for manually controlled quadcopters. The trim positioning adjustment is done manually by doing a zero-positioning trim for the forward–backward channel. The thrust data calibration results were obtained using a data logger system for three speed values (i.e., 1200 rpm, 1500 rpm, and 2200 rpm). For each speed level, 50 thrust data acquired during 50 s sampling times are obtained. The corresponding data is then averaged to analyze the quadcopter's position during hovering at the considered speed level, as shown in Figure 6. The trim control adjustment only helps to set the zero point. The thrust difference is smaller than 0.5 N for the considered speed values and the thrust data obtained during the 50 s sampling time is quite stable.

Having the thrust data acquired, the adjustment process of the control system on the quadcopter can be started, especially for the pitch and roll angles. The quadcopter tends to move forward and slightly to the right. Based on the rate of error, the proportional gain in the PID for the pitch position must be greater than the one corresponding to the roll position. The data logger can be of assistance to notice the changes that occur by modifying each variable gain in PID. By manually adjusting all of the PID's coefficients, the final value for each coefficient can be seen in Table 2.

An experiment of thrust data retrieval using a data logger was conducted during a sampling time of 118 s, increasing the speed every 10 s. To ensure the accuracy and stability of the thrust data readings, the experiments were performed three times, and the thrust behavior of each rotor can be seen in this result. All of the results for the PID's coefficient adjustment process can be seen in Figure 7. Using Equation (8), the total thrust and torque on each axis can be calculated. This quadcopter model produces a total thrust of 26 N.

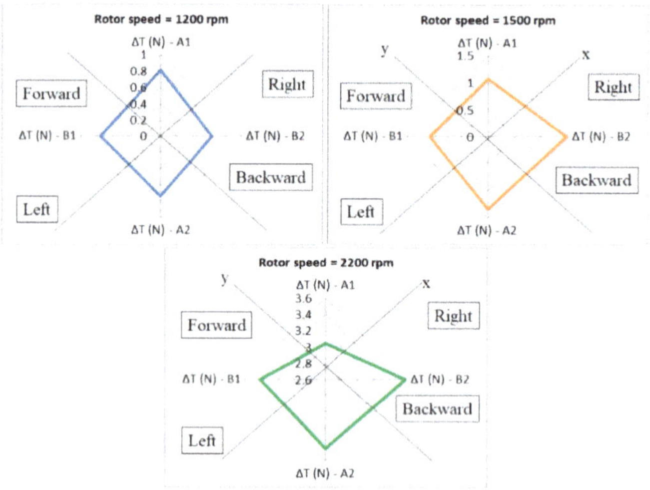

Figure 6. Thrust value with trim adjustment.

Table 2. Proportional integral derivative (PID) coefficient adjustment for roll and pitch position.

Parameter	Rate Roll	Rate Pitch
Proportional Gain	0.215	0.235
Integral Gain	0.140	0.140
Derivative Gain	0.010	0.047

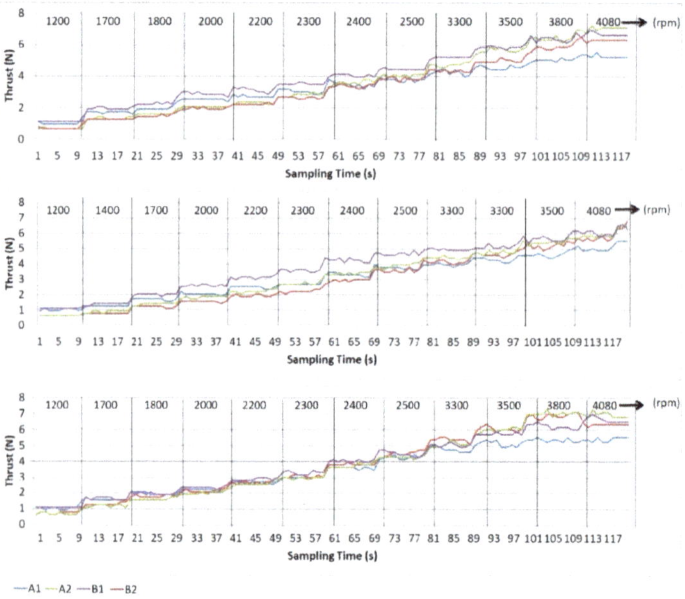

Figure 7. Thrust value with trim adjustment.

In a repetitive and continuous acceleration, it is difficult to obtain the exact angular velocity of a rotor for the same iteration of the sample acquisition. To get an exact angular velocity of each repetition, a discontinuous process can be done for each different angular velocity. However, to demonstrate the reliability and the stability of the data reading with a thrust data logger, this research chose a continuous process with a 10% error tolerance for data verification. The first verification was done to ensure the accuracy of the data thrust reading generated by the same speed parameter. Repeating the sampling process showed that the thrust data logger produced data with less than a 10% difference. This is linear with the error tolerance for the rotor velocity reading. From the 10 s of sampling time for each rotor speed, the stability generated by the data logger could be seen in data output. The flat line shows that the read data is the same, and the spike occurs when the rotor speed increases continuously through the remote control. From the perspective of the controller adjustment with trim and PID, the data showed that the speeds lower than 3000 rpm or the throttle less than 80% of the maximum speed, resulting in the thrust produced in the same axis, has a difference between 0.3–0.5 N.

This proves that the control system can improve the stability of the resulting thrust. For speeds greater than 3000 rpm, the thrust produced by each rotor have a significant value difference, with a maximum of 2 N in the same axis. This can happen because of the turbulence on the airframe caused by a high airflow on each propeller and also because of the vibration that occurs in the quadcopter frame. In the real test conditions, the vibration in the quadcopter can occur in high speed movement and stability can be achieved by adjusting the control system.

The result of the adjustment made by using the data logger can be validated during a field test, where the quadcopter position can be acquired by the means of an orientation sensor. Data acquisition during field tests have been done using telemetry and the Mission Planner open source software [33]. In this case, telemetry was used as a wireless data transmission between the ground station and the quadcopter [34], while Mission Planner was used to reading the data [35–37]. The data retrieval result can be seen in Figure 8.

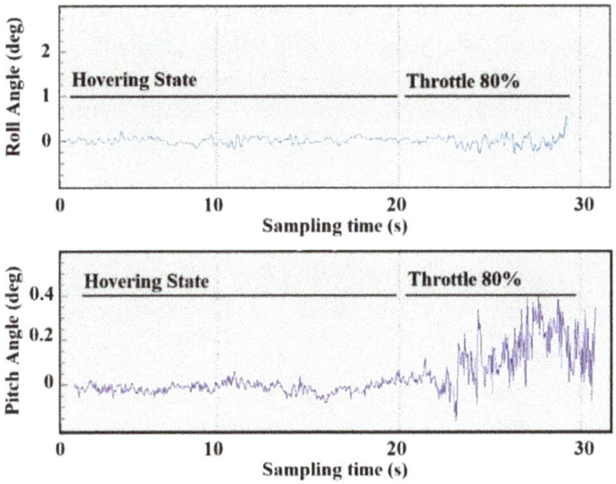

Figure 8. Thrust value with trim adjustment.

Details of the position changes occurring during the hover state and throttle state can be seen by comparing the desired position with the actual one. The result showed a linear variation of the output, with the data indicated by a data logger, where the throttle with a high-speed level resulted in instability in the roll position. While hovering, the pitch position showed more stable data compared to the roll position. An angular change for both of the angles during hovering were less than 0.2°

and both angles were oscillating at an angle of 0.5° at higher speeds. The orientation sensor with PID control in no-wind conditions seeks to stabilize the position of the quadcopter in a zero level, where high velocity oscillations may cause a vibration on the quadcopter's frame. This proves that there is a linear correlation between the orientation sensor data and thrust data logger.

5. Conclusions

A data logger for measuring two-axis thrust on a quadcopter was successfully developed, with a sensitivity level of 0.7 mV/kg, a response time of 1 s, and a maximum thrust of 40 N. This model works at a 125 kHz sampling rate and can be applied only on a symmetrically-structured quadcopter. It was integrated with the quadcopter to measure the thrust value simultaneously on all rotors–propellers and able to perform a calibration process of the flight control system in real time. The results were validated by repeating the 118 s sampling time process to measure quadcopter's thrust, using the same parameter values for each repetition. The maximum error percentage for each measurement was 10%. This value is given by the rotor's speed tolerance for each repetition, which allows for a maximum of 10%. The tolerance level and linear thrust differences indicate that the data logger has a good accuracy level. This was proved by the equally generated thrust at a steady speed, after the quadcopter was set by the data logger. To improve the stability of the quadcopter, the system is used for trim and PID adjustment, for a maximum speed of 4080 rpm and total thrust of 26 N. Knowing the thrust value of each rotor makes it easier to manually set the coefficient value of the PID. In the final adjustment, with the quadcopter having the speed below 3000 rpm, the results have shown that the difference between the rotors is less than 0.5 N. From this experiment, it can be seen that high-velocity rotors can cause vibrations, so that the resulting thrust becomes unstable. Overall, it can be concluded that this low-cost equipment can be statically connected with the quadcopter in adjusting the value of the control system for the quadcopter's dynamic movement. Data processing using a two-axis data logger is enough to obtain the positioning values of the quadcopter expressed in the pitch, roll, and yaw angles as well as the thrust condition during hovering and throttling.

In this study, the thrust value received from the data logger can assist the user in setting the quadcopter control system. For future research, the thrust data logger can be developed, making it able to connect with the flight controller for an automatic adjustment of the rotor speed. The designed data logger should also be able to measure thrust under windy conditions and assist in the flight control tuning process to reduce the imbalance effect of the propeller.

Author Contributions: E.K.: investigation, writing (original draft). I.T.: supervision, writing (review and editing). S.D.: data curation, formal analysis, and methodology. I.D.: formal analysis, funding acquisition, and validation. R.T.: project administration, resources, supervision, and validation.

Funding: This research was funded by European Commission, grant number 2014-0855/001-001, LEADERS—Erasmus Mundus Grant and Unitatea Executivă pentru Finantarea Învățământului Superior, a Cercetării, Dezvoltării și Inovării (UEFISCDI) Romania, grant number 47BG/2016, PNCDI III Programme P2—Transfer of knowledge to the economic operator. The APC was funded by R & D center "Cercetare Dezvoltare Agora Oradea".

Conflicts of Interest: The authors declare no conflict of interest.

References

1. Ajmera, J.; Sankaranarayanan, V. Point to Point Control of a Quadrotor: Theory and Experiment. *IFAC Pap. Online* **2016**, *49*, 401–406. [CrossRef]
2. Schopp, P.; Klingbeil, L.; Peters, C.; Buhmann, A.; Manoli, Y. Sensor Fusion Algorithm and Calibration for a Gyroscope-free IMU. *Procedia Chem.* **2009**, *1*, 1323–1326. [CrossRef]
3. Benzerrouk, H.; Nebylov, A.; Li, M. Multi-UAV Doppler Information Fusion for Target Based on Distributed High Degrees Information Filters. *Aerospace* **2018**, *5*, 28. [CrossRef]
4. Czyba, R.; Szafranski, G. Control Structure Impact on the Flying Performance of the Multi-Rotor VTOL Platform-Design, Analysis and Experimental Validation. *Int. J. Adv. Robot. Syst.* **2013**, *10*, 62. [CrossRef]

5. Hrishikeshavan, V.; Chopra, I. Performance, Flight Testing of Shrouded Rotor Micro Air Vehicle in Edgewise Gusts. *J. Aircr.* **2012**, *49*, 193–205. [CrossRef]
6. Jayakrishnan, H.J. Position and Attitude Control of a Quadrotor UAV using Super Twisting Sliding Mode. *IFAC Pap. Online* **2016**, *49*, 284–289. [CrossRef]
7. Doukhi, O.; Fayjie, A.R.; Lee, D.J. Intelligent Controller Design for Quad-Rotor Stabilization in Presence of Parameter Variations. *J. Adv. Transp.* **2017**, *2017*, 1–10. [CrossRef]
8. Tesch, D.A.; Eckhard, D.; Guarienti, W.C. Pitch and Roll Control of a Quadcopter using Cascade Iterative Feedback Tuning. *IFAC Pap. Online* **2016**, *49*, 30–35. [CrossRef]
9. Yushu, Y.; Ding, X. A Quadrotor Test Bench for Six Degree of Freedom Flight. *J. Intell. Robot. Syst.* **2012**, *68*, 323–337. [CrossRef]
10. Malandrakis, K.; Dixon, R.; Savvaris, A.; Tsourdos, A. Design and Development of a Novel Spherical UAV. *IFAC Pap. Online* **2016**, *49*, 320–325. [CrossRef]
11. Liu, T.; Inoue, Y.; Shibata, K. Wearable Force Sensor with Parallel Structure for Measurement of Ground Reaction Force. *Measurement* **2007**, *40*, 644–653. [CrossRef]
12. Aghili, F. Design of a Load Cell with Large Overload Capacity. *Trans. Can. Soc. Mech. Eng.* **2010**, *34*, 449–461. [CrossRef]
13. Maranzano, B.J.; Hancock, B.C. Quantitative Analysis of Impact Measurements using Dynamic Load Cells. *Sens. Biosens. Res.* **2016**, *7*, 31–37. [CrossRef]
14. Casas, O.V.; Dalazen, R.; Balbinot, A. 3D Load Cell for Measure force in Bicycle Crank. *Measurement* **2016**, *93*, 189–201. [CrossRef]
15. Richiedei, D.; Trevisani, A. Shaper-Based Filters for the Compensation of The Load Cell Response in Dynamic Mass Measurement. *Mech. Syst. Signal Process.* **2017**, *98*, 281–291. [CrossRef]
16. Huang, S.C. Development and Implementation of Load Cell in Weight Measurement Application for Shear Force. *Int. J. Electron. Electr. Eng.* **2017**, *5*, 240–244. [CrossRef]
17. Lee, W.K.; Yoon, H.; Han, C.; Joo, K.M.; Park, K.S. Physiological Signal Monitoring Bed for Infants Based on Load-Cell Sensors. *Sensors* **2016**, *16*, 409. [CrossRef] [PubMed]
18. Tinkerforge. Available online: http://www.tinkerforge.com/en/shop/load-cell-1kg-czl635.html (accessed on 21 January 2018).
19. Ma, J.; Song, A.; Pan, D. Dynamic Compensation for Two-Axis Robot Wrist Force Sensors. *J. Sens.* **2013**. [CrossRef]
20. Mohammed, A.A.S.; Moussa, W.A.; Lou, E. High Sensitivity MEMS Strain Sensor: Design and Simulation. *Sensors* **2008**, *8*, 2642–2661. [CrossRef] [PubMed]
21. Ballo, F.; Gobbi, M.; Mastinu, G.; Previati, G.A. Six Load Cell for the Analysis of The Dynamic Impact Response of a Hybrid III Dummy. *Meas. J.* **2016**, *90*, 309–317. [CrossRef]
22. Adamo, F.; Andria, G.; Di Nisio, A.; Carducci, C.; Guarnieri, C.; Lay-Ekuakille, A.; Mattencini, G.; Spadavecchia, M. Designing and Prototyping a Sensor Head for Test and Certification of UAV Components. *Int. J. Smart Sens. Intell. Syst.* **2017**, *10*, 646–672. [CrossRef]
23. Kluger, J.M.; Sapsis, T.P.; Slocum, A.H. A High-Resolution and Large Force-Range Load Cell by Means of Nonlinear Cantilever Beams. *Precisi. Eng.* **2016**, *43*, 241–256. [CrossRef]
24. Boschetti, G.; Caracciolo, R.; Richiedei, D.; Trevisani, A. Model-based Dynamic Compensation of Load Cell response in weighing machines Affected by Environtmental Vibrations. *Mech. Syst. Signal Process.* **2013**, *34*, 116–130. [CrossRef]
25. Cetinsoy, E.; Dikyar, S.; Hançer, C.; Oner, K.T.; Sirimoglu, E.; Unel, M.; Aksit, M.F. Design and Construction of a Novel Quad Tilt-Wing UAV. *Mechatoronics* **2012**, *22*, 723–745. [CrossRef]
26. Bouzid, Y.; Siguerdidjane, H.; Bestaoui, Y. Nonlinear Internal Model Control Applied to VTOL Multi-Rotors UAV. *Mechatronics* **2017**, *47*, 49–66. [CrossRef]
27. Huang, H.; Hoffmann, G.; Waslander, S.; Tomlin, C.J. Aerodynamics and Control of Autonomous Quadcopters in Aggresive Maneuvering. In Proceedings of the 2009 IEEE International Conference on Robotics and Automation, Kobe, Japan, 12–17 May 2009; pp. 3277–3282. [CrossRef]
28. Magnussen, O.; Ottestad, M.; Hovland, G. Experimental Validation of a Quaternion-based Attitude Estimation with Direct Input to a Quadcopter Control System. In Proceedings of the 2013 International Conference on Unmanned Aircraft Systems (ICUAS), Atlanta, GA, USA, 28–31 May 2013; pp. 480–485. [CrossRef]

29. Patel, K.; Barve, J. Modeling, Simulation, and Control Study for the Quadcopter UAV. In Proceedings of the 2014 9th International Conference on Industrial and Information Systems (ICIIS), Gwalior, India, 15–17 December 2014; pp. 1–6. [CrossRef]
30. Kuantama, E.; Vesselenyi, T.; Dzitac, S.; Tarca, R. PID and Fuzzy-PID Control Model for Quadcopter Attitude with Disturbance Parameter. *Int. J. Comput. Commun. Control* **2017**, *12*, 519–532. [CrossRef]
31. Moghaddam, H.F.; Vasegh, N. Robust PID Stabilization of Linear Neutral Time Delay System. *Int. J. Comput. Commun. Control* **2014**, *9*, 201–208. [CrossRef]
32. Bolandi, H.; Rezaei, M.; Mohsenipour, R.; Nemati, H.; Smailzadeh, S.M. Attitude Control of a Quadrotor with Optimized PID Controller. *Intell. Control Autom. J.* **2013**, *4*, 335–342. [CrossRef]
33. Ardupilot—Ground Station (GCS) Software. Available online: http://ardupilot.org/planner/index.html (accessed on 18 October 2017).
34. Hanafi, D.; Qetkeaw, M.; Ghazali, R.; Than, M.N.M.; Utomo, W.M.; Omar, R. Simple GUI Wireless Controller of Quadcopter. *Int. J. Commun. Netw. Syst. Sci.* **2013**, *6*, 52–59. [CrossRef]
35. Romano, E.; Todeschini, M.G.; Vigano, G.P.; Sacco, M. DroneAGE: An Advamced Graphic Environment for Planning and Control of Drone Missions. In Proceedings of the Conference and Exhibition of the European Association of Virtual and Augmented Reality, Bremen, Germany, 8–10 December 2014. [CrossRef]
36. Sanchez-Lopez, J.L. A Vision Based Aerial Robot Solution for the Mission 7 of the International Aerial Robotics Competition. In Proceedings of the 2015 International Conference on Unmanned Aircraft Systems (ICUAS), Denver, CO, USA, 9–12 June 2015; pp. 1391–1400. [CrossRef]
37. Obome, M. Mission Planner Software and Data Sheet. Available online: http://www.ardupilot.org/planner/docs/common-install-mission-planner.html (accessed on 25 September 2017).

MDPI
St. Alban-Anlage 66
4052 Basel
Switzerland
Tel. +41 61 683 77 34
Fax +41 61 302 89 18
www.mdpi.com

Symmetry Editorial Office
E-mail: symmetry@mdpi.com
www.mdpi.com/journal/symmetry

www.ingramcontent.com/pod-product-compliance
Lightning Source LLC
Chambersburg PA
CBHW051851210326
41597CB00033B/5852
9783039210060